AF540723

ETHNIC GROUPS AND HEALTH DIMENSIONS

ETHNIC GROUPS AND HEALTH DIMENSIONS

By

Dr. Rajesh Khanna

Manager (Development Sector)

Mott MacDonald Group

Noida (U.P.)

&

Dr. Anup K. Kapoor

Professor

Dept. of Anthropology

University of Delhi

Delhi–110 007

DISCOVERY PUBLISHING HOUSE

NEW DELHI-110002

First Published-2007

ISBN 978-81-8356-213-3

Published by

DISCOVERY PUBLISHING HOUSE
4831/24, Ansari Road, Prahlad Street,
Darya Ganj, New Delhi-110002 (India)
Phone: 23279245 • Fax: 91-11-23253475
E-mail:dphbooks@rediffmail.com
dphtemp@indiatimes.com

Printed at:

Sachin Printers, Delhi-

Preface

Among all those enrolled in a basic university course in anthropology, only a few will end up as professional anthropologists. But in going on to become human scientists, medical doctors, biologists, health experts, demographers, development experts and policy makers, a far greater number will need knowledge of, and facility with, anthropology, which is an indispensable tool for their work. There is more than one road to anthropology and a realistic first course must take that fact into account.

It must recognize, as well, the importance of a feeling of accomplishment and satisfaction in the student. Too often teachers in other disciplines have reason to look upon anthropology as a course which they could teach better; too often do we find that our students do not grasp concepts and methods which we know are not that hard. Some of the causes of this unfortunate situation are within our control and can be removed. A main purpose of this book is to make the general health anthropology course a more satisfactory learning experience for the student, and to accomplish this, several guidelines have been followed.

The anthropological level and sophistication of the book increases steadily from beginning to end. It has been our experiences that students find such a rise to be compatible with their development and their interests.

Ethnic groups and health dimensions are presented in such a way that the student can see their relevance. We have attempted to present our material in such a manner that it is within the intellectual horizon of a reader who has holistic understanding of man. Moreover we have also highlighted the relevance of fieldwork among ethnic groups of Himalayas. Each chapter has its own greater complexities, and we hope that some readers will use this book as a springboard for deeper studies. Beyond some people, other readers with an interest in the subjects discussed here may find in it some materials upon which to base their views as citizens.

We have tried to give all or latest references on the various topics of research which are usually not available to the students or scholars.

To acknowledge adequately all the help we have had would require a chapter in itself, but some debts are so great that they must be mentioned here. The basic idea of this type of book was conceived in the Dist. Kinnaur, Himachal Pradesh itself. As no health anthropological studies have been carried out so far on these ethnic groups and adjoining Himalayan areas.

We are grateful to Prof. Dr. S.C. Tiwari, former Head and Dean, Faculty of Science (1984-1987), Late Prof. Dr. J. S. Bhandari, Head (1990-1993), and Prof. Dr. P. K. Seth, Head (1996-1999), for their constant support and encouragement. We are thankful to Prof. Dr. R. N. Vashisht, Prof. Dr. B. G. Banerjee, Dept. of Anthropology, Punjab University, Chandigarh, and Prof. Dr. J. K. Pundir, Dept. of Sociology, Ch. Charan Singh University, Meerut (UP) for their suggestions and comments on various chapters. We also get the assistance and support of Dr. Manoj Singh, Mr. Bhaskar Singh, Dr. Achan Mungleng, Dr. Sudeep Ghosh, Dr. Rajeev Kamal Kumar, Mr. Sanjay Suman and Dr. P. K. Patra, who are at different places in the world, in the form of views and informations for the successful accomplishment of the present work.

This being a research with intensive fieldwork covering one of the very remote and difficult area, the help received from various officers is very valuable. Apart from these officers, we have also been helped by the local people to a great extent. We would like to put on record the appreciation for all of them for their help.

We have been acknowledging together as mentioned above when it comes to family, we would like to acknowledge individually.

Rajesh Khanna wishes to express his sincere thanks to his parents and wife, Poonam who constantly encouraged and provided the much needed moral support from time to time. He is also thankful to his brother, Deepak and his wife; sister Rachana and her husband for constant support and encouragement during the study.

A.K. Kapoor's wife, who teaches in the same department, the reader can just imagine how much she would have helped. A special and warm thanks to her for managing the household during my absence.

Finally, this book is dedicated to Rajesh Khanna's uncle and aunty, Mr. Hari Singh Verma and Dr. Premlata Verma, as an inadequate, if sincere, expression of filial gratitude.

Rajesh Khanna

A. K. Kapoor

Contents

List of Abbreviations Used

Abbreviation	*Details*
ACDPO	Additional Child Development Project Officer
ADMO	Additional District Medical Officer
ANC	Ante-Natal Care
ANM	Auxiliary Nurse Midwife
AWC	Anganwadi Centre
AWW	Anganwadi Worker
BDO	Block Development Officer
BMO	Block Medical Officer
BMS	Basic Minimum Services
CD	Community Development
CMO	Chief Medical Officer
CDPO	Child Development Project Officer
CHC	Community Health Centre
CMHO	Chief Medical and Health Officer
CNA	Community Need Assessment
DFW	Department of Family Welfare
DPO	District Project Officer
DWCD	Director of Women and Child Development
FRU	First Referral Unit
FW	Family Welfare
GOI	Government of India
HQ	Head Quarter

HP	Himachal Pradesh
HRD	Human Resource Department
ICDS	Integrated Child Development Service
ICPD	International Conference on Population and Development
IEC	Information Education and Communication
IFA	Iron and Folic Acid
ITDP	Integrated Tribal Development Programme
IUD	Intra Uterine Device
INHP	Integrated Nutrition and Health Programme
MNP	Minimum Need Programme
MPW	Multipurpose Worker
NGO	Non-Government Organisation
NHP	National Health Policy
NPP	National Population Policy
OPD	Out Patient Department
PHC	Primary Health Centre
PNC	Post-Natal Care
POL	Petrol-Oil-Lubricant
PPC	Post Pertem Centre
RMP	Registered Medical Practitioner
SC	Sub Centre
SC	Scheduled Caste
ST	Scheduled Tribe
SDH	Sub-Divisional Hospital
Sq. Km	Square Kilometer
TBA	Traditional Birth Attendant
UDC	Upper Division Clerk
WHO	World Health Organization

1

Introduction

Individuals and societies have long considered various definitions of health. In defining health, researcher usually fell into three areas. The first, the perception of health, is either seen as a subjective or objective phenomenon, and in terms of whether it extends beyond the physical domain. The second includes the means of improving and maintaining health. The third, considers the value and aim of health, i.e. how it allows one to function. These three areas are usually considered together in historical and contemporary definitions.

It has long been recognized that there is a close interaction between a healthy mind and a healthy body. Furthermore, health was considered in antiquity a beneficial asset and one that required action by the individual to preserve it. Thus, in the hippocratic writings it is said that "a wise man ought to realize that health is his most valuable possession and learn to treat his illnesses by his own judgement" (Rosen, 1976). During classical times and all through the middle ages, the struggle for survival, a struggle that aimed to overcome common epidemics, childhood infections and the hazards of childbirth, all limited populations' abilities to improve their physical health. Under such conditions, strong emphasis was given to mental, social and spiritual dimensions of health.

It is reported that by the end of the eighteenth century, public health and medical advances had laid the basis for people to believe that a long life could be obtained through societal actions, as well as through the actions of individuals. In fact, at the turn of the last century it was expected that all the challenges to health and disease would be conquered in a short period of time. Midway through the twentieth century, and with the start of the use of antibiotics and vaccines, this belief endured and remained in some ways an impediment to the full realization of a broader definition of health.

Prior to the Second World War, Sigerist, a well known public health professional, expressed the view that "health is, therefore, not simply the absence of disease; it is something positive, a joyful attitude to life, and a cheerful acceptance of the responsibilities that life puts upon the individual. A healthy individual is a man who is well balanced bodily and mentally, and well adjusted to his physical and social environment". (Sigerist, 1941)

Development of the notion of social responsibility for health and the duty of individuals for the care of their health was espoused by Dr Andrija Stampar, who was to become President of the First World Health Assembly of WHO. It was he who played a crucial role in drafting the definition of health that was to be incorporated into the first paragraph of the preamble to the WHO Constitution and subsequently into the International Covenant on Economic, Social and Cultural Rights.

Over half a century ago, the founders of the World Health Organization defined health as "a state of complete physical, mental and social well-being and not merely the absence of disease or infirmity". The Constitution further recognized "the enjoyment of the highest attainable standard of health ... as one of the fundamental rights of every human being" (World Health Organization, 1948). This "right to health", as it became expressed in an abbreviated version in many subsequent documents, includes the right to adequate food, water, clothing, housing, health care, education, security in the event of unemployment, sickness, disability, old age or lack of livelihood in circumstances beyond an individual's control.

Greater emphasis in the definition to equity and social justice was given when the 30th World Health Assembly decided in 1977 that the main social targets of governments and WHO in the coming decades should be "the attainment of all citizens of the world by the year 2000 of a level of health that would permit them to lead socially and economically productive lives". (World Health Organization, 1977) This statement is important in that it specifies both what level of health is needed and what will be accomplished at that level.

Optimal health is defined as a balance of physical, emotional, social, spiritual, and intellectual health. Lifestyle change can be facilitated through a combination of efforts to enhance awareness, change behaviour and create environments that support good health practices. Of the three, supportive environments will probably have the greatest impact in producing a lasting change.

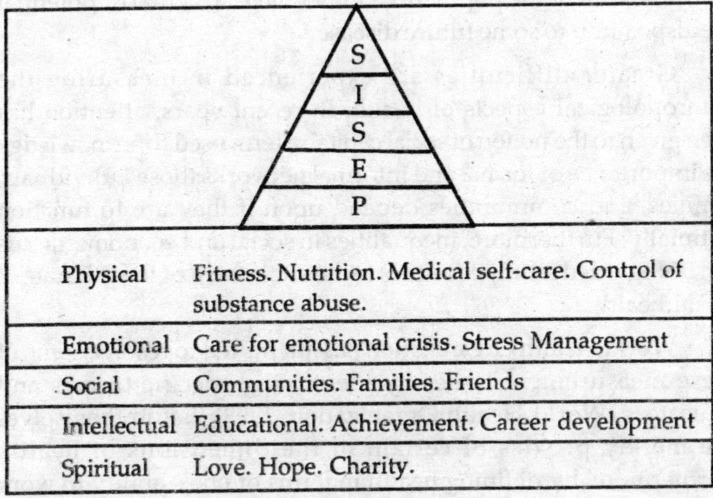

Physical	Fitness. Nutrition. Medical self-care. Control of substance abuse.
Emotional	Care for emotional crisis. Stress Management
Social	Communities. Families. Friends
Intellectual	Educational. Achievement. Career development
Spiritual	Love. Hope. Charity.

Fig. 1.1: Heath Dimensions

Over the decades, there have been many criticisms of the definition of health and of the shorthand version of "health as a human right". Some considered the definition too inclusive and should focus rather on the physical domain of health, the rationale being that health and its achievement was best left to health professionals and to the application of specific health and medical interventions. There are others who felt the definition excluded important dimensions, such as the spiritual and ethical dimensions of health. The third concern was that many felt that it was unrealistic to believe that all could be healthy. Protagonists of this view point out that there are genetic impediments to the attainment of health by all; that there are limits to the availability of resources available to ensure that all can attain the highest level of health; and that our scientific knowledge remains incomplete with regard to the true determinants of health and effectiveness of interventions.

Additional difficulties follow from these definitions and relate to the problem of measuring health and implementing programmes that would improve the health of individuals and populations. The physical dimension of health could be measured in terms of life expectancy, the infant mortality rate and other relatively objective measures. However, with advances in technology, particularly in the fields of imaging and genetic screening, now one can recognize that almost all of the populations either have an actual or potential predisposition to some future disease.

Greater difficulties are experienced in measuring the anthropological aspects of health. In recent years, attention has been given to the notion of social capital, a term used to acknowledge the importance of formal and informal networks those individuals, families, and communities depend upon if they are to function optimally. Furthermore, inequalities in social and economic status can be regarded as impediments to the attainment of the full state of social health.

World Health Organization has attempted to resolve some of these measurement issues by identifying specific targets and indicators (World Health Organization, 1998) that in themselves are merely proxies of certain of the dimensions of health. Furthermore, by defining health in terms of one's ability to work productively and participate fully in social life, a greater degree of operationalization is attainable.

Now one needs to recognize that health is everyone's business and that the ultimate responsibility for its attainment is shared between individuals, families, communities, governments at all levels and intergovernmental agencies.

Globally, there has been a substantial decline in the infant mortality rate, maternal deaths have remained unacceptably common, the level of childhood immunization has increased dramatically and access to primary health care, including water and sanitation, continues to improve, albeit at a slower rate in the last decade than in previous decades (World Health Organization,

- A government sector that provides publicly financed and managed curative and preventive health services from primary to tertiary level, throughout the country and free of cost to the consumer (these account for about 18% of the overall health spending and 0.9% of the GDP), and
- A fee-levying private sector that plays a dominant role in the provision of individual curative care through ambulatory services and accounts for about 82 per cent of the overall health expenditure and 4.2 per cent of the GDP. Nationwide health care utilization rates show that private health services are directed mainly at providing primary health care and financed from private resources, which could place a disproportionate burden on the poor.

The provision of health care by the public sector is a responsibility shared by state, central and local governments, although it is effectively a state responsibility in terms of service delivery. State and local governments incur about three-quarters and the center about one-quarter of public spending on health. The responsibility for health is at three levels. First, health is primarily a state responsibility. Second, the center is responsible for health services in union territories without a legislature and is also responsible for developing and monitoring national standards and regulations, linking the states with funding agencies, and sponsoring numerous schemes for implementation by state governments. Third, both the center and the states have a joint responsibility for programmes listed under the concurrent list. Goals and strategies for the public sector in health care are established through a consultative process involving all levels of government through the Central Council for Health and Family Welfare.

The outcomes from meetings of the Central Council for Health and Family Welfare have provided a thrust to various sub sectors within the health sector. The private and voluntary sectors have emerged as an important arm of the health sector. From 1 April 1996 a change has been effected in the family welfare services with targets for contraceptive methods being replaced by a target-free approach. A huge campaign to eradicate poliomyelitis through pulse polio immunization (PPI) was initiated in 1995. The traditional system of medicine is now playing a more significant role due to escalating costs of health care. State health systems/projects have been formulated to improve efficiency in the allocation and use of health resources through policy and institutional development. Specific efforts have been made to consolidate and strengthen the PHC infrastructure, under the minimum needs programme, by providing enhanced assistance to regions with severe health problems, supporting voluntary organizations, improving IEC activities, etc. The convergence of services to provide a holistic approach to population control has also been promoted. In March 1995 a separate Department of Indian System of Medicine and Homoeopathy (ISM & H) was created within the Ministry of Health and Family Welfare.

At the beginning of the First Five Year Plan, there were 725 Primary Health Centres, functioning in the country. Subsequently over the past-56 years, the health services organization and infrastructure have undergone extensive changes extension in stages following review by a number of Expert committees namely.; Mudaliar Committee in 1962, Chadha Committee in 1963, Mukherjee Committee in 1966, Jungalwala committee in 1967, Kartar Singh Committee in 1973 and Srivastava Committee in 1975. Progressive changes have been introduced in the programme over the 6th and 7th Five Year Plan period when national norms for population coverage were adopted. During the 8th Plan, the emphasis is mainly on consolidation of the existing health infrastructure rather than expansion. The thrust given to qualitative improvement in health services through strengthening of physical facilities like provision of essential equipments, supply of essential drugs and consumables, construction of buildings and staff quarters, filling up vacant posts of medical and para-medical staff and in-service training of staff.

Much before the Alma Ata Conference the Constitution of India provided (Under Directive Principle of State Policy: Article 36-51 Part IV) as follows: " The State shall in particular direct the policy towards securing that the health and strength of workers, men and women and the tender age of children are not abused and that citizens are not forced by economic necessity to enter avocations unsuited to their age or strength, that childhood and youth are protected against exploitation and against moral and material abandonment. The State within the limits of its economic capacity and development make effective provision for securing the right to work, to education and to public assistance in case of unemployment, old age, sickness and disablement and in other cases of underserved want. The State shall make provision for securing just and humane conditions of work and maternity relief. The State shall regard the raising of standard of living of its people and the improvement of public health as among the primary duties."

In the 30th World Health Assembly in 1977 referred to Health for all slogan as " a level of health that will permit people to live a socially and economically productive life". India as a signatory to the Alma Ata Declaration is committed to the achievement of Health for all by 2000 AD. India is also signatory to the South-East Asian Charter on Health. Against this background, the current objective of state and national health plans in the health sectors is to organize and provide universal primary health care to all sections of the society with special attention to the needs of those living in the tribal, hilly and remote area. Greater equity in the access to health care, better utilization of limited resources, appropriate policy development, active participation of the community in achieving self reliance in the health and sustaining the health services and inter- sectoral action are the key components of the Global Strategy of health for all.

The Indian Health Care System operates at the national level by Ministry of Health and Family Welfare, under which all the states and union territories come. States control the district health institutions. Chief Medical Officer/Chief Health and Medical Officer is the head of the district health institution, sub-district hospitals come which controls to the community health centres which is present in 1/1,00,000 population. Under community health centres, primary health centres come which present 1/30,000 population for plains and 1/20,000 for tribal and hilly region. Primary health centres (PHC) are present in block. Under PHC, sub centres come which are present 1/5000 population plain and 1/3000 population for tribal and hill region. Keeping in view that most of the tribal habitation is concentrated in far flung areas, forest land, hills and remote villages, and in order to remove the imbalances and provide better health care and family welfare services to Scheduled Castes/ Scheduled Tribes, the population coverage norms of establishment of rural infrastructure have been relaxed (Kumar and Chakrapani, 1994).

The First United Nations Advisory Mission, which visited India in 1955, to make recommendation for improvement in the family planning programme, has observed. In 1966, with the appointment of a Commissioner of Family Planning with the rank of Director-General, this lacuna was filled to a certain extent. At the state level, the State Family Planning Bureau is headed by the Joint Director, Family Planning and Maternal and Child Health and an officer of the rank of a secretary/deputy secretary heads the cell in the Secretariat. Each district is entitled to its own Family Planning Bureau with a Class 1 Officer in charge. The District Bureau has divisions for administration, education and information, field operation and evaluation: mobile units for IUD and sterilisation are also attached to the District Bureau. The headquarters of the rural family planning organisation is the Primary Health Centre, with sub-centres attached to it. For the urban family welfare planning centres, different patterns of staffing are provided, according to the size of the population to be covered. Each city with a population above 2 lakhs is entitled to its own Family Planning Bureau. All cities with a population of more than 10 lakhs are allowed the pattern of the District Family Planning Bureau. Different pattern are laid down for the family planning units of public sector undertakings and those of the Ministry of Railways, Post and Telegraphs Department and Ministry of Defence depending on the size of the population to be covered.

A separate Tribal Development Planning Cell has been functioning under the Ministry of Health and Family Welfare, Directorate General of Health Services since 1981 to co-ordinate the policy, planning, monitoring, evaluation etc. of the Health Care Schemes for welfare and development of Scheduled Tribes and Scheduled Castes.

Various Public Health Programmes are being implemented in the country and SCs/STs are deriving full benefit of the same. However, Programme Officers have been directed to ensure that plan funds to the extent of 8.1 per cent for Tribal Sub Plan and 16.5 per cent for Special Component Plan are allocated in proportion to the total population.

National Population Policy

Population policy in India has had a long and somewhat chequered history. Starting in the 1950s as a largely urban clinic-based programme, the family planning programme increasingly became, in the 1960s, rural in its focus and more community oriented in its approach. By the late 1960s, however, contraceptive method-—specific targets had become a key instrument in the programme, and in the 1970s, the programme and the policy within which it was embedded became increasingly target-driven, centralized and top-down. It operated largely in isolation from other sectors, such as education, which might plausibly have been argued to have some impact on demographic behaviour and patterns. Indeed, population and health policies were usually only nominally linked to each other even though both had a home in the Ministry of Health and Family Welfare.

The Population Policy of India underwent a major shift with the Programme of Action Adopted at the International Conference on Population and Development (ICPD) held at Cairo in September 1994. The Programme of Action placed the population in the context of development .It was recommended that instead of population control programme focusing on demographic targets, the emphasis should be on the needs of individuals. It may be reiterated that at the ICPD, the nations of the world agreed that the government give special attention to the education of the girls, the empowerment of the women. It was also recommended that comprehensive productive health services should be provided to enable couples to achieve their reproductive goals, and to determine freely responsively the number and spacing of their children. Thus the ICPD Programme of Action endorsed the concepts of reproductive and sexual health and rights and emphasized the need for providing services to achieve these goals.

The National Population Policy 2000 (NPP 2000) recognises the immense potential of Social Marketing in expanding the outreach and coverage of health care products and services, and emphasises the need to formulate and implement social marketing schemes for provisioning products and services, through partnerships between the voluntary sector, non–government organisations, the private corporate sector, Government, Panchayati Raj Institutions and the community. It recognises that all of this will accelerate achievement of the national socio-demographic goals.

National Health Policy

A National Health Policy was last formulated in 1983, and since then there have been marked changes in the determinant factors relating to the health sector. Some of the policy initiatives outlined in the NHP-1983 have yielded results, while, in several other areas, the outcome has not been as expected. The NHP-1983 gave a general exposition of the policies which required recommendation in the circumstances then prevailing in the health sector. The noteworthy initiatives under that policy were:

1. A phased, time-bound programme for setting up a well-dispersed network of comprehensive primary health care services, linked with extension and health education, designed in the context of the ground reality that elementary health problems can be resolved by the people themselves;
2. Intermediation through 'Health volunteers' having appropriate knowledge, simple skills and requisite technologies;
3. Establishment of a well-worked out referral system to ensure that patient load at the higher levels of the hierarchy is not needlessly burdened by those who can be treated at the decentralized level;
4. An integrated net-work of evenly spread speciality and super-speciality services; encouragement of such facilities through private investments for patients who can pay, so that the draw on the Government's facilities is limited to those entitled to free use.

While noting that the public health initiatives over the years have contributed significantly to the improvement of these health indicators, it is to be acknowledged that public health indicators/ disease-burden statistics are the outcome of several complementary initiatives under the wider umbrella of the developmental sector, covering Rural Development, Agriculture, Food Production, Sanitation, Drinking Water Supply, Education, etc. Despite the impressive public health gains, there is no gainsaying the fact that the morbidity and mortality levels in the country are still unacceptably high. These unsatisfactory health indices are, in turn, an indication of the limited success of the public health system in meeting the preventive and curative requirements of the general population. The period after the announcement of NHP-83 has also seen an increase in mortality through 'life-style' diseases- diabetes, cancer and cardiovascular diseases. The increase in life expectancy has increased the requirement for geriatric care. Similarly, the increasing burden of trauma cases is also a significant public health problem. Another area of grave concern in the public health domain is the persistent incidence of macro and micro nutrient deficiencies, especially among women and children.

NHP-1983, in a spirit of optimistic empathy for the health needs of the people, particularly the poor and under-privileged, had hoped to provide 'Health For All by the year 2000 AD', through the universal provision of comprehensive primary health care services. In retrospect, it is observed that the financial resources and public health administrative capacity which it was possible to marshal, was far short of that necessary to achieve such an ambitious and holistic goal.

Against this backdrop, it is felt that it would be appropriate to pitch NHP-2002 at a level consistent with our realistic expectations about financial resources, and about the likely increase in Public Health administrative capacity.

The main objective of the new National Health Policy, 2002 is to achieve an acceptable standard of good health amongst the general population of the country. The approach would be to increase access to the decentralized public health system by establishing the infrastructure in deficient areas, and by upgrading the infrastructure in the existing institutions. Primacy will be give to preventive and first line curative initiatives at the primary health level through increased sectoral share of the allocation. The increased outlay will be utilized for strengthening existing facilities and opening additional public health services outlets consistent with the norms for such facilities. The recommendation of this Policy will attempt to maximize the broad based availability of health services to the citizens of the country on the basis of realistic consideration of capacity. Under the broad macro policy prescriptions contained in this policy, State Governments will have the flexibility to design separate schemes, tailor made to the health needs of different socio-economic sections of society including the tribals.

The recommendations of NHP-2002 will, therefore, attempt to maximize the broad-based availability of health services to the citizenry of the country on the basis of realistic considerations of capacity. The changed circumstances relating to the health sector of the country since 1983 have generated a situation in which it is now necessary to review the field, and to formulate a new policy framework as the National Health Policy-2002. NHP-2002 will attempt to set out a new policy framework for the accelerated achievement of Public health goals in the socio-economic circumstances currently prevailing in the country. (National Health Policy, 2002).

ECOLOGY AND HEALTH

Himalaya means the "Adobe of snow" and the Great Poet Kalidas had depicted it as an ensouled Divinity and all along in India, it is supposed to be the eternal abode of gods, sages and ascetics. Thus we have innumerable myths, legends and mysteries associated with the Himalayas. Many sacred and holy centre with temples and shrines in them also attract thousand and thousand of people to these places from the hoary past, who visit and revisit these shrines and ashrams on certain specific auspicious dates, on which large fairs and pompous festivals are help at these places. Many sacred rivers have sprung out of the Himalayas—the water of which is considered as Amrita (nectre) by the Hindus.

The Himalayas form the greater mountains range of the world. Making the natural northern boundary of India, it displays a series of range from Peshawar in the West to NEFA in the East. It is 'Him-Alya'—Home of Snow and is full of peaks, valleys, passes, rivers, lakes and glaciers. The southernmost part—the Siwalik ranges—is the area of human habitation. (Chadha, 1989; Melkania and Melkania, 1988).

Peculiar to Himalayan ecology, altitude, gradient, snow, rainfall, wind and temperature are some noteworthy factors and their composite or individual effect create an impact—positive and negative, among the inhabitants. They feel constant insecure due to dangerous tracks, forest full of lurking beasts, varieties of natural upheavals, climatic extremes, crop failure, disease and death etc. different factors of their habitat. These elements of insecurity at every steps have made Himalayans to be God-Fearing. They are so scared of their ecological setting that even the sound produced by the breeze on the dry leaves in the lonely forests in enough to accept the presence of some of malevolent spirit. In this manner the Himalayans ought to perceive a number of deities and spirits, on behind each of the ecological happenings and place them in their pantheon. (Kapoor, 1993; Kapoor, 1994; Kapoor, 1996) Likewise liturgy, shrines, religious performance, specialists and ideology etc. various aspects of religion in Kinnaur region reflect ecological bearing.

Until the first half of the twentieth century, ecology was a buzz word; even in academic circles it had nothing to do with modern dimensions, problems and predicaments. Ecologist, Botanists, Zoologists and others involved in the study of biotic, abiotic and symbotic relationships concentrated on issues like population dynamics of insects and pests in a particular surrounding, habitat or, in terms of specific flora, fauna, humidity, temperature, etc. and visualized their impact on the food resources base and the inter- species struggle for survival. Furthermore, their explorations were of a piecemeal nature and were limited to the animal and plant world, these had hardly anything to do with the overall environmental perspective in terms of human existence.

Ecology as one of the sharper methodological tools in anthropology has contributed an additional dimension to its already existing theoretical baggage. Cultural ecologist speaks for an intimate relationship between a culture and its environment. Such an interaction is taken as natural. In cultural ecology, efforts are also directed to the study of adaptation. For Marshall D. Sahlins (1969), " The truism that culture are ways of life, taken in a new light, is the ground premise—cultures are human adaptations. Culture, as a design for society's continuity, stipulates its environment. By its mode of production, by the material requirements of its social structure, its standardized perceptions, a culture assigns relevance to particular external conditions. A culture moulds itself to significant external conditions to maximize the life chances. There is an interchange between culture and environment, perhaps continuous dialectic interchange. The environment–culture relation need not be one to one, by environment is never, thereby a powerless term. Societies are typically set in fields of cultural influence as well as fields of natural influence. They are subjected to both. They adapt to both". The cultural ecologist search for cultural variation in the adaptation of societies for cultural variation is explained in searching a cultural trait as adaptive in a specific physical and / or social environment. " For cultural ecologist the relationship between culture and environment is one of mutual interaction; a society's mode of ordering natural features, whether cognitively or through physical means, frequently, leads to changes in the milieu, thereby confronting the society with new challenges to adaptation to environment have consequences for other areas of life". (Honigmann, 1976).

According to Steward (1955), "The method of cultural ecology involved the analysis of (a) The interrelationship between environment and exploitation or productive technology (b) The interrelation between 'behaviour' patterns and exploitative technology and (c) The extent to which those behaviour patterns affect other sectors of culture". That the culture and environment are not the separate spheres and they are involved in dialectic interplay.... or what is called feedback or reciprocal causality", are considered as the most important contributions of Julian Steward.

Even today, with all the furore of an ecological catastrophe, 'ecocide' or 'ecological harakiri' and pollution holocaust, the strides that ecological studies have made are limited in many respects. The full implications of the word, especially in its socio-economic perspective, are still beyond the comprehension of the masses, whose involvement is necessary for any corrective effort to be successful. And the few who know the implications still continue to ignore the warning in the hope of nature's self-corrective propensity or form lack of knowledge of the upper limits of population or their distributive effects in space and times or perhaps, in order to sub serve their deluded self interest. (Gupta and Bansal, 1998).

Ecology requires to be understood in terms of the concept of eco-system as:

1. The eco-system has historical aspects; the present is related to the past, and the future to the present.

2. Man, the dominant organism on earth, is strongly dependent on its resource base and is inextricably tied to his environment.
3. Environment problems in reality are social and economic in nature. Any change in the socio-economic perspective, outlook and politics is bound to have an impact on the resource base, and in turn on the ecology. It would matter whether or not the socio-economic system is conditioned by the *laissez- faire* perspective and spurious or faulty criteria of economic development. The latter, in particular, lay emphasis on unidimensional and unidirectional change, the end product of which is the growth of the national or the per capita output. It assumes that development is costless process. Such facts as development in the forms of the concomitant environmental pollution and ecological destruction are ignored. Not only this. The optimum rate of development and growth is myopically defined in relation to the present only; and in a bid to resort to such an optimisation, the well being of the future generations is endangered through excessive exploitation of the exhaustible resources and the destruction of the supportive bio- system.

The fundamental base of the social – culture lies in man to man interaction; the material culture, on the other hand, is essentially an outcome of man's interaction with environment. Man to man interaction, in this case, occupies a secondary place. Man's capacity and equipment to exploit the environment determine the level of technology, a major ingredient of material culture. The nature of material objects and material way of living, possessed by a group of people are also, many times, considerably directed by the environments around. The arguments are true in case of physically and socially less exposed situations.

Exposed groups have, through culture contact, imbibed some material culture traits of the outsiders. Tribal, and for that matter even the non- tribal communities, inhabiting interior and outlying areas, have greater relationship of dependence on local environments. Their material ways of living represent their adaptability to native surroundings. Innumerable examples, in this context, can be cited from various parts of the country, and even globe, posing varied environments and variety of ethnic groups. Tools, implements, containers, food, dress and decoration, house type etc., establish relationship, many times, with the environments, around.

To investigate into the biological affects of high altitudes on man, the precise definition of adaptability needs further clarification. Harrison (1966) described it as the capacity of the individual to respond to changes in the environment in ways, which facilitate survival. The most pertinent aspects of adaptability are that it is concerned with the totality of the human responses at the level of both the individual and the population—to the totality of the environment.

Altitude being one of the significant determining factors the life and living of the people of these areas distinctly differ from the rest of Indian living in the plain. In almost all respects including roads communications and other infrastructural facilities these areas are deficient and are consequently lagging behind the average level of development of the rest of the country.

Recently there is being spurt of interest in the ecology of this country. This awakening of phenomenal interest is the work of social reformers, ecologists and environmentalist who feel deeply concerned at the unprecedented damage done to the environmental resources of the country. The perspective on health would be very close to what Meyar Fortes (1976) calls the state of "the individual in the day to day process of living". That is, it examines how the communities or the individuals meet the challenges of environment and seasons. It also attempts to understand the causation of illness as embodies in the ideas and beliefs of the people. The focus is more on understanding the parameters of health in relation to ecology. It however, does not treat health as dependent variable of ecology but as an independent variable and as a factor of production process including social relationship of production and bearing ecology has on health and how ethnic relations are mediated through ecology and health. (Behera, 1996).

Health Ecology is a broad, interdisciplinary approach to health and wellness. It is based on the dynamic interplay among diverse environmental, political, spiritual, philosophical, psychological, and personal factors and it incorporates an ecological perspective. Among, the tribal communities it is difficult to study the problems of health isolated from other aspects of their culture. Health is considered as the expression of harmony with the Universe (Basu 1985). A person in tribal society is not healthy unless his environment is healthy and that, which causes disease and illness, may also cause failure of crops, mishap in family and misfortune to village. (Behera, 1995).

ROLE OF NON GOVERNMENT ORGANISATIONS IN HEALTH SECTOR

The past five decades have witnessed the difficult problems encountered in providing health care services to poor people, the majority of whom live in more than half-a-million villages and in the proliferating slums of cities. Charitable and voluntary organizations since time immemorial have been contributing significantly towards the health care of the community. With the passage of time, Non-Governmental Organizations (NGOs) have equipped themselves adequately and come up enthusiastically in providing services like relief to the blind, the disabled and disadvantaged and helping the government in mother and child health care including family planning programmes. As a result, now all concerned have realized the potentiality of NGOs and their considerable merit compared to the public/private health sectors because of their staffs motivation, dedication and sympathy for the deprived sections of society and their personalized approach towards the solution of the problems.

As per the National Population Policy (NPP) 2000 and National Health Policy (NHP) 2002, there should be greater involvement' of NGOs in the implementation of different health and family welfare programmes in the country. In order to utilize the high motivational skills of NGOs on an increasing scale, it has even directed that the national health programmes should earmark a definite portion of the budget in respect of identified programme components, to be exclusively implemented through the NGOs. Consequently, the perception of strengthening linkages between the Government and NGOs has become very crucial for India's health sector reforms, especially for the decision-making, planning and management procedures.

It is very interesting and encouraging to note the evolution of NGOs' activities in the health sector in India. In recognition of the crucial role played by them, Government of India started granting financial aids to NGOs for various schemes. The laudable role played by the various national and regional level NGOs is briefly documented in the various documents, where special mention has been made of such organizations like All India Blind Relief Society, Family Planning Association of India (FPAI), Indian Medical Association, Indian Red Cross Society, National Society for the Prevention of Blindness, Sant Parmanand Blind Relief Mission, T.B. Association of India, Bombay Mothers and Children Welfare Society; to name a few. After the Alma Ata Declaration in 1978, greater role for the NGOs was seen to ensure Health for All through the primary health care approach. Their role was also considered as most crucial to translate the concept of 'People's Health in People's Hands' into action. (Kapoor, 2000).

Planning Commission and the Central Council of Health and Family Welfare emphasized the role of NGOs in health care delivery system. Subsequently a very large number of NGOs took the health and family welfare field and made notable contributions all over the country. There are a large number of voluntary organizations, like Voluntary Health Association of India, Hind Kusht Nivaran Sangh, Christian Medical Association of India, Paediatrics Association of India, Federation of Obstetrician and Gynaecologist of India, Lions Club, Rotary Club, Family Planning Foundation of India (which are mainly funding research and training), that have shown their sustainable contribution to the community. Some of these organizations are working for control of specific diseases such as Tuberculosis, Leprosy, Blindness, Family Planning, while many others are providing Maternal and Child Health and General Health Care services. Of late, the Government of India is encouraging NGOs to take up contraceptive and immunization services in a big way including training of nurses, ANMs, AWWs, Dais etc. The government has tried to streamline the system by identifying National NGOs, Mother NGOs and Field NGOs with a view to facilitate flow of funds and supervision of work output.

It is evident that NGOs would now have to play a key role for the 'health sector reform'. 'Reform' simply denotes "a sustained process of fundamental change in policy and institutional arrangements, guided by the government, designed to improve the functioning" and 'performance of the health sector and ultimately the health status of the population". Thus, the essence of reform is "sustained, purposeful change to improve the efficiency, equity and effectiveness of the health sector". As a result, for the health sector reform, the role of NGOs is directed towards the most vital components of health and family welfare programmes, i.e. population stabilization and primary health care.

However, a few studies show that for some health and family welfare activities, the NGOs given responsibilities of health centres could not perform well as compared to the government health centres. For example, there were less number of cases reporting for antenatal care and deliveries to NGO centres than government run centres. Similarly post natal care and child health care are less well organized in the NGO sector. This may be partly due to lack of understanding of their appropriate role regarding such services. (Kapoor and Singh, 1997; Khanna and Kapoor, 2004).

The functioning of the grassroot level workers in the NGO sector needs to be related to definitive health and family welfare action plans. For translation of 'medical' and 'health' approaches and knowledge into practical action, they require to know how to deal with problem situation by involving the use of simple and inexpensive interventions. All these can be readily implemented by NGO personnel who have undergone short course trainings on these issues. However, quality of training would be of crucial importance for the success of the approach that has been repeatedly advocated i.e., shift from 'medical' to 'health' and 'individual' to 'community' orientation of health and family welfare service provision.

ETHNIC GROUP: DEFINITION AND PERSPECTIVE

The word ethnic comes from the Greek ethnos, originally meaning nation. In its earliest English usage, about 1470 B.C., the word referred to culturally different heathen countries or nations (those not Christian or Jewish). Apparently, the first usage of ethnic group to denote national origin developed in the period of heavy immigration from southern and eastern European nations to the United States in the early twentieth century. Since the 1930s and

1940s, a number of prominent social scientists have suggested that the narrower the definition of ethnic group, the more in line with the original Greek meaning of nationality, makes the term more useful.

The term ethnic means of or pertaining to a group of people recognized as a class on the basis of certain distinctive characteristics such as religion, language, ancestry, culture or national origin. The ethnic group is a set of individuals whose identity as such is distinctive in terms of common cultural traditions or heritage.

The term ethnic group has been used by social scientists in two different senses, one narrow and one broad. Some definitions of the term are broad enough to include socially defined racial groups. For example, in broad definition, an ethnic group is a social group distinguished by race, religion, or national origin. Like the definition of racial group, this definition contains the notion of set-apartness. But here the distinctive characteristics can be physical or cultural, and language and religion are seen as critical markers or signs of ethnicity even where there is no physical distinctiveness. Further, ethnic group has been defined exclusively as:

A single family of social identities—a family which, in addition to races and ethnic groups, includes religions, language groups and all of which can be included in the most general term, ethnic groups, groups defined by descent, real or mythical, and sharing a common history and experience.

Other scholars prefer a narrower definition of ethnic group, one that omits groups defined substantially in terms of physical characteristics (those called racial groups) and is limited to groups distinguished primarily on the basis of cultural or national-origin characteristics. Cultural characteristics include language; national origin refers to the country (and national culture) from which the person or his or her ancestors came.

Definitions of racial group and ethnic group, which emphasize their social meaning and construction directly, reject the biological determinism that views such groups as self-evident with unchanging physical or intellectual characteristics. People themselves, both outside and inside racial and ethnic groups, determine when certain physical or cultural characteristics are important enough to single out a group for social purposes, whether for good or for ill. One should note that racial group and ethnic group are only two of the terms used in research on racial and ethnic relations. Among the other terms are majority group and minority group. A minority group is defined in terms of its subordinate position: A group of people who, because of their physical or cultural characteristics, are singled out from others in the society in which they live for differential and unequal treatment and who therefore regard themselves as objects of collective discrimination.

In this book, the broader term ethnic group is extensively used. An ethnic group is a group of people who identify with one another, or are so identified by others, on the basis of either presumed cultural or biological similarities, or both. Like race and nation. Initially, the notion of ethnicity developed in the context of European colonial expansion, when mercantilism and capitalism were promoting global movements of populations at the same time that state boundaries were being more clearly and rigidly defined. In the nineteenth century, modern states generally sought legitimacy through their claim to represent "nations."

Nation-states, however, invariably include indigenous populations that were excluded from the nation-building project, or recruit labourers from outside their borders; such people typically constitute ethnic groups.

Members of ethnic groups, consequently, often understand their own identity in terms of something outside of the history of the nation-state — either an alternate history, or in historical terms, or in terms of connection to another nation-state. Such identity often expresses itself through various "traditions" which, although often of recent invention, appeal to some notion of the past. Sometimes ethnic groups are subject to prejudicial attitudes and actions by the state or its constituents. (Gupt, 1997).

In the twentieth century, people began to argue that conflicts among ethnic groups or between members of an ethnic group and the state can and should be resolved in one of two ways. Some, scholars argued that the legitimacy of modern states must be based on a notion of political rights of autonomous individual subjects.

According to this view the state ought not to acknowledge ethnic, national or racial identity and should instead enforce political and legal equality of all individuals. Others argued that the notion of the autonomous individual is itself a cultural construct, and that it is neither possible nor right to treat people as autonomous individuals. According to this view, states must recognize ethnic identity and develop processes through which the particular needs of ethnic groups can be accommodated within the boundaries of the nation-state.

The biological race too, is usually taken into account, and some consider it as a basic platform on which cultural heritages can be preserved and continued via a genetical perpetuation. This concept is however adverse by those who believe that the ethnic group can be accessed also by spontaneous choice or—more commonly—marriage (allowed exogamy), and is not closed to new adherents.

Besides, some authors suggest that an element of voluntarity should always exist in the individuals, in the sense that the appurtenance to an ethnic group, if considered as a co division of culture, necessarily has to be backed by the individual's acceptation. Also, an ethnic group ordinarily expresses its social character by evidencing common behaviours in common forms, like ritual or conventional collective actions (even in familiar habits), which are easy to be eventually abandoned. In most systems, the eventual punishment for those individuals who wouldn't follow the "ethnic rules" is only the expulsion from the group so, it is said, any individual is usually free enough to abandon it at will. Opponents stress that the familiar education (quite generally recognized as a distinctive element) often makes it difficult to abandon the original heritage, especially if racial or geographical or linguistic elements force the group to develop an internal solidarity to protect members from external aggression and/or isolation (racism, xenophobe). Sometimes, it is said, there wouldn't really be an alternative.

At a political regard, an ethnic group is a social entity, which has to be respected as a minoritarian yet relevant component of the whole society. In case the ethnic group represents the vast majority (better if near to the totality) of the population within a given state territory, it becomes a nation, which expresses its common culture with self-governmental powers and with international acknowledgement. In this sense, an ethnic group is a minoritarian social group, which lives in a system by which it suffers for sometimes relevant differences (like in economy, for instance), in a relationship (when the ethnic group is recognized and accepted) made of tolerance rather than of co division. When the differences become hard to sustain, from both sides, the majority tends to absorb the ethnic group in itself, often by trying to annihilate its cultural heritage. On the other side, the ethnic group might develop, among its individuals, extremist positions which sometimes explode with violence or terrorism, which can be as dangerous as harder the state's opposition becomes.

The definition of an ethnic group is often referred to that of ethnicity, but the particular characters, which can identify a human group, as an ethnic entity is relevant for legal aspects too. The attempt to delete an ethnic group as such (genocide) is usually punished by most legal systems. In general, it could be resumed that an ethnic group is a community of human beings who:

- Live as a minority inside the state they are in, and claim for the respect for their social entity;
- Use a common language (which might be different from the official language of the state in which they live);
- Have common familiar and social habits and conventions, often the same religion;
- Come, by ancestry, from the same places, usually implying that at some moment a migration has happened and it is probable that a racial element can be in common too, among members.

Ancestry is important to the concept of ethnic group whether it is defined in a narrow or a broad sense. Perception of a common ancestry, real or mythical, has been part of outsiders' definitions and of ethnic groups' self-definitions. In addition to a sense of common ancestry, a consciousness of shared experiences and of shared cultural patterns is important in shaping a group's identity.

The aggregation of biological and socio-cultural characteristics constitutes an ethnic group. Within a category of ethnic group, caste, scheduled caste, tribe and communities are included. The anthropological studies of such ethnic groups as well as communities[1] have been reported in India. (Ghurey, 1924; Sinha 1960; Sinha 1965; Mandelbaum, 1972; Kapoor 1998; Kapoor 2000 (a & b); Kapoor 2003; Misra and Kapoor 2002-2003; Kumar and Kapoor 2004; Kapoor et al. 2004; Kapoor and Kapoor 2005; Misra and Kapoor 2006).

In this book, the usual meaning of ethnic group will be the narrower one—a group socially distinguished or set apart, by others or by itself, primarily on the basis of cultural characteristics. Such set-apart groups, usually develop a strong sense of a common cultural heritage and a common ancestry. In the context of Kinnaur, the population is divided into four major ethnic groups, i.e., the Kanet, the Koli, the Badhi, the Lohar. Mainly there are two divisions; one of Kanet and other includes rest of the ethnic groups. The Kanet are considered to be superior of high-class people, other groups are considered to be socially of low class and looked down by the Kanet. This division is based on endogenous groupism with respect to their own social entity; have different social habits and conventions; come, by ancestry, from the different ancestral group.

Scheduled Tribes

The term 'Scheduled Tribes' first appeared in the Constitution of India. Article 366 (25) defined scheduled tribes as "such tribes or tribal communities or parts of or groups within such tribes or tribal communities as are deemed under Article 342 to be Scheduled Tribes for the purposes of this constitution". According to Article 342 of the Constitution, the Scheduled Tribes are the tribes or tribal communities or part of or groups within these tribes and tribal communities which have been declared as such by the President through a public notification. As per the 1991 Census, the Scheduled Tribes account for 67.76 million representing 8.08 per cent of the country's population. Scheduled Tribes are spread across the country mainly in forest and hilly regions. The essential characteristics of these communities are:

- Primitive Traits
- Geographical isolation

1 Community is referred to a group of people who may have occupational, linguistic religious or regional characteristics.

- Distinct culture
- Shy of contact with community at large
- Economically backward

The word 'tribe' is generally used for a "Socially cohesive unit, associated with a territory, the members of which regard themselves as politically autonomous" (Mitchell, 1972). The ideal type of tribe can be characterized as a socially homogenous unit having its own dialect, political and cultural institutions and territory, which isolate it from the outside influences.

There is no agreed definition of a tribe. The word, as such, has dictionary meaning of "a race or family descended from the same ancestor: an aggregate of families forming a community usually under the government of a chief." The Hindi equivalent 'Adivasi' has a clearer connotation. It means those who are the earliest inhabitants of the country.

The objective of the public policy in India since independence has been the promotion of rapid and balance economic development with equity and justice. This is necessitated because India has large traditionally disadvantaged groups of scheduled caste (SCs) and scheduled tribes (STs) as well as large backward areas. The gap between traditionally disadvantaged groups of SCs/STs and rest of the population was very large. (Burman, 1997). Therefore, the need was to reduce this gap and enable them to contribute to the country's development. The Constitution of India provides for protective discrimination in favour of STs and SCs for removing age-old social disparities from which these classes suffer. This is the most important instrument for upliftment of SCs/STs in India. The Constitution of India provides in Articles 15(4) that the state can make" special provision for advancement of any socially and educationally backward classes of citizens or for the schedules caste and scheduled tribe". The Directive Principles of State Policy also lay down in Article 46 the need for "promotion of educational and economic interest of scheduled castes, scheduled tribes and other weaker sections". The Indian Constitution also makes special mention in Article 244 with regard to administration of Scheduled areas and tribal areas. In addition, there are Special Component Plan and Tribal Sum Plan for SCs and STs, respectively. The objective of these instruments has been to improve the standard of living of these traditionally disadvantaged groups of SCs/STs thus reducing the gap between them and the rest of the population.

The Constitution of India incorporates several special provisions for the promotion of educational and economic interest of Scheduled Tribes and their protection from social injustice and all forms of exploitation. These objectives are sought to be achieved through a strategy known as the Tribal Sub-Plan (TSP) strategy, which was adopted at the beginning of the Fifth Five Year Plan. The strategy seeks to ensure adequate flow of funds for tribal development form the State Plan allocations, schemes/programmes of Central Ministries/Departments, financial and Developmental Institutions. The cornerstone of this strategy has been to ensure earmarking of funds for TSP by States/UTs in proportion to the ST population in those State/UTs. Besides the efforts of the States/UTs and the Central Ministries/Departments to formulate and implement Tribal Sub-Plan for achieving socio-economic development of STs, the Ministry of Tribal Affairs is implementing several schemes and programmes for the benefits of STs.

There are now 194 Integrated Tribal Development Projects (ITDPs) in the country, where the ST population is more than 50 per cent of the total population of the blocks or groups of block. During the Sixth Plan, pockets outside ITDP areas, having a total population of 10,000 with at least 5,000 scheduled tribes were covered under the Tribal Sub-Plan under Modified Area Development Approach (MADA). So far 252 MADA pockets have been identified in the country. In addition, 79 clusters with a total population of 5,000 of which 50 per cent are schedule tribes have been identified.

In order to give more focused attention to the development of Scheduled Tribes, a separate Ministry, known as the Ministry of Tribal Affairs was constituted in October 1999. The new Ministry carved out of the Ministry of Social Justice and Empowerment, is the nodal Ministry for overall policy, planning and coordination of programmes and schemes for the development of Scheduled Tribes.

The mandate of the Ministry includes social security and social insurance with respect to the Scheduled Tribes, tribal welfare planning, project formulation research and training, promotion and development of voluntary efforts on tribal welfare and certain matters relating to administration of the Scheduled Areas. In regard to sectoral programmes and development of these communities, the policy, planning, monitoring, evaluation as also their coordination is the responsibility of the concerned central Ministries/ Departments, State Governments and UT Administrations. Each Central Ministry/Department will be the nodal Ministry of Department concerning its sector. Ministry of Tribal Affairs supports and supplements the efforts of State Governments/U.T. Administrations and the various Central Ministries/Departments for the holistic development of these communities.

Recently, Ministry of Tribal Affairs declared their National Policy on Scheduled Tribes. The National Policy on Scheduled Tribes seeks to promote the modern health care system and also a synthesis of the Indian systems of medicine like ayurveda and siddha with the tribal system (National Policy for Schedule Tribes, 2004). It also seeks to:

- Strengthen the allopathy system of medicine in tribal areas with the extension of the three-tier system of village health workers, auxiliary nurse mid-wife and primary health centres.
- Expand the number of hospitals in tune with tribal population.
- Validate identified tribal remedies (folk claims) used in different tribal areas.
- Encourage, document and patent tribals' traditional medicines.
- Promote cultivation of medicinal plants related to value addition strategies through imparting training to youth.
- Encourage qualified doctors from tribal communities to serve tribal areas.
- Promote the formation of a strong force of tribal village health guides through regular training-cum-orientation courses.
- Formulate area-specific strategies to improve access to and utilisation of health services.

- Strengthen research into diseases affecting tribals and initiate action programmes.
- Eradicate endemic diseases on a war footing.

(i) Tribal Development Programmes

With independence the hiatus between tribes and castes is sanctified by the Constitution. And over the years, due to political compulsions, internal ethnic divisions and phonetic variations, the number of scheduled tribes has increased from 212 in 1950 to more than 698 now and they constitute about 8.08 per cent of the Indian population. In short, the scheduled tribe in India is basically a politico-administrative category, and has little objective socio-cultural or economic connotation. But the anthropologist, who until recently were *de facto* tribal researchers, have uncritically accepted the administrative category. For all the tribal researchers of today, a tribe, which is included in the official list of scheduled tribes. A few years back, there were some attempts to define tribe, but the definitions were not only dissimilar but also even contradictory to one another. This could not have been otherwise for the definitions were made on the basis of one or two communities already notified as tribes. If the common indicators are only taken into consideration, then excluded a dozen or so small communities, the rest of those listed under the scheduled tribes category cannot be called tribes.

The notion communes with regards to the tribes have been that they are (a) isolated and closed entities due to their unique historical and cultural setting; (b) a historic and static societies, surviving as a 'cultural lag'; (c) homogenous and unstructured units of production and consumption; (d) being backward, are exploited by the non- tribals; and (e) because of, every tribe expenditure of the state extended towards the tribals. (Pathy, 1976).

The concept of the tribal development is based on twin approaches firstly, the protection of interest of the tribal people through legal-administrative support and secondly to implement development programme to promote their standard and level of living. Presently the main instrument to such development is tribal sub-plan launched in 1974-75 the sub–plans implemented in different States and Union Territories have benefited a large number of tribals. These sub- Plans are implemented through the State Plan, Special Central Assistance, Centrally Sponsored programmes and institutional finance. Under the revised 20 Point Programme, a high priority has been given to the development of the Scheduled Castes and Scheduled Tribes.

After the promulgation of the Constitution of India, which provides a number of the safeguards for the welfare and development of the Backward people, particularly the Scheduled Caste and Scheduled Tribes, the Government of India and also the State Government and Union Territory administrations started looking after the interest of the tribal people. A large number of schemes for the overall development of the Scheduled Castes and Scheduled Tribes of India have been launched and a good amount of money has been spent for the Successful implementation of these schemes. Not only their representation in the both Houses of Parliament and State Legislative Assemblies are guaranteed by the Constitutional provisions, but they also enjoy the facility of the reservation of the services in the government and semi government organisations. The Tribal advisory Council in State is meant for rendering advice on such matters pertaining to welfare and upliftment of these people. The Commissioner for the Scheduled Tribes and Scheduled Castes is to look after their interest.

In order to organise a special administration particularly to check alienation of lands for the primitive tribes that would be more or less the same throughout the tribal territories, the Scheduled Districts Act was passed 1974 by the Indian Legislature. This specified the tribal areas all over the country into Scheduled Tracts. In spite of these legislative safeguards, alienation of the tribal lands could not be altogether checked. Even after these enactments the aboriginal populations suffered enormous disabilities from the enforcement of the rather ill-conceived forest and game laws and interference with tribal festivals on moral ground. The result was, as pointed out by Elwin (1957), an extraordinary increase in incidences of drunkenness, which augmented, excise revenues and profit of the contractors and wine dealers at the expensed of the bewildered aboriginals.

The second phase in the administration of the tribal began with the passing of the Government of India Act of 1919 when the tribal areas were reconstituted under section 52-A of that Act into backward tracts after removing them either fully or partially from the jurisdiction of the administrative ministries. There were altogether 19 areas thus classified, of which four were entirely beyond the pale of the legislatures. With respect to the others, the legislatures were empowered to pass laws but in that case, too, the operation was dependent on the decision of the Government General.

In the Government of India Act of 1935, more stringent provision for special treatment of tribal areas was made by covering the backward tracts into Areas of Total Exclusion or Partial Exclusion. In the Sixth Schedule of the said Act, four areas under the first head and 11 areas under the second were at first proposed. It may be noted here that the Wazirs of Lahul and Spiti were exceptions in this respect as though declared as excluded areas, they were not considered as such as backward tracts. Their inclusion was, however, due solely to the political and administrative reasons. These enactments were primarily aimed at protecting the aboriginal people from exploitation and for allowing them freedom to pursuit their traditional method of living.

(ii) Tribal Sub Plan

The Constitution of India in its very preamble pledges to secure to all citizens justice, social, economic and political. To redeem this pledge, Article 38 enjoins the state to try to promote the welfare of the people by securing and protecting as effectively as it may, a social order in which justice, social, economic and political shall inform all the institutions of the national life. Article-46 of the Constitution further lays down that the state shall take special care of the educational and economic interests of weaker sections and in particular of Scheduled Castes and Scheduled Tribes and to protect them from social injustice and all forms of exploitation.

In pursuance of the above Constitutional directive, the first systematic effort for the development of the tribal area was initiated in 1955 in the form of Special Multipurpose Tribal Development Blocks. The modified version of this programme was taken up on a larger scale during the 2nd Five Year Plan when the Tribal Development Blocks were stared. This programme was further expended during the Third Five Year Plan and all areas with more than 2/3rd Tribal concentration were covered by the end of this period. Although it was agreed, in principal, to extend the programme to all those area, which had more than 50 per cent, tribal population, but it was not possible to do so during the Fourth Five Year Plan. The development effort in the then existing Tribal Development Blocks was consolidated by extending the period of their operation. The area coverage therefore, remained practically unchanged upto the end of the Fourth Five Year Plan. In the mean time this programme was reviewed on a number of occasions notably by a Study Team appointed by the Planning Commission under the chairmanship of Shri. Shilu Ao. It was also found that the development effort from the general sector programme was not adequate in these areas particularly, in the context of their comparatively lower economic base. It was, therefore, decided to evolve a new strategy for the development of the tribal areas from the beginning of the fifth Five Year Plan. The problem of tribal development was broadly classified into two categories:

1. Area being more than 50 per cent tribal concentration; and
2. Dispersed tribals.

In respect of the former, it was decided to accept an area development approach with focus on tribals. For dispersed tribals, family oriented programmes were decided for being taken up. Thus, the strategy of Tribal Sub Plan for areas of tribal concentration was evolved in the Fifth Five Year Plan beginning from 1974-75. The Tribal Sub-Plan strategy adopted comprised

1. Identification of development block in the state where tribal population was in majority and their constitution into Integrated Tribal Development Programme (ITDPs) with a view to adopting there in an integrated and project based approach for development;
2. Earmarking of funds for the Tribal Sub Plan and ensuring flow of funds from the State and Central Plan sectoral outlays, Special Central Assistance and from Financial Institution; and
3. Creation of appropriate administrative structure in tribal areas and adoption of appropriate personnel policy.

In the Sixth plan emphasis shifted from welfare family and beneficiary oriented development scheme within the general framework of socio-economic programme specifically directed at and designed for the benefit of the scheduled tribes. The ambit of tribal sub-plan was widened in the Sixth Plan and Modified Area Development Approach (MADA) was adopted for over smaller contiguous area of tribal concentration having 10,000 population and of which 50 per cent or more were tribals. During the Seventh Plan, the tribals for beneficiary oriented programmes. The planning strategy for tribal development during this Plan continued to be a mix of beneficiary oriented and infrastructure and human development programmes. Special emphasis was placed on minor irrigation, social and water conservation, cooperation, rural roads and land reforms in the infrastructure sector drinking water supply, general education, technical education and health in the social services sector and agriculture, horticulture, animal husbandry, dairy development fisheries, forest and small village and cottage industries in the production sector.

Eighth Plan: The Tribal Sub Plan Strategy adopted since fifth Five Year Plan has yielded results and has proved beneficial for the socio- economic development of tribals and tribal areas. The strategy has generally helped in focusing the attention of the planners and implements on special needs of the tribal society and tribal areas and adopting a more integrated approach for their development. As a result of this, there has been a spurt in investment in tribal areas. Besides this, a fundamental change in the process of formulating the Tribal Sub Plan on Maharashtra Model has been introduced at the end of the Eighth Five Year Plan. The past practice of planning process from top to bottom exercise has been reversed and the decentralised planning process with ITDP as basis has been started. In such an arrangement the Tribal Development Department is able to decide priorities of the schemes to be implemented in tribal areas and will also be in a position them to their felt needs.

Ninth Plan: The Ninth Five Year Plan has been commenced on schedule from 1st April, 1997 covering the years 1997-98 to 2000-2002 and the year 1999- 2000 shall be the 3rd year of the Ninth Five Year Plan. As per guidelines of the Planning Commission, Govt. of India and Union Welfare Ministry the, Ninth Five Year Plan lays greater emphasis on accelerated growth in employment, provision of basic minimum services to the people, eradication of poverty, provision of food security. In order to ensure that the quality of life of the people is enhanced, seven basic minimum services like drinking water, primary health services, primary education, housing in the shelter less, mid day meal in primary education, Rural roads connecting the villages and public distribution system have been given high priority as focussed in the Chief Ministers conference organised by the Planning Commission. The Ninth Plan has given high priority for the welfare of Scheduled Castes and Scheduled Tribes. Agriculture and rural Development has been given special attention. Besides this, major efforts shall be made to reduction in population growth through intensified family welfare programme so that the social and economic benefits are ensured to maximum. For achieving these goals an outlay of Rs. 496 crores in the Ninth Five Tribal Sub Plan 1997-2002 has been approved. (Tribal Sub Plan, 1999-2000; Tribal Sub Plan, 2000-2001).

(iii) Tribal Area

Tribal areas generally mean areas having preponderance of tribal population. However, the Constitution of India refers tribal areas within the States of Assam, Meghalaya, Tripura and Mizoram, as those areas specified in Parts I, II, IIA and III of the table appended to paragraph 20 of the Sixth Schedule. In other words, areas where provisions of Sixth Schedule are applicable are known as Tribal Areas. In relation to these areas Autonomous District Councils, each having not more than thirty members has been set up. These councils serve as an instrument of self-management and have powers of legislation and administration of justice apart from executive, developmental and financial responsibilities.

Scheduled Caste

According to Berreman (1972), 'Caste systems are living environments to those who comprise them. Yet there is a tendency among those who study and analyze them to idealize or intellectualize caste, and in the process to squeeze the life out of it. Caste is people, and especially people interacting in characteristic ways and thinking in characteristic ways. Thus, in addition to being a structure, a caste system is a pattern of human relationships and it is a state of mind.'

The Indian society is highly stratified and is divided into schedule castes, scheduled tribes etc. Hindu caste system is highly complex institutions, though social institutions resembling caste in one respect or another are not difficult to find elsewhere, but caste as we know it in India, is an exclusively Indian phenomenon. The word caste comes from the Portuguese word 'casta' signifying breed, race or kind. Risley (1915) defines it as ' a collection of families or groups of families bearing a common name; claiming a common descent from a mythical ancestor, human divine; professing to follow the same hereditary calling; and regarded by those who are competent to give an opinion as forming a single homogenous community' is generally associated with a specific occupation and that a caste is invariably endogamous, but is further divided as a rule, into a smaller of smaller circles each of which is endogamous. (called Jati). The internal exogamous division of the endogamous caste is Gotra. The main features of caste are hierarchy, endogamy and hyergamy (male of higher caste marrying a female of lower caste) occupational association; consciousness of caste membership and restriction on food, drink and smoking; distinction in dress and speech and confirmation to peculiar customs of particular caste ritual and other privileges and disability; caste organisation and caste mobility.

The essence of the caste is the arrangement of socio-economic hereditary group than hierarchy. Each caste is associated with hereditary occupation and had a limited monopoly over it. However generally speaking most practised agriculture along with their traditional occupation. Even agriculture as a single occupation cannot be associated with castes, as agriculture also means number of things; land ownership, tenancy and labour. Often the artisan and serving castes do not earn enough from traditional occupation, so they augment their income by working as casual labourers or tenants on land.

The President of India by a special order scheduled particular caste among Hindu and Sikhs in particular area for special treatment that also applies to tribe irrespective of their religious persuasions. The Scheduled Caste and Schedule Tribe have been specified by 15 Presidential Orders issued under the provision of Articles 341 and 342 of the constitution. They are listed in Schedule Castes and Schedule Tribes Orders (Amendment) Act 1976.

TRIBES OF HIMACHAL PRADESH

The Kinnaur and Lahaul-Spiti districts, in their entirely and Pangi and Bharmour (now tehsil Bharmour and Sub Tehsil Holi) Sub-Division of Chamba district constitute the tribal areas of Himachal Pradesh, fulfilling the minimum criterion of 50 per cent S. T population concentration in a C.D Block. These are situated in the north and north east of the Himachal Pradesh forming a contiguous belt in the far hinterland behind high mountain passes and are amongst the remotest and most inaccessible areas in the state with average altitude being 3281 metre above mean sea level. The distinguishing mark of the tribal areas in the state is that they cover very vast area but extremely small in population with the result that per unit cost of infrastructural activity is very exorbitant.

These areas have also been declared as Scheduled Areas under the Fifth Scheduled of the Constitution by the President of India as per the Scheduled Areas (Himachal Pradesh) Order , 1975 (CO 102) dated 31st November 1975. The five ITDPs are Kinnaur; Lahaul; Spiti; Pangi and Bharmour. Except Kinnaur, which is, spread over 3 C. D Blocks, rest of the ITDPs comprise 1 CD Block each.

Area and Population of Himachal Pradesh

Geographical area of tribal area continues to be in same in 1981 but the population has increase from 1,33,847 in 1981 to 1,51,433 in 1991. Sex ratio has improved from 976 in 1991 to 970 in 2001. In tribal areas, Density of Population per sq. km continues to remain constant at 6. The decimal growth rate has continued to be lower than that for the state as a whole.

The tribal area constitutes 42.49 per cent of the State's geographical area and represents 2.93 per cent of the total population of the State. Of the total population, 69 per cent are scheduled tribes, 18 per cent scheduled castes and rest are others. Males and females are in ratio of 54: 46. The entire population in the tribal belt continues to be rural. In Himachal Pradesh, two MADA pockets were identified in the Chamba district. Together, these covered an area of 891 sq. km. and population of 38,870; the proportion of ST population in these pockets was 55 per cent (1991). Coupled with tribal areas, 100 per cent of ST population was covered under sub plan treatment (1991). 13 per cent of the Scheduled Tribe Population in the State was dispersed outside the tribal area and tribal pockets (1991). The ultimate objective of Sub Plan strategy being 100 per cent coverage of ST population under its treatment, the Union Welfare Ministry came out with SCA supplementation for such dispersed tribes in 1986-87 and, in this way, 100 per cent ST Population in the State came under sub plan ambit.

Mechanism for Implementation of Tribal Sub Plan

The concept of incorporating Tribal Sub Plan in the Annual Plans of the state was first introduced by the Planning Commission, Govt. of India on the eve of 5th Five Year Plan. Comprehensive development of Tribal Areas focusing particularly on the Welfare of Individual tribal families was the main objective of the tribal sub plan. The procedure followed in the state till 1995-96 for the formulation of Tribal Sub Plan of the State was briefly that State Planning department used to allocate plan outlays to different administrative departments in consultation with Tribal Development Department. The department then used to curve out outlays for Tribal Sub Plan as per their own discretion and priorities. The concerned department were also deciding which of the scheme, programme and development works were to be taken up from the funds set aside from the Tribal Sub Plan. There was, therefore, a feeling that the tribal Sub Plan was merely agglomeration of the State Plan schemes taken up in the Tribal Area and emphasis given mainly arithmetical figures rather than the scheme really benefiting tribal families. There was no attempt to formulate the scheme in consultation with the Integrated Tribal Development Project level Officers. Consequently, the mechanism of re-appropriation and diversion of outlays had to contribute till the end of the financial year. Keeping in view the above lapses and shortcomings in the formulation of Tribal Sub Plan, the State Government decided to introduce fundamental change in the process of formulation Tribal Sub Plan at the directions of Ministry of Welfare, Govt. of India from 1996-97 onwards. Under this new system, the State Planning department shall communicate 9 per cent ceiling of the total State Plan outlays to the Tribal Development Department who shall in turn, allocate the divisible outlays to each of ITDP viz. Kinnaur, Lahaul, Spiti, Pangi and Bharmour. The indivisible outlays in the nature of grant-in-aid etc shall be conveyed to the Administrative department. Each ITDP has its own need and requirement as such each ITDP shall be free to determine its own priorities and allocates funds only to those schemes which are relevant to the area. Each

ITDP shall prepare its plan in consultation with the concerned Project Advisory Committee headed by the respective Honourable MLA of the Area.

The Tribal Sub Plan in respect of ITDP prepared in consultation with the Project Advisory Committee shall be compiled by the Tribal Development Department and dovetail the same in the main Tribal Sub Plan. The Heads of Departments simultaneously furnish draft plan both to the Planning Department as also to the Tribal Development Department. The Draft plan Planning Board and after their approval, the State Plan is submitted to the Planning Commission and the Union Welfare Ministry where the General Plan is discussed in the working groups set up by the Planning Commission, discussion on the Tribal Sub Plan takes place in the Union Welfare Ministry a day earlier to the one fixed for the general plan in the Planning Commission. The State Government has been leading with the Planning Commission that the State Sub Plan Should also be simultaneously considered by the working groups. The main responsibility of finalizing the Tribal Sub Plan within the ceiling so indicated now rests with Tribal Development Department and do not with the Administrative department as was the previous practice. The outlays for different schemes as now to be finally decided by the Tribal Development Department, keeping in view the actual benefits accruing the tribal people. The schemes are now being scrutinized very carefully by the Tribal Development Department.

For equitable flow of fund to the 5 ITDPs, the state has evolved an objective formula based on 40 per cent population, 20 per cent area and 40 per cent relative economic backwardness of each ITDP and, based on this formula, the share of each ITDP is as given in Table 1.1.

Table 1.1: Share of Each ITDP

ITDP	*Share of ITDP*
Kinnaur	30%
Lahaul–Spiti	34%
Lahaul and Spiti	18% and 16% respectively
Pangi and Bharmour	36%
Pangi and Bharmour	17% and 19% respectively

In Himachal Pradesh, 9 per cent of the State Plan flow has been earmarked to the Tribal Sub Plan. It may further be stated that such flow to the Tribal Sub Plan has always been above the per starting with 3.65 per cent in 1974-75, the level has reached 9 per cent for 1993-94.

Administrative Set-up for the Tribal Sub Plan

The tribal areas in the state are well-defined administrative units. The ITDP Kinnaur comprises the whole district; the ITDP Lahaul comprises Tehsil Lahaul and Sub Tehsil Udaipur and the rest of the Three ITDPs are by the name of Spiti, Pangi and Bharmour (Now Sub Tehsil Holi and Tehsil Bharmour) comprise Tehsils by the same name. In terms of Community Development (CD) blocks, ITDP Kinnaur consists of three Blocks namely Kalpa, Puh and Nichar and the rest of the four ITDPs are constituted of one CD Block each by the name of Lahaul, Spiti, Pangi and Bharmour respectively.

The pattern of administration in the tribal areas as also in the rest of the Pradesh had been identical except that in April 1986, the ITDP Pangi was put under the charge of an officer of the rank of Resident Commissioner and all officers there were merged with his office and he was the epitome of ultimate authority of the ITDP on the one side and the Government on the other. He exercised powers of Commissioner for revenue matters and that of District Magistrate in Pangi Sub- Division. ACRs initiated/reviewed by him were to be sent by him direct to the concerned administrative Secretary in respect of gazetted officers and he was the final accepting authority in relation to non-gazetted establishment. He was made disciplinary authority in respect of all non-gazetted establishments and also vested with powers of imposing minor penalty with respect to gazetted officers. He was also delegated full powers to accord administrative approval and financial sanction for all works. This experiment was a great success and there was demand from the public representatives for introduction of such type of administration in other ITDPs also and accordingly, single-line administration has now come to prevail in all the ITDPs alike with effect from 15th April 1988. This arrangement has cut down delays and improved the delivery system.

Project Advisory Committee have been constituted for each of the five Integrated Tribal Development Project headed by the Local

MLA and on which Members of Parliament representing the area, Chairman, Panchayat Samiti (s) Members of T.A.C., from the area, and Project Level Heads of Offices are represented. The District Commissioner/Assistant District Commissioner is the Vice-Chairman of the Committee. The Committee looks after formulation as well as implementation of the Sub-Plan at the Project Level and also the dispensation under nucleus budget funds.

HEALTH TRENDS: AN OVERVIEW

Anthropology and Demography

In anthropology, due emphasis is given to demographic analysis in health studies of any particular society. Demography is considered as the interdisciplinary science and policy oriented concerned with understanding and measuring population change in any society. Demography is a professional and global interdisciplinary science during 1930s. The practitioners of demography are originated from diverse field such as sociology, economic and statistic.

Anthropology is considered as the holistic study of man irrespective of time, space and culture and in demographic research, fertility, mortality and migration of human beings are studied in detail. Therefore, both demography and anthropology are concerned with the study of human beings in various dimensions. (Howell, 1986).

Various renewed anthropologists borrow the research interest from demographers in family formation and dissolution, health, migration and fertility and then phrase these issues as kinship, birth practices, creation of affinal ties etc. The Euro-American anthropologists practise parallel tracts along which anthropology and demography often run.

Not much of anthropological attention is generally given on fertility and mortality whereas the demographic attention always widely focuses on mortality and morbidity. Therefore, demographic-anthropology is considered as a more recent specialised subject. Before 1970s, without taking any training in formal demographic methods, some anthropologists studied the population change. These studied populations were usually too small for formal demographic analysis. (Caldwell et al, 1987).

Depending on second hand collected data, demographers reach for conclusion without personal involvement in data collection process whereas anthropologists focus on understanding the issues related to communities. Demographers are criticized by anthropologists for their rational economic models of demographic behaviour. In this sector, only demographers as well as anthropologists are not able to develop their adequate and accurate research activities, they co-operatively depend on each other and then demographic anthropology originated.

The basic difference between demographer and anthropologist is that former generally adopt large scale and quick data collection methods for national census survey whereas latter prefer small scale and intensive data collection field methods for the research of a particular area or community.

In anthropological sense, one type of qualitative data embody the own ideas of the local respondents and it is referred to as 'emic' data and another type of qualitative data is based on own constructs and categories of researchers and it is also referred to as 'etic' data (Pike, 1954). Demographers generally follow 'etic' method for data collection. The formal surveys generally collect the as usual data where qualitative methods specially collect the better quality data (Bleek, 1987).

Beside this, qualitative methods are allowed the researchers for probing the different nuances of meaning and solve all evident contradictions in responses. Qualitative data is sometimes used to generate the explanations for quantitative demographic differentials (Ford & Arcury, 1984). To get the reasonable explanations by retrospective as well as time-sequence data, anthropologists always try to combine the qualitative and quantitative methods. The background of the events like-illness, deaths, etc is probed by case studies and in-depth interview methods. In historical demography, the various scholars like Swedlund and Armelagos (1976), Hassan (1979) and Willigan and Lynch (1982) have published many literatures and these literatures are very helpful to anthropological researchers. Dyke and Morrill (1980) conducted a survey on the field of genealogical demography and wrote their report in a volume where both techniques and results were demonstrated skillfully. Bruch (1980) and Netting, et al., (1983) targeted on the household and Skinner (1977) followed at the implications for families. According to Wilson (1975), it has been proposed that biological theory in anthropology generates the positive focus upon demographic-anthropology.

Demographic anthropology is an important and substantial for understanding the theoretical and methodological orientation. The researchers of anthropology and demography emerge the new consideration of the related literature, apply the suitable method for data collection and then prepare the survey report. (Choudhary and Kumar, 1976; Gogoi, 1990).

(i) Fertility

Due to exponential growth of population, production, consumption, waste and land occupancy are increasing. The significant aspects of demographic studies are considered as fertility, mortality and migration of human beings. In demographic studies, fertility occupies the central position. It is considered that population growth is depended on fertility. The term 'fertility' is to understand the actual reproductive performance of women. In simple word 'fertility' is considered as the capacity to produce both living and death offspring.

Benjamin defined (1965), 'Fertility measures the rate at which a population adds to itself by births and is normally assessed by relating the number of births to the size of some section of population, such as the number of married couples of the numbers of women of child bearing age, i.e. an appropriate yard stick of potential fertility'. Thompson and Lewis (1965),'Fertility is generally used to indicate the actual reproductive performance of a woman or groups of women. The crude birth rate (number of births per 1000 population per year) is only one measure of fertility'.

There is always debut about the difference between fertility and fecundity. Many scholar considered that there is no difference between fertility and fecundity, it may be used as synonymous terms. The word 'fecundity' deals with capacity of a man and woman or a couple to take part in reproduction and by which produce living offspring. According to Thompson and Lewis (1965), 'Fecundity is a biological potential—the physiological capacity to participate in reproduction. The absence of this potential is known as fecundity or sterility'.

The fertility is basically a biological factor, which is governed by social rules and regulations. The universal psychological factors are directly linked with fertility. The fertility rate is affected and influenced by various biological (like diseases, food habits etc) and social factors (early age at marriage or child marriage, polygamy etc). The various contraceptives like—oral pills, loop, condom and other unnatural measures like abortion infanticide, etc. directly regulate the child birth.

Fertility indicates the number of children, which were produced by the women. Various attempts have been made to study the association of culture with fertility behaviour among certain tribes through social and demographic variables. It has been observed that among tribals, every aspect of life from birth to death is being influenced by the prevalence of customs, beliefs and notions. Though fertility is a biological phenomenon there are a number of other factors influencing the levels and differentials of fertility among tribals. Demographers usually measure the fertility differentials by taking into account women's income, occupation, education, family type, age at menarche, age at marriage, etc. (Dandekar & Dandekar, 1953; Sen, 1956; Dandekar, 1959; Roy Burman, 1961; Banerjee and Mukherjee, 1961; Nag, 1962; Hussain 1970a; Hussain 1970b; Das, 1973; Vidyarthi & Rai, 1977; Sahu, 1983; Kapoor and Kapoor, 1985; Banerjee, 1988; Basu & Kshatriya, 1989; Choudary, et al., 1994; Adak and Chaudhari, 1996; Kapoor and Kshatriya 2000; Kapoor 2001; Biswas et al, 2001; Kapoor et al. 2002; Misra and Kapoor 2005).

Bhasin and Bhasin (1990) studied communities of Sikkim and Himachal Pradesh. It has been found that biological, sociological and economic factors always influence the fertility. The social and economical well status of the people is well established by producing the large number of children. The preference for male child leads to high fertility. After bearing sons, they make themselves as socially eligible for inheriting the family-properties.

Kshatriya et al (1993) reported that total fertility rate of Bison Horn Madia tribe of Madhya Pradesh has been estimated as 5.46. The total fertility rate of Indian National population has been calculated as 4.55, which is low as compare to total fertility rate of Bison Horn Madia. Socio-economics, cultural, environmental, sanitation, awareness of health and health care practices, etc. are considered as the important factors, which influence their high fertility. Chaudhary et al.(1994) defines that different fertility rate have been measured of Mirdha (of Orissa) mother whose age group was varied form 15 to 50 years. The percentage of female live birth (49.49%) is high as compared with male live birth. Due to worst environmental condition, poverty, illiteracy, unawareness of family planning, etc. have been specified as responsible factors, which influence high fertility among Mirdha tribe.

Pandey (1994) studied tribal and non tribal groups in Jabalpur district of Madhya Pradesh. According to him, the average age at marriage of tribal males and females have been accounted as 18.6 years and 15.7 years respectively. The average age at marriage of non-tribal males and females is 18.9 years and 16.7 years respectively. The fertility rate is affected by age at marriage, duration of fertile age, potentiality of age specific fertility rate of female, etc.

Knowledge, attitudes, practice of birth control devices, etc. are indirectly related with the education and occupation of mothers (Caldwell, 1979). For taking up a final decision, it is very much required to develop self confidence and acquire knowledge. Occupation of mothers not only provide exposure but also communicate an interaction with the outsiders for acquiring knowledge and growing self confidence (Buvinic et al,1972).

Khongsdier and Ghosh (1995) studied that fertility rate of War Khasi is quite low as compared with the other tribal population of India. Most important causes of their low fertility are that 43.37 per cent unmarried women among all have been followed whose age is varied from 15 to 19 years (fertile age). Moreover, the concentration of those unmarried females has not been found among children group. Most of them have been included in fertile age group but in unmarried condition. It has been estimated that total fertility rate is more or less equal with the number of member of completed family. So, their study indicates that it is a steady fertility rate but sometimes with the changing period of time, only its trends have the chances to be declined.

Sengupta and Chakravarty (1995) conducted a study among Ahom tribe of Assam and observed that fertility is influenced by various socio-cultural factors which determine the type of family. Low fertility has been registered in joint family whereas comparatively high fertility has been registered in nuclear family. Similarly, fertility has been estimated as higher of rural women than urban women.

Pandey and Tiwari (1996) observed that average number of child born per mother is 1.9 among the Hill Korwa tribe of Madhya Pradesh. Due to their extreme poverty, low number of population involved in agriculture, low literacy rate (3%), even low age at marriage of female, etc., their large number of child birth is performed. But after birth, very less number of children is survived among them. Both fertility and child mortality of this tribe is defined as very high and as result, the number of survival off springs are very low (1.9 per mother). Guha (1997) reported that fertility rate per mother was measured (1.77) as very low among Bhotia women of Uttaranchal. Further it was found that in the lowest age group, sex ratio is very low whereas in the highest age group, sex ratio is followed as high. This is meant that from lower to higher age groups, the female death is higher among them. Higher fertility is affected due to higher female death specially who attended in reproductive age groups.

Pandey and Goel (1999) mentioned that fertility rate (39.9) of Abjhumaria tribe of Madhya Pradesh is very high as compared with others groups. This study specifies that maximum fertile women are included within 20 to 24 years age groups and these women perform large number of childbirth. Among them, family planning methods is not adopted by anybody due to extreme poverty and illiteracy.

According to the study of Yadav et al. (2001), crude birth rate, general growth rate, child-women ratio, total fertility rate and average number of children born of Lohar. Gadiyas of Madhya Pradesh have been found to be 128.9, 76.17, 1622.22, 466.66 and 4.13 respectively. This study explained that due to very low acceptance of family planning devices, their fertility has been recorded as very high. Besides this, due to lack of proper medical facility and family welfare services, their child mortality is high and in these circumstances, the couples are in suspense to the child death.

Biswas and Kapoor (2003) reported that in Saharia tribe of Madhya Pradesh, illiteracy, low socio-economic status, early age at marriage, poor health status of women, unavailability of proper health care facility, still birth, infant and child mortality, non-acceptance of family planning devices etc are the attributed reasons which influence their high child women ratio, crude birth rate, general fertility rate, age specific fertility rate, total fertility rate, gross reproduction rate women in wed lock.

(ii) Mortality

The mortality is another important demographic indicator. The United Nations (UN) and World Health Organization (WHO) (1970) define death as 'the permanent disappearance of all evidence of life at any time after birth has taken place (post natal) cessation of vital functions without capacity of resuscitation'. Mortality indicates permanent extinction of all signs of life from living body and it is occurred at any time after birth.

According to Hauser and Duncan (1959), 'Death prior to complete expulsion or extraction from its mother of a product of conception, irrespective of the duration of pregnancy, the death is indicated by the fact that after such separation the foetus does not breathe or show any other evidence of life, such as breathing of heat, pulsation of umbilical cord or definite movement of voluntary muscles'. According to the World Health Organization, most of the deaths are occurred by the effect of various diseases. It is observed that there is non-formality in death rate in all countries. Even death rate is varied from one section to another section of a society and even from one state to another state of country.

It is discriminated that mortality may differentiate between time and war, different social classes and within a nation, developed and less developed countries, etc. Due to loneliness and depression of widow, divorce and unmarried life, death is more proven as compared with those who lead a happy, affectionate and married (familial) life. Mortality is also influenced by different environmental effects (Heer & Smith, 1968).

According to Thompson and Lewis (1965), 'Marriage is selective as regards both physical constitution and social adaptability'. Due to physical as well as mental problem, some persons are restricted not to marry and in that condition, such persons always sick death for die in a shorter period of life. In other side, those married couples are understandingly passing the regularity of living; their death rate is very low. Occupation and mortality also have close linkage with each other. It is observed that the underground mining work and marshy and mosquitoes affected working places enforces to high death rate.

In India, infant mortality is considered as serious problem. In infant condition and also above the age of 55 years, generally the death rate is very high. It is observed that the maximum attention is always provided to infant whereas those persons who attained 55 years age and above received less attention. The reason for this can be attributed to understanding related to infant mortality as significantly thought as a whole whereas old age mortality is considered as natural death. The various causes like—lack of medical facility, inexperienced nurse, congenital, mal transformation, immature birth, temperature variation, illiteracy of parents regarding negligence of child care, pollution, poverty, effecting nutritional food, unexpected sex of the child, ill health of mother, shorter birth interval between two children, lack of breast milk, non-availability of nutritional food during pregnancy, system of early age at marriage, etc. influence the mortality rate (Thompson & Lewis, 1965). It has been reported that the tribal mortality is quite upsetting, especially when seen in the context of Indian national population.

The poor food habits contribute the malnutrition especially among children and pregnant mothers which leads to increased susceptibility to morbid conditions. In India, maternal care is largely neglected and the expectant mother to a great extent is not inoculated against tetanus. The consumption of iron, calcium and vitamins during pregnancy is poor. From the inception of pregnancy to its termination, no specific nutritious diet is consumed by a woman. Vaccination and immunization of infant and children are inadequate. In addition, extremes of magico- religious beliefs and taboos tend to aggravate the problems which influence high mortality (Dandekar & Dandekar, 1953; Dandekar, 1959; Ghosh, 1970; Sharma, 1978; Sirajuddin & Basu, 1984; Murthy, 1987; Chetlapalli et al, 1991; Khongsdier, 1992; Khongsdier, 1995; Adak, 1994).

Kshatriya et al. (1993) reported 1.011 child mortality rate, as highest and 0.876 child mortality rate, as lowest have been measured of 35 to 39 years and from 15 to 19 years respectively aged women of Bison Horn Madia tribe of Baster district. The tribal are fully depended on one crop cultivation and more or less forest products. The poor quality of diet leads to malnutrition, which creates diseases specially among children and pregnant tribal women. Due to non immunization of both ante-natal and infants, instantiation conditions of household compound due to cattle domestication at the same place etc, the tribal are affected by malaria, unspecified fever, diarrhoea, respiratory problem, infections, skin diseases, etc. which lead death.

Adak (1994) reported that highest and lowest number of infants were died by water borne (diarrhoea, dysentery, jaundice, etc.) and unknown diseases respectively among Khasi tribe of Meghalaya. The infant mortality is high due to socio-cultural factor such as illiterate mother, early marriage of mother, involved themselves in miscellaneous works, delivery attended by neighbours or relatives, less lactation, non-immunization and lack of medical check-up during pregnancy etc.

Khongsdier (1995) reported infant mortality of Christian Khasi and non-Christian Khasi were 6.89 and 8.55 respectively among War Khasi of Meghalaya. The difference is attributed to the fact that Christian Kashi are more educated and financially sound. Due to this, they have knowledge about importance of better medical care and utilise better medical facilities to protect their offspring loss. On the other hand, non-Christian Khasi due to illiteracy, poverty, non-availability of proper medical facility, etc have high infant mortality rate.

Sabat and Dash (1996) reported that crude death rate, infant mortality rate are 23.41 and 153.8 respectively among Kandh tribe of Orissa due to acute scarcity of potable water, insanitation, very poor access of modem health care facilities, unfavourable environment, etc.

Sharma and Sharma (1999) reported that crude death rate, infant mortality rate and highest age specific deaths rate (age above 50 years) is measured as 40.08, 242 and 107.14 among Gond tribe of Madhya Pradesh. The high age specific death rate among people with more than 50 years age is due to cough, senility, digestive disorder, etc. Infant deaths are occurred by tetanus, whopping-cough, diphtheria, small pox, cholera, polio, etc.

Pandey and Goel (1999) reported that infant mortality rate is 140 among Abujhmaria of Madhya Pradesh due to illiteracy, poor medical facility and poor financial conditions.

Pandey et al. (2001) reported infant mortality rate as 126.5 among Kamar tribe of Madhya Pradesh. The infant mortality of Kamar is not only depended on biological factors but also it varies in accordance with the health status of mothers, types of household occupation, income, etc. Moreover, their study defines that age at marriage of mothers, awareness about maternal and child health care, etc. are so much indebted for influencing the infant mortality.

In their paper, Biswas and Kapoor (2003) highlighted the reasons for high mortality rate in Saharia tribe of Madhya Pradesh. It has been observed that some of the mortality indicates have been studied in various tribes which show variablility with reference to health facilities, ignorance and awareness (Biswas and Kapoor 2003 & 2004; Basu et al. 2004; Kshatriya and Kapoor 2005; Misra and Kapoor 2005).

(iii) Occupation

It has been observed that there is variation in economic condition of various ethnic groups. Majumder and Madan (1986) have classified Indian tribes on the basis of their economy i.e followed as food gathering economy, agriculture, shifting cultivation, crafts, pastoralism, labouring, etc. Further Vidyarthy and Rai (1977) also classified the tribal economy as forest hunting type, hill cultivation, plain agriculture type, simple artisan type, pastoral type, folk artisan type, service and trade, etc.

Bhowmick (1963) studied Lodha tribe of West Bengal and specified their various professions. Bhanu and Saheb (1983) conducted a study among lrulas of Tamil Nadu and found out that traditional occupation of the tribe was hunting and trappering of porcupines, rats and snakes. Som (1993) conducted a study among various communities of Orissa. It was reported that approximate 75 per cent population of the village (only males) practised various occupations and among them, 41 per cent, 17 per cent, 6 per cent and 0.50 per cent were engaged in agriculture, labour, business and service respectively as primary sources of income. Apart from this, 15 per cent and 18 per cent practised agriculture and labours respectively as secondary occupation.

Kapoor (1996) and Patra (2001) reported that agriculture, animal husbandry, tailoring are primary economic occupations of Raji tribe of Kumaun Himalaya. Simultaneously traditional hunting and gathering is also practised.

Pandey and Tiwari (1996) studied various aspects of life of Hill Korwa tribe of Madhya Pradesh. It was reported that most of the Hill Korwa tribal groups earned through daily labour. 32 per cent, 66 per cent and 0.50 per cent of Hill Korwas practised agriculture labour and service respectively. Sharma (1998) observed that Bharia tribe of Patalkot mainly practised agriculture, farm-labours and pastoral. Basketry, leaf plates-making, broom making, rope-making activities for their subsistence.

(iv) Household Income

Household income indicates the total income of the earning member of a particular family. The household income determines the socio-economic status of the household. It has been reported that fertility is affected by socio-economic factors whereas socio-economics factors are generally depended on quantity of income and sources of income. Tiwari (1984) conducted a study among Hill Korwa tribe of Madhya Pradesh and reported that 55 per cent tribes were landless. His study comprises that among them, land alienation is considered as the most acute problems. Fernandes (1993) studied the communal right on shifting cultivated land has been specified among Kond tribe of Orissa and Gond tribe of Andhra Pradesh. Sastry (1993) conducted a study among Chenchu tribal community of Andhra Pradesh and reported that the most of Chenchus were landless.

Sabat and Dash (1996) studied Khond tribe of Orissa and estimated their monthly earned amount of various sources. It has been estimated that below Rs. 500/-, Rs. 501 to 750/-, Rs. 751 to 1000/-, Rs. 1001 to 1500/-, Rs. 1501 to 2000/-, Rs. 2001 to 2500/- and Rs. 3000/- and above per month were earned by 36 per cent, 41 per cent, 13 per cent, 1 per cent, 4 per cent, 1 per cent and 1 per cent of families respectively.

Basu (1999) conducted her study among Santal and Lodha tribes of West Bengal and calculated their monthly income. According to her study, 80.00 per cent, 16.40 per cent and 3.60 per cent, tribes have been categorised whose income was varied form 500 to 900/- rupees, 1000 to 1999/- rupees and above 2000/- rupees respectively. Due to lack of available works opportunity, sometimes they are not offered to practise daily wages and in this situation, they earn very little amount. This earned amount is not sufficient for purchasing their livelihood.

Biswas (2002) reported that most of Saharia families of Madhya Pradesh earn very little amount and it is very difficult to maintain their family budgets. So for their maintaining family budgets either they have to sell any properties otherwise borrow from the money-lenders or contractors with high interest.

(v) Education

Education plays an important role for the development society. Education is considered as a development indicator which also effects the demographic behaviours of any particular population group. In earlier studies, it was reported that non-availability of schools and lack of basic infrastructure is common phenomenon in tribal areas. Tiwari (1984) carried out study among Hill Korwa tribe of Madhya Pradesh and reported that scholarship was considered as the incentive attraction for increasing the number of enrolment of the children of Hill Korwas specially in various pre-primary and primary schools. Patnaik (1989) studied among Saora tribe of Orissa and observed that tribal prefer to learn through formal education. Singh (1995) conducted a study among Taro tribe of Manipur and focussed on their education. It was reported that male and female literacy rate accounts as 26 per cent and 22 per cent respectively.

According to Biswas and Kapoor (2003). The Saharia educational status is no doubt deplorably low. Even their female literacy rate is very negligible and inadequate as compared to male. Saharia are very poor. Unemployment, inequality, regional socio-economic imbalances, etc. are still existed among them. The economic growth and development, social development and development of human resources are greatly lagged behind. For overall development, education takes place a most influential role in determining their social and demographic behaviour. Educational characteristic of Saharia is deemed essential for the study of their population dynamics. It is believed to affect reproductive behaviour, use of contraceptives, health of the children, proper hygienic practices and status of women. Their educational characteristics also provide insight into the 'output' of educational system prevailing through the area. Since educational upliftment is a long term activity and its planning should have a long term perspective. Hence there is required to evolve economic and alternative long educational planning strategies which will not only enhance their household income but also provide strategies which will not only enhance their household income but also provide them educational facility. It may be expected that financial betterment and educational improvement under long term policy will motivate them in understanding their own health hygiene and birth control behaviour and then they will acquaint themselves as socio-economically as well as demographically advanced Saharia.

(vi) Age at Marriage

In various anthropological studies and writing, early age at marriage is reported among the tribal communities. Further the age at marriage is linked with the demography of any population group. The lower fertility is presented by those women who married late (Majumdar, 1962; Agarwala, 1966; Nag, 1974; Khan, 1976). Mutharayappa (1993) organised a comparative study among Jenu Kuraba and Kadu Kuruba tribe of Karnataka. According to this study, it has been observed that age at marriage of males and females varied from 17-21 years and 12-16 years respectively. The average age at marriage of females among Jenue Kuruba and Kadu Kuruba is reported to be 14.1 and 14.2 years respectively. Similarly, the average age at marriage of males was also estimated as 18.6 years and 19.6 years among Jenu Kuruba and Kadu Kuruba respectively.

Roy (1995) studied Kokna tribe of Maharashtra and reported that very early age at marriage was practised among the tribal. Earlier, the preferred age of marriage is in between 10-12 years and 8-12 years for boys and girls respectively. Samal et al (1996) reported that among Jaunsari tribe of Central Himalaya, the average age of marriage of tribe girls is 13.86 years. According to him, among Jaunsari tribe, 4 years is lowest and 30 years is highest age of marriage. The reasons for early age at marriage are associated with their traditional socio-cultural barriers as late age at marriage is concerned with evil.

Chachra and Bhasin (1998) studied the age at marriage of Bhotia tribal in Uttaranchal (erstwhile Uttar Pradesh). It recorded that estimated 19.20 years and 23.20 years is average age at marriage of females and males respectively. They also expressed that the average age at marriage of Lepcha females and males of this state were also measured as 18.08 years and 22.90 years respectively. According to the study of Singh (1995), it has been observed that that among Khond tribe of Orissa, the age at marriage of girls was varied from 14 to 16 years and age at marriage of boys was varied from 17 to 19 years.

(vii) Family Structure

The family structure/size indicate the number of total family members who inhabit in a common residence and share the common food from the same kitchen. Size of the family is varied from one community to another.

Banerjee and Bhatia (1988) studied among Gond tribe of Madhaya Pradesh and observed that family structure of Gond tribe has been categorised as small family (below the number of 4 members), medium family (From 5 to 8 members) and large family (9 members and above). In majority of cases, their family size is depended on individual desire and idealness.

The study conducted by Vithal (1992) on Chenchu tribe in Andhra Pradesh reflects that family size of maximum respondents of Chenchu tribe is varied from lower (2-5) to medium (5-7) and Joint families are rarely found.

Sabat and Dash (1996) studied Kandh tribe of Orissa and observed that four types of families structure i.e. is prevailed among the tribal group i.e Small family below 3 members; medium family between 4 to 6 members; large 7 to 9 members and very large above 10. The average size of family is estimated to be 4.5 per household.

In study conducted by Sharma (1998) on Bharia tribal groups of Madhya Pradesh highlights that both joint families and nuclear families are performed among the tribal group. The average family size varied from 4 to 13 members. The family structure and work spirit are correlated with each other.

(viii) Age at Menarche

Menarche is considered as the first regular flow of blood from wound of girl and it is continued periodically upto the specific age. Age at menarche means the age at first menstruation, which is started from a particular age of girls. Various researchers have reported that the average age at menarche of Indian girls varies from 12 to 14 years. Onset of menarche is not only governed by the physiological changes in the body but various socio-economic factors also influence it.

A few comparative studies indicate that with the progress of civilization the trend of menarcheal age is becoming lower. It has been reported that characteristics like caste; sub-caste; nature of the area studies, i.e. rural, semi-rural and urban; socio-economic status (like family income, occupation of the head of the family and education family size), birth rank and climatic conditions of the geographical area also influence the menarcheal age.

Satwanti et al. (1983) studied the three groups of Indian females: Sportswomen, women from high altitude and women from plains for age at menarche. The women from plains has earliest menarche (x= 13.3 years) as compared to sportswomen (x=15.7 years) and women from high altitude (x= 16.5 years). The difference between ages at menarche in all the three groups was significant. In this paper, the mean age at menarche for these three groups has been compared with that of the female belonging to different population groups.

Bhasin and Bhasin (1990) studied Lepcha and Bhutia communities of Sikkim and estimated age at menarche as 15.55 years and 15.62 years respectively. It has also been reported that heavy body weight indicates the acceptance of better nutritional food items and bearing of continuous good health and in this condition, girls attain menarche at very early age. Their studies confirm that the age at menarche is depended on climatical conditions of particular area. The moral and psychic environment in which and individual lives may also have some significant effect on the mean age of menarche (Shah, 1958; Banerjee & Mukharjee, 1961; Biswas, 1967; Chattopahdyay & Khullar, 1969; Mukherjee, 1972; Ghosh & Kumari, 1973; Mukherjee, 1974; Beall, 1983;

Malik and Hauspie, 1986; Singhal et al, 1994; Baura, 1996). Balgir (1994) measured that mean age at menarche of eight exogamous tribal groups of Assam varied from 12.22 to 13.80 years. According to him, age at menarche is not only influenced by genetic factor but also affected by various occupation, educational level, family size, food habit, living condition, birth rank, climate, etc.

The age at menarche is influenced by environmental factors and biological factors Singhal etal. (1994) conducted their studies on non-vegetarian and vegetarian female among and inferred that 14.08 and 13.67 years were the average age at menarche of vegetarian females and no-vegetarian females respectively. So, it has been explained that age at menarche is also depended on various food items. Choudhary et al. (1994) conducted their study in Mirdha tribe of Orissa and inferred that 1.63 per cent, 66.30 per cent, 31.52 per cent and 0.55 per cent women have been measured who menstruated at the age of 12 years, 13 years, 14 years and 15 years respectively. Through the comparative analysis with the menarcheal age of women of other castes, it has been specified that mean menarcheal age of tribal women is higher than the mean menarcheal age of women of Higher caste. Due to close affinity with higher caste groups, the mean menarcheal age of Mirdha women has been accounted as high as compared with other tribal women.

Kapoor and Kapoor (1986) studied the age at menarche and menopause of the three groups of Bhotia female living at high altitude, Himalayan region-Uttaranchal (erstwhile Uttar Pradesh). It has been reported that Johari Bhotias women has earliest menarche (X= 15.4 years) as compared to Rang Bhotias, settled (X= 15.6) and Rang Bhotias, migratory (X= 16.0). The differences between all these three groups for age at menarche were significant. A trend towards increase in age at menarche with an increase in altitude has been observed but the total fertility period in the three groups remained similar as early menarche has been found to be associated with early onset of menopause and late menarche with late menopause.

Sharma and Chowdhury (1995) measured mean menarcheal age of Gond women of Maharastra as 13.69 years. Menarcheal age and fertility are related with parabolic form. Sengupta and Rajkhowa (1996) conducted a study among Ahom tribe of Assam and measured 12.51 years as the mean menarcheal age. According to them, age at menarche is depended on nutrition, socio-economic status environment, altitude, climate, heredity, food habits, rural-urban residence, family size, etc. (Kalita and Sengupta, 1997).

(ix) Age at Menopause

Menopause is regarded as the symbolic end of womanhood. The menstruation and Ovulation stopped and ceased the reproductive function of women. Age at menopause is an important biological phenomenon plays a great role in reproductive performance of women. Menopause marks the end point of a transition of up to several years during which female fecundity gradually approaches zero.

According to various researchers mean age at menopause of the Indian populations is between 40 and 50 years age. However, the mean age at menopause ranges from 44 to 50 years. The age at menopause is influenced by nutrition, physical environmental factors, etc. (Chatterjee, 1994).

Bhasin and Bhasin (1990) reported that the menopausal age of Lepcha (total), Bhutia (total) and Buddhists of Sikkim were 47.00 years, 42.46 years and 44.33 years respectively. Further, they have measured that menopausal age of Gaddis (total) of Himachal Pradesh was 44.99 years comparatively higher than Lepcha, Bhutia and Buddhists. It was inferred that age at menopause is not only influenced by hot climatic condition but also it is varied in accordance with the body weight of women.

It has been reported that among tribals, age at menopause may play an important role, specially in the absence of any voluntary (fertility control methods) or involuntary (sterility, illness) regulation of fertility (Agarwala, 1966; Gogoi, 1972; Gunasundaramma, 1980; Beall, 1983; Sidhu and Sidhu, 1985; Mastana, 1987; Biswas and Kapoor, 2005).

Majumdar (2001) defined that the age at menopause varied between 40 to 48 years age of women of Mahatos of West Bengal. Due to various factors like environmental, malnutrition, etc., the Mahato women attain low age at menopause which controls their fertility. Sengupta and Rajkhowa (1996) estimated 46.32 years as the mean age at menopause of women of Ahom tribe of Assam. According to them, the age at menopause of women is depended on nutrition, socio-economics status, environment, altitude, climate, heredity, food habit, rural-urban residence, family size, Kalita and Sengupta (1997), the mean age at menopause of women of Sonowal tribe of Assam is 47.22 years. It is considered to be lower than other Mongoloid population of Assam. The age at menopause is affected by food habits, socio-cultural, environmental aspect etc.

(x) Place of Residence

It has been reported that tribal of India generally prefer to reside near to their livelihood. Tiwari (1998) stated that Kawar tribe of Madhya Pradesh prefer plain area for their settlement purpose with complex type of residence. Recently, the Kawars are residing among multi-ethnic villages. Settlement pattern of the villages are found as linear on plain land. Whether they are settling among multi-ethnic villages, their location of cluster is separated from other caste groups. Kawars do not allow to reside with the neighbours who belong to low caste groups as compared with them. According to them, if they reside with low caste groups, then they are also considered as the groups of low social status.

Soni (1998) reported that Bhil tribe of Madhya Pradesh construct their residence on higher ground to keep eye on cultivated fields. The settlement at high place relates to privacy and higher ground considered as the abode of Gods and Goddess. Bhasin and Bhasin (1990) studies specify that specially in Sikkim state, the houses of communities are found to be very neat and clean. Their high populated villages are usually settled on the slope of various hills. The settlement of houses in Himachal Pradesh is not well planned and people build their houses at any available suitable place.

Anthropology and Health

Anthropology of health is of recent origin. As subject for specialized anthropological study, health and ill health is as yet only in its adolescence. Anthropology of Health is application of anthropology to the health field. (Aandot, 1978; Susanne, 1982) The anthropology has its genesis in certain fundamental aspects of human life whereas health and health care services possess certain distinctive characteristics. (Febrega, 1971) According to the Herbert E. Klarman, the inherent characteristics of health and medical care

are: uneven and unpredictable incidence of illness; external effects, as in the case of infectious diseases; health and medical care being a need, the consumer's inability to evaluate the effects of choices before him, thanks to lack of knowledge; the mixture of consumption and investment elements; large components of personal service and non-profit motive. (Klarman, 1965).

Anthropologist emphasis that cultural influences every man's activities by biological and non-biological. Culture determines to a large extent: (a) The type and frequency of diseases in a population; (b) the way people explain and treat diseases (c) The manner in which person respond to the delivery of modern medicine (Logan and Hunt, 1978).

Anthropologist interest in understanding the ways in which human behaviours effects the maintenance of health and the occurrence and control of disease has been used as an aspect of applied anthropology. The overall impact of culture on diseases are culture patterns diseases; some culture produce a higher incidence of given psychiatric disorders through certain child rearing practices; culture produce personality types, especially vulnerable to certain kinds of illnesses; culture may serve certain diseases through proper sanctions and structure on acceptable behaviours; culture may produce psychiatric disorders differently in given segments of the population through certain stressful roles; culture may perpetuate malfunctioning by regarding it in certain prestigious roles; culture may effect breeding pattern selectivity; culture through patterns of faulty hygiene can produce toxic and nutrients deficiencies influencing the health conditions of the people. Chaudhary (1986), while discussing about the health problems of tribal, explained the areas of interest of anthropologist.

- Health and culture including the traditional belief in supernatural concerning diseases;
- Health, food habit and environment—covering the sanitation, water supply, settlement pattern, the total physical environment affecting health food habits and food during socio-religious occasions;
- Medicine, health and community—the traditional and modern health practitioners, their position in the society, concept and treatment of diseases, nature and use of medicines, traditional and modern;

- Fertility and mortality among the tribal, variation and reasons, the population problems of small tribes, use of traditional and modern practices of birth control;
- Interaction of traditional and modern system of medicine at various levels, reasons for non adoption of modern practices;
- Traditional tribal medicine—its use and application with certain development and modification, study of indigenous methods of treatment.

Hasan and Prasad (1959) used the term "Medical Anthropology" in an important article published in the journal of Indian Medical Association. The pioneering definition of Hasan and Prasad pointed out that medical anthropology studies biological and cultural aspect of man from the point of view of understanding the medical, medico-social, medico-historical, medico-legal and public health programmes of human being. Further Hasan and Prasad (1959) listed a number of these areas including nutrition and growth, and the correlation of body build and a wide variety of diseases such as arthritis, ulcers, anaemia and diabetes. The role of anthropologist in health study is very prominent as anthropologist can explain about traditional beliefs and practices; how social factors influence health care decision; and as to how health and disease are aspects of total cultural pattern, which change only in the context of broader and more comprehensive socio-cultural changes.

The important of ecological aspects besides socio-cultural was advocated by Livingstone (1958), Wiesenfeld (1967), Dunn (2000), Alland (1970), and Mc Cracken (1971) and others. The ecological orientation is concerned with dimensions of diseases that is how do factors of biology, culture and environmental pressure influence the process or distribution of diseases. It also sees what are the socio-cultural, including the cognitive, consequences and concomitants of a given disease in particular groups. The root source of ecological orientation was the scientific revolution in evolutionary biology that erupted along a broad front of biological disciplines during the 1940s and laid the necessary theoretical foundations for dealing with human evolution and adaptation as the complex interaction of cultural and biological factors under given environmental conditions.

Drawing on the synthetic theory of evaluation, Allands (1970) explained the interrelatedness of culture, biology, environment and disease in the adoptive process. "In general, the incidence of disease is related to genetic and non-genetic factors. Any change in the behavioural system is likely to have medical consequences, some of which will produce changes in the genetic system. On the other hand, disease induced changes in the genetic structure can effect the behavioural system. Such effects may be the result of population restructuring or the emergence of new immunological patterns which alter the possibilities for natural alterations in the environmental field and provide new selective pressures relating to health and disease which must be met through a combination of somatic and non-somatic adaptations". The ecological model conceptualized health disease as measures of the effectiveness with which human groups, combining biological and cultural resources adapt to their environments. The model also views health and disease in their feedback effects on culture, biology and response to environment.

The place of the medical system and the adaptive equation varies evolutionary. Alland (1970) notes that in the primitive and technologically simple societies, past and present, their medical theories and specific therapeutic procedures had and have less direct impact on the control of disease than those customs and behaviours outside the medical systems and that whatever may be the rational they serve to minimize disease through positive feedback from the environment.

The domain of ethnomedicine is "those beliefs and practises relating to disease which are the products indigenous cultural development and are not explicitly derived from the conceptual framework of ethnomedicine" (Hughes, 1968). It is the lineal descendent of the early interests of anthropologists in non-western medical systems. It is also referred to in the literature as "folk medicine", popular health culture". "Ethnoiatry" and "ethnoiatrics" (Huard, 1969). Ethnomedicine, the contemporary term for the vast body of knowledge that has resulted from the curiosity and the research methods used in adding to it, is of interest to anthropologists for both theoretical and practical reasons. On a theoretical level, medical beliefs and practises constitute a major element in every culture. Consequently, they are interesting in their own right and also for the insights they give into other aspects of the culture they are part. On a particular level, knowledge of indigenous medical beliefs and practises is important in planning the health programmes for and in delivering health services, to traditional people.

Disease classification is one of the important areas of study in ethnomedicine. (Bhasin and Srivastava, 1991) Modern medicine classifies diseases in terms of single taxonomy of universal categories. Hence from the standpoint of this taxonomic system, a recognized disease retains its identity wherever it occurs, regardless of the cultural context. Whereas the disease classifications of indigenous medical systems tend to be confined within cultural boundaries, in ethnomedicine, there is often a marked variation in disease entities recognized from culture to culture. Foster proposes disease classification into two categories, one is personalistic and the other naturalistic. Personalistic system illness is believed to be caused by the active purposeful intervention of a sensate agent who may be a supernatural being luke a deity or god, a non-human being such as a ghost, ancestor, or evil spirit of a human being such as witch or sorcerer. In naturalistic medical system, illness is explained in impersonal, systematic terms.

In ethnomedicine the study of disease classification has an important place. Anthropologists take into consideration the natives point of view adopting the "emic" approach to know how and by what criteria the people classify the diseases. Modern medicine classifies diseases in terms of single taxonomy of universal categories, wherein disease retains its identity wherever it occurs, regardless of the cultural content. Whereas the classification of diseases in indigenous medical systems is directly related to their culture and ecological systems in which they service. So, indigenous classification of disease changes from culture to culture. What is illness in one culture is not considered as illness in the other culture. Each culture has its own therapy, which includes magico–religious, mechanical and chemical procedures. These procedures comprise an impressive series of practises that demonstrate empirical therapeutic knowledge, including trephining, bone setting, removal of ovaries, obstetrics, including caesarean section, leparotony, urulectomy, comparative anatomy, autopsy, cautry, inoculation, baths, poulticer, inhalations, laxativer, enemas, ointments and cupping. (Ackerknecht, 1942; Laughlin, 1963 and Huard 1969).

The pharmacopoeia of ethnomedicine is copious and includes such proven drugs as quinine, opium and coca, cinchona, copaiba, curare, chaulmoogra, oil, ephedrine, and ranworfia (Lieban, 1977)". Even when a mechanical or a chemical therapy is employed, magico-religious elements may form an essential part of the prescription. The treatment may be regarded as incomplete without attention to mystical factors involved in the etiology of the illness. Many prophylactic practises are widely prevalent in the various indigenous medical systems. These include both mechanical and magico-religious measures such as bathing, massage and rapid rewarding to prevent hypothermia, dietary restrictions, surgery, inoculation, incantations, amulets and prayers at shrines.

When illness occurs, it may be ignored or treated without the help of a specialist. If treatment is sought from medical practitioners various types of specialists may be available, in particular culture including herbalist's diviners, shamans, midwives and masseurs. Therapists may specialize in only one type of skill or calling or they may combine several in their practice (Nurge, 1958; Polgar, 1962 and Lieban, 1977). Qualifications for indigenous medical roles vary considerably. In some cases, no formal training may be required for practitioners; in others, a long apprenticeship may be customary. Some curers acquire the healing techniques only as a gift of supernatural powers (Metzer and Williams, 1963).

In India, the research work on health has been carried out by various anthropologist and social scientists like Khare 1957; Hasan, 1981; Sinha, 1990; Chaudhary, 1990; Basu, 1990; Basu, Jindal and Kshatriya, 1990; Kapoor 1992; Basu, 1992; Kar, 1993; Basu (ed), 1993; Basu, 1993; Bagade et al 1995; Maulik and Mohapatra, 1995; Basu, 1996; Bhasin, 1997; Barua and Phukan, 1998. According to Kshatriya (2000),'The following strategies, if actively pursued could go a long way towards improving health and overall development of the tribal population of India:

1. Formulation of realistic developmental plans based on needs of specific tribal group;
2. Positive tribal culture/values, traditional skills should be encouraged and inducted into the mainstream of life;
3. Most of the tribal communities have a wealth of folklore related to health. Documentation of this folklore available

in different socio-cultural system could provide the model for promoting appropriate health and sanitary practices in given eco-system;

4. Encouraging the development of horticulture with emphasis on local fruits;
5. Study of nutritional status and physical growth among tribal children in different ecological setting should be carried out for an effective strategy to plan nutrition intervention programmes;
6. Hundred per cent immunisation for pregnant mothers and children below five years must be targeted;
7. Attempts should be made to introduce oral re-hydration therapy among different tribal population through their culturally accepted food habits;
8. Simple kits should be provided at PHC level and staff to be trained for genetic disorder tests like sickling, G-6-PD enzyme deficiency etc.;
9. Health education should be imparted by the local people (preferably women) with guidelines provided by health functionaries. It can also be imparted through distribution of leaf-lets and playing of audio and where possible video cassettes, preferably in local dialects at weekly markets, ghotuls, schools etc. Health Education through community participation;
10. Organisation of short-term orientation courses on tribal culture for health workers at district and sub-divisional headquarters;
11. Training of tribal girls as nurses, midwives to generate better response;
12. In a difficult tribal/hilly areas, Mobile Health Teams should be formed to provide more coverage. Tribal "HAATS" can be taken as the focal point of the activity;
13. Awareness about STD and AIDS must be further enhanced;
14. Local community leaders, clan chiefs must be involved in the decision making process, in which women must also be included;

15. Efforts should be directed towards achieving a respectable literacy rate for women'.

(i) Maternal and Child Care

The mothers and children's health have been considered as most important factors for the improvement of national health status. For this cause, maternal and child health aspects have been included in family welfare programme of India. Nowadays, these health service activities have been continued as child survival programme and safe motherhood programme. (Kantikar, 1979).

Bhasin and Basin (1990) reported that during pregnancy, some special foods are consumed by the women. Pregnant women avoids nutritional and fatty substances due to fear of abnormally increase the size of baby among the various communities in Sikkim and Himachal Pradesh. Due to social taboo, the pregnant women do not consume certain fruits and meat of animals. After delivery, the new born mothers are initiated with strong alcoholic drink which is prepared by boiled milk and rice. The mothers' breast feed their babies for minimum three months and then they slowly initiate the supplementary foods.

Mishra (1993) reported that the maximum mothers (34 per cent) and minimum mothers (2 per cent) initiated breast feeding within the birth of 49 to 72 hours and 73 to 96 hours respectively among various communities of Orissa. It has also been reported that educational level of mothers influence the frequency of breast feeding to the children. Most of the rural mothers (70.00 per cent) practised breast-feeding during above 24 months whereas very less percentage of mothers (3.50 per cent) fed breast milk to the babies below 6 months of age. After birth, sweet water was initiated as first food by highest percentage of mothers (31.5 per cent) whereas only 5.5 per cent mothers fed breast milk directly (colostrums).

Basu et al. (1994) reported that maternal care and child care are not satisfactory among Kheria tribe of Orissa. According to the study, 84 per cent women did not change their diet during pregnancy and after their delivery whereas 10 per cent reduced their quality of food intake and the remaining women did not give any proper information. Further it was reported that during pregnancy, 67 per cent women were immunized against tetanus and 94 per cent of deliveries were performed at home with the help of help of senior women. 69 per cent, 74.7 per cent and 71 per cent children upto the age of three years were immunized against BCG, DPT and Polio respectively.

Karmakar et al. (1995) reported that during pregnancy, most of the tribal women of Singhbhum district in Jharkand (erstwhile Bihar) do not take any additional/nutritious diet. Low birth weight of the babies was recorded and resulted pregnancy complications due to poor diet intake, ignorance of early marriage and unhygienic practices, etc.

Pandey et al. (1994) reported that during pregnancy, Khairwar women of Madhaya Pradesh were not benefited by accepting any ante natal care and even during the pregnancy, they continue their daily routine work like arrangement of water, cutting the fuel wood, etc. Most of the deliveries were conducted by local aged women in the community and complicated deliveries were left at the mercy of God. The trained medical personnel did not play any role in delivery in the community.

Nanda and Niranjan (1999) reported that ante natal care of a particular community is depended on their place of residence, occupational status, educational level, medical services, etc among various scheduled tribe of India. The ante natal care found to be higher in educated families than non-educated families. The non-working women got so much ante natal care as compared with the women who were engaged in agriculture. The ante natal care was not availed due to against their customary law, financial problem, inconvenient, no time to visit health institutions etc.

Prasad and Nagaraj (2001) reported that 44.3 per cent and 55.7 per cent deliverers took place in home and institution respectively among various communities of Andhra Pradesh. Whereas, 40.3 per cent and 59.7 per cent deliveries were carried out with the help of trained medical personnel and untrained personnel respectively. Pregnancy and child birth are taken as the cultural and spiritual phenomenon in most of the communities and it has the great influence on their behaviour.

Nayak and Babu (2001) reported that maximum deliveries of Scheduled Caste and Scheduled Tribe women were performed at home among Scheduled Caste and Scheduled Tribe of Orissa. In tribal groups, 36.5 per cent and 41.7 per cent deliveries were assisted by traditional birth attendants and relatives respectively. Doctors and nurses attended only 3.4 per cent and 3.9 per cent deliveries respectively. These studies indicate that 45.4 per cent (Scheduled Caste) and 56.2 per cent (Scheduled Tribe) mothers did not receive any TT doses, Iron and folic acid tablets were received by 47.7 per cent (Scheduled Caste) and 52.9 per cent Scheduled Tribe) women respectively.

(ii) Health Seeking Behaviour

In a particular environment, health and disease are depended on the resources of biological and cultural factors. It has been reported that in tribal societies, health and disease is totally related with socio-cultural factors.

Pandey and Saxena (1988) reported that population density, female literacy rate, growth rate, achievement of maternal and child health care, etc. in the population of tribal area are very low as compared with the population of non-tribal area in Madhya Pradesh. In accordance with socio-cultural differentiation among tribal groups, health seeking behaviour is also varied from one to other tribal groups.

Bhatia (1990) reported that for getting free check-up and free medicine, initially the sickness persons take the health care services from their local primary health centre among tribal area specially of Dang district in Gujarat. In case of non-recovery from government health centre, they start availing the private health care facility. This shift from public health care to private health care can be attributed to non satisfactory supply of medicine, slow process of recovery etc. Due to such problem, they are demotivated from public health centre based treatment and take decision to return to their traditional healing practices.

Rajyalakshmi and Greevani (1992) reported that health problem of four tribal groups (Jatapu, Savara, Gadaba and Kondadora) of South India are caused by mainly scabies and dermatitis. The significant number of these tribal people is suffering from gastric disorder, bronchitis, B-complex deficiency, anaemia and other diseases. The poverty and environmental hygiene are indebted to create skin disease of large number of tribal people.

Kar and Gogoi (1993) reported that according to the symptoms, all the diseases have been categorized and mentioned that disease is caused by spirit, deities and also other supernatural power among Nocte tribe of Arunachal Pradesh. For healing from diseases, they arrange worship programme of various deities and sacrifice the coconut, hen and others. Occasionally they also prefer natural herbal medicine for the treatment of their diseases. In case of failure in these methods, they make shift to modern medical treatment.

Chattopadhyay and Prasad (1995) reported that according to their health status, Nicobarese boys and girls were classified into five categories. The maximum percentage of boys (57.45 per cent) and girls (49.12 per cent) were categorised in 'poor' body build. The lowest percentage of only boys (2.13 per cent) was come under 'good' category of body build. Nicobarese are ignorant in regard to quality and quantity of food items for their children in accordance with various stages of their physical growth. Through the year, Nicobarese children suffer from gastro intestinal disorders, bronchial disorders, malaria, scabies and other diseases.

Bhagwat and Cirmulay (1997) reported that rural Karnataka is a typical example for traditional or indigenous health care practices. For easily feedable medicines and quickly relief, allopathic treatment is initially preferred by rural people. Only traditional medicines are valued for treatment of some certain particular diseases. The traditional medical system is applied through various techniques like as herbal and animal, kitchen materials used, magico-religious treatments, acupuncture, acupressure and massages according to nature of diseases.

Dutta (1999) highlighted the traditional healing practices of Singpho in Arunachal Pradesh. According to their perception, the modem process of treatment is also unable to cure all the diseases. Firstly, they take the help of village head (Gaon Bura) and then worship Lord Buddha due to fail of treatment of the hospital. Finally, they consult magician (Shamon). In spite of advancement in medical system, they still believe evil spirits and witchcrafts. Magician usually arranges worship and offerings to the particular spirit or deity. Further it is being reported that before conversion to Buddhism and establishing hospital, their survival was totally depended on nature and natural sources like medicinal plants etc.

Pandey and Tiwary (2001) reported that most of the Pandos of Madhya Pradesh are illiterate and inhabiting in socio-culturally-backward position. For the treatment of adult persons, usually 'Gunia' is selected as healer. After diagnosis the patient, 'Gunia' serves the herbal medicines without taking any cash. In accordance with the severity of diseases, 'Gunia' is offered by fowl, coconut, locally made wine, etc. In case of the treatment of infant, they may select either traditional or modem medicine men. This study expresses that about health hygiene, ante natal care, post natal care and family planning, they fully depend on their own perception.

Sharma (2001) reported that Gond of Madhya Pradesh is accepting the contaminated food and water and their living standards are unhygienic. Due to this, they are frequently suffering from the diseases of malaria, asthma, conjunctivitis, ear- nose and throat infections, itches, malnutrition, etc.

Some of the study highlighted the health seeking behaviour among tribes with focus to their socio-cultural dimensions, nutritional dynamics, religious beliefs, and educational levels (Kumar and Kapoor, 2005; Kapoor, 2006; Misra and Kapoor 2006-2007; Tungdim et al., 2007).

(iii) Family Planning

Family planning indicates to suitable methods for family control and by adopting these, the undesired births are come under control. The main objectives of the family planning are to decrease births rate of less than 39; to influence the couples for organising small family size in their society; to develop their personal knowledge about various family planning methods; and to supply the various necessary family planning devices and equipments.

It is being reported in various studies that birth rate and death rate in tribal population are very high as compared with general population. High death rare is directly linked with high birth rate. For this cause, health services and birth control devices are much required for these tribal people. Bhasin and Bhasin (1990) reported that 11.96 per cent Lepchas and 13.38 per cent Bhutias of Sikkim have been estimated who adopted family planning devices. Beside this, 12.47 per çent Gaddis of Himachal Pradesh adopted various family panning devices. Among family planning acceptors, the percentage of males is very low as compared with females. Highest family planning adopted persons have been specified whose education was in middle school level. They are the members of traditional society and changing pattern of this traditional society is dead slow. Most of them are ready to limit their fertility by using any birth control devices. Their studies reflect that for controlling fertility, there, no any social guidance has been provided but ecological condition saves their high population growth.

Piplai (1993) reported that Oraon and Tamang—two tea garden tribal communities, in Jalpaiguri district of West Bengal, are entitled to get food items at subsidized, free water supply, free health services, free educational facility and others. Still, number of family planning users is far less common among them. The acceptance of family planning method is 7.23 per cent couples and 19.00 per cent couples among Oraon and Tamang tribes respectively. Significant number of couples was sterilized through vasectomy and tubectomy. It is being reported that very less number of women used loop. The oral pills and condom users were not existed among both the tribal groups.

Basu et al. (1994) reported that family planning methods were used by only 16.06 per cent respondents among Dudh Kharia tribe of Orissa. In the context of family size, 44.90 per cent respondents were concerned to procreate four children whereas three children were desired by 32.56 per cent respondents. Among Dudh Kharia, 84.04 per cent respondents denied to use any family planning methods. Among non-user groups, 61.9 per cent respondents believed that birth of children were considered as the gift of God.

Chachra and Bhasin (1998) reported that knowledge of contraceptive devices are almost universal among Kumauni and Bhotia communities of Uttar Pradesh but yet the number of users are quite low. Among contraceptive acceptors, 49.58 per cent have been identified who used any methods whereas 9.2 per cent males and 27.39 per cent females were sterilized. Beside this, 13.57 per cent females used IUD whereas 1.41 per cent males used condom. Pills and other methods were adopted by 1.06 per cent and 2.00 per cent tribal women respectively. Among contraceptive non-user groups, 45.95 per cent respondents clarified that they wanted male offsprings whereas 34.68 per cent respondents also clarified that they wanted more children and for these reasons, they did not adopt any contraceptive method.

Mahapatro et al. (1999) reported that the basic modern family planning methods are classified into five types i.e. physical, physiological, psychological, biological and social among Bhattara Tribe of Madhya Pradesh. Under this physical practices, condom, intra uterine device, Pills and others are defined as important methods whereas rhythm is included in physiological method. Coitus interrupts is regarded as psychological one. Jelly, creams, oral pills, etc. are come under bio-chemical type whereas abstinence and ritual taboos are considers as the family planning methods of social practices. Among them, 95.0 per cent respondents had experience about knowledge of family planning methods whereas 16.1 per cent respondents had positive attitude about the devices. Whereas only 8.0 per cent respondents used birth control devices among the tribal.

Sharma and Sharma (1999) reported that due to religious belief, Gond tribe of Maharashtra do not have any desire to avoid conception and reproduction because it is considered as the sin and against behaviour of God. Since most of the Gond people are agriculturist and production of agricultural crops, requires more manpower. Therefore, the reproduction of many offspring is not considered as bad. High mortality rate among the tribal is considered as an important factor which influences (them) not to adopt family planning devices.

Pandey and Tiwary (2001) reported that 87 per cent of Pando of Madhya Pradesh were aware about family planning methods. Among them, 61 per cent and 6 per cent respondents had the knowledge of vasectomy and laparoscopy respectively. Only 18.5 per cent and 16.7 per cent respondents utilized natural methods and nirodh respectively. Their study showed that two couples both husbands and wives adopted vasectomy and tubectomy respectively. Besides these Pando had knowledge of several temporary family planning methods like copper- T (5.60 per cent), nirodh (16.7 per cent) and oral pills (10.5 per cent). Due to poor health service facility, very less number of respondents of Pando used family planning methods which has its the direct impact on demographic structure.

Biswas et al,.2001) have observed that the Kamar tribe of Madhya Pradesh population had been increasing from 1911 to 1971 at different rates. But during 1971-81, there was sudden sliding down of the population may be due to large scale family planning programmes, proper enumeration, out – migration and infant and old age mortality rate.

OBJECTIVES OF THE STUDY

The main aims and objectives of the present study are as follows:

1. To study the socio- cultural dimensions of Kinnauris;
2. To study the health profile and health care system of four ethnic groups of district Kinnaur, Himachal Pradesh;
3. To study the drug logistic system and health development in district Kinnaur and its impact;
4. To suggest the appropriate strategy for improvement of health status in district Kinnaur.

SCOPE OF THE STUDY

There is an urgent need to investigate into the health issues in order to help the planners and administrators in planning and administering health related programmes for the community. Moreover health planner and administrators need feedback of operational information for decision making. The present study explore into health profile of four ethnic groups, drug logistic system and health development in the district and to suggest appropriate strategy to be followed for improvement of health status of Kinnauris.

STUDY AREA

The Himalayan represent the youngest mountain range, having been uplifted about 60-70 million years ago. It is also geographically a very complex and its origin is not yet clearly understood with the development of the means of communication in the region, many of its remote parts have now become accessible and are being systematically investigated and habituated due to pressure of population (Basu, 1998). These developmental activities have interfered with the natural ecosystem of this range and deterioration in the environment by aggravating the natural dynamic process of the earth. These dynamic processes are already in a more active state in the region because of certain geological, geo-dynamical, watering, erosion and seismic processes.

The Himalayan mountain area is approximately 5,23,000 sq km (16.13% of the land surface of India). The Himalayan sediments areas i.e. the areas which are directly fed by the Himalayan rivers and which form the Indo-Gangetic and Brahmaputra plains, including 7,26,000 sq km with 22.26 per cent of the area. Thus, the grand total of the Himalayan region is 1,294,000 sq. km. And 38.4 per cent of the land resources area of India. As per an estimate, in the Himalayans and its foot hill live about 150 million people and a total of about 300 million people live in the Himalayan mountain area and Himalayan sediment area.

The state of Himachal Pradesh situated in the north western part of India is a land of lush green forests, deep river valleys, beautiful plateaus and snow-capped lofty mountains. It is bordered with Jammu & Kashmir in the North, Punjab in the West and South-West, Haryana in the South, Uttaranchal in the South-East and with China in the East. The area as a whole is hilly, mountainous and the major part is inaccessible, physiographically complex, snow covered and forest clad. The altitude varies from 450 to 6500 Mts. above sea level in Himachal Pradesh. According to the census report of 2001, the total population of Himachal Pradesh is a little more than six million (6,077,248) people.

For the present study, the Kinnaur district of Himachal Pradesh has been selected. The geographical location of Kinnaur is between 31-05′-20″ and 32-05′-20″ north altitude and between 77-45′-0′ and 79-00′-50″ east longitude. The entire district is spread over the Himalayan mountainous terrain. The Kinnaur valley is stretched over about 80 kilometers in length and 64 kilometers in breadth. The area lying within the district territory is nearly 6679 sq. kilometers. This is 11.7 per cent of the total area of Himachal Pradesh. The Kinnaur district is divided into three blocks, Kalpa, Nicher and Puh. The tehsil in Kalpa, Nichar and Puh blocks are Kalpa and Sangla, Nichar and Morang and Puh respectively. (Hangrang is sub teshil in Puh Block).

The inhabitants of Kinnaur district; i.e. Kinnauri has been spelt out with many variants such as Konawri, Koonauri, Kanauri,

Kannauri, Kinnaura and Kinnauri (presently in use). The ethnic population of Kinnaur district, Kinnauri are divided into five groups i.e. Kanet, Koli, Badhi, Lohar and Nagloo. The entire population is classified as Scheduled Caste or Scheduled Tribe. The population of the district comprises of five distinct classes of Kanets (Rajputs), Chamangs {(Kolis, preparing shoes, weaving, tailoring and music drum beating)} and Domangs {Lohar (Blacksmith) and Badhi (carpentry), Nagloo (Basket makers)}. The Kanet constitute the majority of Kinnaur population. They are declared as Scheduled tribe people. The other four ethnic groups i.e Koli, Badhi, Lohar and Nagloo constitute the scheduled caste population of district.

The Koli, Badhi, Lohar and Nagloo mainly comprises of the artisan class and are considered untouchables whether they be Koli having the occupation of preparing shoes, weaving, tailoring and music drum beating, the Badhi having the carpentry and mason, Lohar – Blacksmith and Nagaloo-Basket making.

Here in this book, the literal meaning of Kinnauris is the inhabitant of district Kinnaur. However, every person residing within the territorial limits of Kinnaur district cannot be termed as Kinnauris. In the ancient literature, the abode of the Kinnaur tribe has been described as Kinner Desa, which has been corrupted and pronounced as Kinnaur.

In the context of present study, four ethnic groups were selected i.e Kanet, Koli, Badhi and Lohar. The rationale for selection of these four ethnic groups is uniform and wider distribution in the district. Nagaloo being a smaller group and confined to particular geographical area was excluded from the preview of study. In the book, further grades/groups with in the ethnic groups are not considered for the matter of data analysis. It is interesting to point out that these ethnic groups are classified as Scheduled Caste and Scheduled Tribe.

2

Area and People

An Ethnographic Live

AREA

Geographical Location

Kinnaur is surrounded by Jammu and Kashmir in the north, Tibet on North/North-East, Uttaranchal in the East/South-east, Haryana in the south and Punjab in the Southwest. The Nagri region of the western Tibet bound Kinnaur on the east, the district is separated from Tibet by the Zaskar mountains. The Dhaula Dhar range of the mountains forms its southern boundary and it separates Uttar Kashi district of Uttaranchal and Rohru tehsil of Shimla district from it. Srikhand Dhar separates the districts from Kullu and Rampur regions in the west The North Kinnaur district is separated from Spiti region of Lahaul and Spiti district by rivers Spiti and Para near the international boundary with Tibet.

The Sutlej valley divides the district equally. The elevation of the peaks in mountain ranges very between 5,180 and 6,770 meters and therefore is covered with snow all the year around. The total geographical area of the district is 5,553 square kms. The geographically condition of the Kinnaur district have created barrier and hence the district is conveniently divided into three culturally barrier which are as follows:

(a) Lower region dominated by Hindu culture of the main land;

(b) Middle region having a mixture of Tibetan and Hindu influence;

(c) Upper region influenced by Tibetan culture.

The above mentioned cultural division correspond to the administrative division of the district into three sub division, namely Puh, Kalpa and Nichar.

Kinnaur can be divisible into three different culture zones. "Zone-I includes most part of Puh sub-division and is dominated by Buddhism. Zone-II covers Nachar sub-division and is dominated by the local Hinduism and Zone-III is sandwiched by the above two zones and includes Kalpa sub-division and part of Puh sub-division. This last zone has a mixed religion both Buddhism and Hinduism (local)" (Raha, 1974).The face of the district presents high hills and low dales rapid and rushing stream and streamlets and is marked by precipitous sky-high mountains with their peaks perpetually covered with snow. The main Sutluj valley almost equally divides the district. (Khan 1996; Harnot and Verma, 2000).

Baillie (1882), "Kunawar is that part of Bischur which embraces all the northern, north-eastern and eastern tracts and lies entirely in and behind the snowy hills, chiefly comprised in the glen of the Sutlej, and running through and beyond these mountains, cutting them in a line, diverging not far from east and west; the Himalaya range here taking a direction from north-west and north-east to west-north-west and south-south-east, while the river runs through in nearly similar course.

The river Sutluj enters the district from Tibet in the north east near the village of Namgia and leaves it at the western end at Chaura near Wangtu bridge.

Administrative Structure

The Kinnaur district is divided into three blocks, Kalpa, Nichar and Puh. The tehsil in Kalpa, Nichar and Puh blocks are Kalpa and Sangla, Nichar and Morang and Puh respectively. (Hangrang is sub teshil in Puh Block). The Deputy Commissioner is the pivot round whom the entire administration revolves in the district. He wields wider administrative and financial powers than any other in the state of Himachal Pradesh. This system is known as Single

Line administration introduced in December, 1963. Under this system the Deputy Commissioner writes the annual confidential reports of all the officers in the district. In 1965 the Deputy Commissioner was delegated the power to transfer within the district class III and IV employees within the district in consolation with the head of the office concerned. This enabled him to ensure the co-ordination of all departments. He is empowered to sanction casual leave and tour programmes of all district level officers. The development schemes of various departments have also to route through the Deputy Commissioner.

In Kinnaur district, number of villages habited and inhabited are 189. The details of number of village habited and inhabited are given in Table 2.1.

Table 2.1: Details of Number of Villages Habited and Inhabited

District/Tehsil/ Sub Tehsil	*No. of Villages*								
	Habited			*Un Habited*			*Total*		
	1971	*1981*	*1991*	*1971*	*1981*	*1991*	*1971*	*1981*	*1991*
Hangrang	8	8	14	-	-	43	8	8	57
Puh	12	12	27	-	-	53	12	12	80
Morang	12	12	39	-	-	103	12	12	142
Kalpa	12	12	36	-	-	48	12	12	84
Sangla	11	11	27	-	-	83	11	11	110
Nichar	22	22	85	-	-	104	22	22	189
Kinnaur	**77**	**77**	**228**	-	-	**434**	**77**	**77**	**662**

Important Places

- RECONG PEO (2670m): Recong Peo is the district Headquarter having a panoramic view of Kinner Kailash. It is located 260 km from Shimla.
- KALPA (2759m): Beyond Recong Peo (14 kms. from Powari) on the link road, is the main village of the District-Kalpa. Across the river, facing Kalpa is the majesty of the Kinner Kailash range.

- NICHAR (2150m): This village is situated between Taranda and Wangtu on the left bank of Sutluj about 5 kms. above Wangtu.
- SUMDO/KAURIK: On the border of Spiti at a distance of 104 kms. and 124 kms. respectively from Kalpa, are the entry points to Spiti valley.
- SANGLA VALLEY: Sangla valley starts 57 kms. short of Kalpa which has been named after a beautiful and populous village Sangla. Sangla is situated on the right bank of Baspa river 17 kms. from Karcham. It is also known as Baspa Valley since Baspa river flows through this area. This is the most charming valley in the entire District of Kinnaur. A famous temple dedicated to Nages god is situated in the valley.
- CHITKUL (3450m): This is the last and highest village in the Baspa valley. It is situated on the right bank of Baspa river. There is a road along the left bank from Karcham. There are 3 temples of local goddess Mathi, the main ones are said to have been constructed about 500 years ago.
- KOTHI: Kothi is also called Koshtampi. It is little below Kalpa, and is overshadowed by the Kinner Kailash peak. The village with its attractive temple, gracious willows green fields and fruit trees makes an altogether lovely landscape. Goddess Shuwang Chandika temple is in the village.
- PUH: Puh is locally pronounced Spuwa, is the tehsil Headquarter. 71 kms. from Recong Peo. It is situated above the National Highway-22 having all modern amenities as well as green fields, vineyards, apricot, almond and grape orchards enhance its beauty. The local god is called Dabla, who neither has any dwelling nor possesses an ark. The only manifestation of the deity is a pole with a small idol set on its upper portion and adorned with Yak tail hair and long pieces of coloured cloth. The whole being called Fobrang, it is occasion brought to the Santhang.

- RAKCHHAM (2900m): Rakchham is situated on the right bank of river Baspa. Its name has been derived from "Rak" a stone and "Chham" a bridge.
- RIBBA (2745m): Ribba or Rirang is another large populous village at a distance of 14 kms. from Morang, the tehsil hqrs. Situated between the villages of Purbani and Rispa. In the local dialect *Ri* stands for Chilgoza and *Rang* means a peak of a mountain.
- LEO: Leo is about 105 kms. from Recong Peo perched on a small rocky eminence, on the right bank of the Spiti river, is the hqrs. of sub-tehsil Hangrang in Puh subdivision. The famous temple of Jamato is situated at the place.
- LIPPA (2438m): Situated near the left bank of Taiti stream. The village can be approached from Kalpa by the old Hindustan Tibet road to Jangi-Lippa-14 kms. Ibex are said to be found near the forest. The three Budhist monasteries here are dedicated to Galdang, Chhoiker Dunguir and Kangyar.
- MORANG (2591): This village is situated 39 kms. away from Kalpa on the left bank of river Sutluj. The location is very beautiful and approach to this picturesque village is through apricot orchards. The local deity is Urmig and there are three structures dedicated to the deity each existing in Thwaring, Garmang and Shilling. Generally these are empty as the ark of the deity remains in the fort. On a sacred day the ark is taken to the above named places. The ark has got 18 'mukh', made of silver, gold and brass. The 18 mukh represents the 18 days of the great epic Mahabharat.
- CHANGO (3058m): At a distance of 122 kms. from Kalpa, is a collection of 4 hamlets in Pargna Shuwa, sub-tehsil, Hangrang on the left bank of river Spiti. It is encircled on every side by high hills which is a witness to the presence of a former lake. Buddhism is generally practised here but there are some local Hindu deities too namely Gyalbo, Dabla and Yalsa.

- NAKO (3662m): Nako is situated about 2 kms. above the Hangrang valley road and is 103 kms. from Kalpa on the western direction of the huge mountain of Pargial. This is the highest village in the valley and the existence of lake formed out of the masses of the ice and snow above adds beauty to the village. Local village deity is Deodum and another Lagang temple with several idols exist here.

Geology

Rocks varying in age from Pre Cambrian to Permo-Carboniforuous are exposed in the Kinnaur district. The district can be sub divided in three main sectors (Between the border of Kinnaur and Shimla district and Jangi, between Jangi and Shipkila, Area north and west of Shipkila) on the basis of geological formation. i.e between the border of Kinnaur and Shimla District and Jangi; between Jangi and Shipkila; and area north and west of Shipkila.

Climate

Kinnaur district experience the cool temperate climate. The climate in major parts of the district is cool and dry and there is wide rainfall variation. The cool temperate zone climate that is found in the district is characterised by heavy clouding, persistent and prolonged long winters from October to May. Snowfall occurs during the winter months.

Temperature in different parts of the district varies depending upon elevation. Temperature starts rising rapidly from the end of February and are the highest in June which is the warmest month in the Lower Kinnaur. In upper Kinnaur, which is an arid tract, the warmest months are July and August. In humid and sub humid sections of the region, the temperatures start decreasing gradually by the end of June with the onset of rains.

In winter, the impact of temperate cyclones is quite visible when the district experiences spells of cold weather and mercury often falls below freezing point. Frost occur from May and October everywhere on higher elevation. The area is free from storms of great intensity such as cyclone. Thunderstorms occur very rarely in this region.

The rainfall in lower Kinnaur upto Wangtu does not vary greatly from rainfall in Shimla i.e. about 200 cm. Beyond that the amount of rainfall decreases progressively. The Kapla and Pangi face the stronger currents of the monsoon. The place, Karchham is the limit of monsoonal reach. Further Puh lies outside the monsoon zone.

Forest

Forest is very important for the Kinnauris. They are dependent on the forest for their day to day need of timber and firewood. The livestock depend on it for fodder and grazing. Besides producing rain (ecological balance), the forest produce varieties of medicinal plants serves as income for the people. Sizeable parts of wooded area are covered by Chilgoza trees which bear edible nuts and fetch handsome price.

According to forest Department statistics, total forest area of the district is 4,85,599.48 hectares. The forest area is decreased to 41 per cent in year 1997-98 and further increased (by 29%) in year 1999-2000.

During the last century, forests were properly demarcated and reserved. Later on these forests were exploited by their owners or the right holders for timber or fuel or heavily lopped for fodder or green manure. This was period (1997-98) of uncontrolled exploitation of reserved forests, when forest crop was destroyed and soil laid to bare through destructive misuse.

Prior to 1850, no records were available of the early history of forests. It seems almost certain areas must have been under forest. In 1850, Indian traders purchased deodar trees from the Raja of Bushahr at 2 annas each. It was only after 1859 that the forests were worked more systematically. Before the merger, the forests of Bushahr state were taken from the Durbar by the Punjab Government on lease. According to the terms of the lease certain forests were kept reserved for the use of ruler. With the merger, the lease was annulled and the management and control of all the forests was transferred to the Himachal Pradesh Administration.

The chief factors affecting the distribution and quality of forest vegetation are rainfall and elevation aspects. On the basis of rainfall, the district can be divided into two zones, wet zone and dry zone. Wet zone gets rainfall due to the monsoon, while the dry zone does not get any rainfall.

Pinus longifolia, Pinus wallichrana, Cedrus deodars, Picea smithiana and Abies pindro, Pinus geradiana form a broad belt of forest along both sides of the Sutluj valley and the side streams between the cliffs of the gorge below and alpine pastures and eternal snow above. From the river side at 3500 feet to alpine pasture at 12,000 feet on the right base consists of grass lands and higher up are the forest belts. On the lower slopes upto 5000 feet Chir pine occurs in pure form and higher up gives way to Quercus incana and Rhodenondron arboretum. On sheltered ravine bank between 5000 to 12000 feet Cedrus deodara and Pinus wallichiana form intervine forests, higher up from 7,000 to 10,000 Picea smithiana with mixture of broad leaf spices predominate. Upper forests consist of Picea smithiana and Abies pinfro which merge with top belt of Quercus semecarpifolia alone which are the alpine pasture.

These include forests of Chini and Kilba Kailesh ranges. In this zone low level pine is the neoza pine which forms a very open forest. Above these, there are deodar forming pure forest belt with light mixture of neoza. Still above, the blue pine and spruce, followed by silver fir between 9000 and 12000 ft, is found.

On the basis of composition, the forests of Himachal are broadly classified into coniferous and broad-leaved forests. Deodar, kail, chil, spruce, and chilogoza pine are the coniferous species. Chilgoza pine, which produce edible nuts, grow in the district of Kinnaur and are the only forests of neoza in India (Gupta 1965).

Among the broad-leaves species, sal, ban, oak, mohru oak, kharsu oak, walnut, maple, bird cherry, horse chestnut, poplar, simal, tun and shisham are the important species which grow in these forests. The forests of Kinnaur are valuable timber forest and source of considerable revenue. The total forest area of Kinnaur district is given in Table 2.2.

Table 2.2: Total Forest Area *(In Hectares)*

Year	*Reserved Forest*	*Non-Classified Forest*	*Total*
1993-94	18820.12	619025.3	637845.42
1994-95	18820.12	619025.3	637845.42
1995-96	18820.12	619025.3	637845.42
1996-97	18820.12	619025.3	637845.42
1997-98	23685.69	353533.28	377218.97
1998-99	23685.69	353533.28	377218.97
1999-00	28203	456636.17	484839.17
2000-01	28203	456636.17	484839.17
2001-02	28203	457396.48	485599.48

Source: District Kinnaur at Glance-2002.

Flora/Fauna

Fauna: No systematic botanical survey has been conducted in the area. Nature has endowed tract with various fauna due to considerable variation in the elevation and climate. The species of alpine fauna found in the district are Blue sheep, Brown bear, Hill fox, Jungle cat, porcupine, European bat, Himalayan langur etc. Out of the mammals that denizens of Kinnaur district, brown bear, mush deer, Tibetan antelope and ibex are better known. Common retile species are the spotted Agama, Indian Chameleon, common krait and the harmless creatures. In amphibian family, frogs are found at the places. Among the lizards, monitor lizards, common house gecko and garden lizard are seen occasionally. The indigenous fish fauna is uniformly distributed in the waters of district. The exotic fish species Salmo fario (brown trout) is found in Baspa river in the Sangla valley.

Flora: The species of the plants found in the district are Berberis Aristata, Berberis Lycium, Berberies Petiolaris, Betula Utilis, Cedrus Deodara, Pinus excelsa, Pinus longofolia, Pinus Gerardiana etc.

Lakes and Springs

There is no large sized lake in the district. There is one lake in Nako situated on the western declivity of the large mountain of Leo Pargial about 1.6 Kms. above the left bank of the Spiti River. Another small lake, locally called Sorang, is situated above the villages of Ramni and Jamni in Nichar tahsil. A small lake is also situated above the Labrang village of Puh tahsil which is locally called Tomchho. Springs mostly come from snow are scattered all over the district. The water of spring is used for drinking purposes as well as for irrigation by constructing kuhls. There are number of hot water springs in Nichar tahsil. One such spring is found at Nathpa three at Tapri and one at Joktiaring hamlet about five kilometres towards east of Nichar.

Rivers and Water Resources

River water is main sources of water and besides this there are two small tanks located in Nichar Block. Several springs and springheads are scatters all over the district. Another source of water is snow. Sutluj, the principal river of the district, has its origin in the Himalayan and has plentiful and perennial sources of water. Within the district the length of the Sutluj is about 130 kilometers. Another river of the district is Spiti or Lee. The Baspa river, another feeder of the Sutluj, rises from Dhaula Dhar mountain of the Himalayan passes through valley bearing its name and meets Sutluj at a place called Karchham after a distance of 72 km.

House

There is remarkable difference in housing pattern of lower and upper Kinnaur. In the lower Kinnaur, the houses are of double storeys but the houses of nobles or landlords comprise of more than two storeys. The roof of the houses is mostly flat and is made of wooden planks covered with tree bark and overlaid with earth. Walls are decorated with picture of men, women, trees, moon, stars, sheep, goats and flowers drawn crudely with a white chalk like material. The doors are often folding and open inwards. In the upper Kinnaur, the houses are usually built of stone. These are flat roofed and covered with earth. These are usually ill build on account of scarcity of wood.

There is one living room on each storey and ground floor is used for cattle shed, except where there is worship place as in case of ironsmiths. The first floor is always used for the living purpose. It has extended wooden balconies. Since each story comprises only one room, therefore there is no separate bathroom or kitchens. For the dual purpose of letting the smoke out and admitting the light, hole is provided in the room. The earth is set in the centre of the room. The teshilwise number of families (scheduled caste and scheduled tribe) and average family size is given in Table 2.3.

Table 2.3: Tehsil-wise Number of Rural Families and Average Family Size (1981)

District/ Teshsil	*No. of Families*			*Average Family Size*		
	Schedule Caste	*Scheduled Tribe*	*Total*	*Schedule Caste*	*Scheduled Tribe*	*Total*
Hangrang	5	701	706	3.8	4.42	4.41
Puh	122	1,049.	1171	5.28	4.85	4.85
Morang	128	1,445	1573	4.5	5.11	1.43
Kalpa	207	1702	1909	5.00	5.07	5.07
Nichar	521	2559	3080	4,76	5.11	5.03
Sangla	293	1391	1684	5.36	5.51	5.43
Kinnaur	**1276**	**8847**	**10123**	**4.96**	**5:08**	**5.92**

Source: District Kinnaur at Glance-2002.

Population

It appears that the aborigines of these hills are today identifiable among the members of various scheduled castes and tribes who form a considerable part of the population. These people are represented by such castes as Kolis, Badhis, Loharis etc. These people form the lowest socio-economic strata of the hill society. They ordinarily come from the original stock of the Kolarian (Kolis) race which once inhibited the whole of the western Himalaya. These classes form in many respects one of the most interesting sections of the Himachal Community. They are non-Aryan origin but through inter-caste marriages, much social interfusion has taken place. The entire population is classified as Scheduled Caste or Scheduled Tribe. The population of the district comprises of four distinct classes of Kanets (Rajputs), Chamangs {(Kolis, preparing shoes, weaving, tailoring and music drum beating)} and Domangs {Lohar (Blacksmith) and Badhi (carpentry), Nagloo (Basket makers)}. The year wise population growth and percentage of Scheduled Tribe and Scheduled Caste population is given in Table 2.4.

Table 2.4: Yearwise Population Growth and Percentage of SC/ ST

Year	*Population*	*Decennial Variation*	*Sex Ratio*	*Population Density*	*Percentage of Schedule Caste to Total Population*	*Percentage of Schedule Tribe to Total Population*
1901	27,232	-	91	-	-	-
1911	28,470	+4.55	935	-	-	-
1921	28,191	-0.98	922	-	-	-
1931	30,455	+8.00	941	-	-	-
1941	33,238	+9.17	910	-	-	-
1951	34,475	+3.72	1070	-	-	-
1961	40,930	+18.87	969	-	-	-
1971	49,835	+21.61	887	8	19.40	68.41
1981	59,547	+19.49	885	9	10.63	74.27
1991	71,270	+19.69	856	11	26.87	55.18
2001	77,007	+17.79	851	13	-	-

Source: District Kinnaur at Glance-2002.

Alexander Gerard was the first person to give a rough estimate about the Kinnaur population. As rough estimate, during his travel in year 1822 and travel log published in 1841, the total population of Kinnaur was 9853 person. (Gerard, 1841).

According to 1901 Census, the total population of Kinnaur was 27,232, which has raised to 77,007 in year 2001. Increase in population always has a great impact on the socio-economic environment of the area. The population of the district has shown an increase of 49,775 during the period of 10 decades. Negative growth rate has been observed during the year 1911-21, the population of Kinnauri has reduced by 279 persons. The decrease can be attributed to influenza epidemic of 1918. (District Census Handbook. 1991; Chib, 1998).

In Kinnaur district, the rise of population growth through vacillating has always been below 10 per cent till 1951. During 1951-61 and 1961-71, the growth rate has been 18.87 per cent and 21.61 per cent respectively (District Kinnaur at Glance-2000). The population growth rate, during the period, was due to in migration in the district. With formation of new district, Kinnaur, the government servants, technical and non-technical persons for construction of infrastructure like roads, bridges etc have entered in the district. For year 1981 and 1991, the population growth rate has stabilised to approx 19 per cent and for year 2001, the decline in growth rate was observed to 17.79 per cent from 19.69 per cent. This decline in population growth can be explained as out migration of Kinnauri people in search of better economic opportunities.

Teshil-wise male female population is given in Table 2.5. It can be observed that the maximum population (34%) is in Nichar Tehsil followed by Kalpa (22%) and Sangla (15%) Tehsil.

Table 2.5: Tehsil-wise Male Female Population (2001)

District/Tehsil	*Population*		
	Male	*Female*	*Total*
Hangrang	2051	2006	4057
Puh	4026	3608	7634
Morang	5169	5080	10249
Kalpa	8967	8057	17024
Sangla	6034	5610	11644
Nichar	14631	11768	26399
Kinnaur	**40878**	36129	77007

Urban Population is not included

Source: District Kinnaur at Glance-2002.

Chib (1998) in his paper stated that growth of population in Kinnaur has led to many far reaching socio-economic implications. Nonetheless, it has its good as well as bad consequences. In case steps are taken to contain and check the attempts at disrupting the ecological balance, directly and implicitly, as also to dispense "distributive justice", the population growth in Kinnaur is not likely to prove detrimental in forceable future excepting the plight of women which warrants early and urgent attention of the social workers, government and the Kinnaura male society itself.

Language

Various Vocabularies have been published by various travellers and writers like Captain A. Gerard (1942), Diack (1996). Apart from Pandit Tika Ram Joshi (1909) there are several works on linguistics aspects of Kinnaurs like Bailey (1909), Konow (1905).

According to Bailey (1909, 1910, 1975), "Kinnauri is spoken over the whole of Kannaur except in the extreme east, where a dialect of Tibetan is current. It has five dialects (1) Lower Kannauri in the West of Kannaur, north of Sutlej, its area is about 12 miles from east to west and 6 miles from north to south (2) Standard Kannauri (3) Chitkhuli, spoken only in two villages in the Bospa valley viz. Chitkhul and Raksham (4) Theborskad, spoken in the east of the state, in the villages of Lippa, Asran, Labran, Kanam, Shunnam and Shaso i.e. between the Lippa river and the Tibetan speaking area of Kanaur".

Eickstedt (1926), "In these region, some remnants of very old languages also survive, the complex pronominalised Himalayan languages. They have in the main Tibeto-Burman character manifest traces of an older substratum of Munda tongues. These linguistic affinities with the aboriginals of central India are to be found in Chamba, Lahaul and Bashahr. Did Munda Peoples, coming from the east, once rules in these regions? Have they any indirect connection with the Khas? We donot yet know".

The Kinnauri dialect, called *Homskad,* is the mother tongue of nearly seventy five per cent of the population of Kinnaur. There are as many as nine different dialects used by various sections of the population. One of them called *Sangnaur* is spoken in solitary village of Tehsil Puh. The Village on the Tibetan border speak Tibetan dialects of the western Tibet. The extent of this spoken Tibetan is limited to the villages of Nesang, Kune and Charang only and adjoining Tibet. *Jangiam* dialect is spoken in Jangi, Lippa and Asrang Villages of Morang Tehsil. The *Shumccho* dialect is spoken in the villages of Kanam, Labrang, Spilo, Shyaso, Tailing and Rush Kalang of Puh Tehsil. A *Kinnauri-Jangram* mixture is the dialect which is closer to that of certain parts of the adjoining areas of Kinnaur.

A number of dialects are spoken by the inhabitants of district Kinnaur which came under 'Kinnauri' or 'Kanauri'. According to the classification of languages, made by the Linguistic Survey of India, 'Kanauri' comes under Tibeto- Chinese Family of languages. It has further been classified as a language belonging to Western Sub-Group of Pronominalized Himalayan Group belonging to Tibeto-Himalayan Branch under Tibeto-Burman Sub Family (Census of India, 1961). Bajpai (1991) has mentioned that there are as many as nine different dialects used by various sections in district Kinnaur. According to him, " one of them Sangnaur is spoken in a solitary village Sangnaur of Tehsil Puh. The villages on the Tibetan border speak Tibetan Dialects of western Tibet. The extent of spoken Tibetan is limited to the village of Nesang, Kunu and Charang adjoining Tibet. Jangiam dialect is spoken in Jangi, Lippa and Asrang villages of Morang Tehsil. The Shumceho dialect is spoken in the villages of Kanam, Labrang, Spilo, Shyaso, tailing and Rushkalang of Puh Tehsil. A Kinnauri-Jangiam mixture is the language used in Rakchham and Chitkul villages of Sangla Tehsil. The Scheduled Caste a language which is closer to that of certain parts of the adjoining district of Kinnaur. Besides these dialects the educated people of Kinnaur can speak Hindi also. Both men and women, especially in Sangla and Kalpa valley, can speak English in addition to their mother tongue and Hindi. (1987)

Steve know (1905) in his article made an attempt to highlight the facts related to Tibeto-Burman Dialect of Kinnaur people. According to him, Kanawari and connected dialects are mixed forms of speech. They all essentially Tibeto-Burman languages. There is also a substratum of languages built up according to the same principles as the Munda family. Further he concluded that Mundas or tribes related to them have once been settled in the Himalayas where traces of their language can still be observed in grammatical features of the dialects spoken at the present day. The non Tibeto-Burman element can be traced in a series of dialects along the ethnographic watershed between Tibetan and Indo-Aryan tribes from Kinnaur in the West, to Nepal in the East.

Food

Kinnauri have three meals a day—in the morning, in the noon and at night. Their staple food includes wheat, *ogla*, *phafra* and barely. Besides this *kangni* (kawing), *cheema*, maize (*chhallia*), koda (*kodru*) *chollaie dhankher* and *bathu* are also used. Peas, black peas, beans, mash and *masur* are the main pulses consumed. Cabbage, turnip, peas, beans, pumpkin, potato and tomato are the chief vegetables. Kinnauri relish meat. Smoking is widely common among the Kinnauri.

Country made liquor forms an integral part of the trial life and culture. Drinking of country beverages is considered a distinct phenomenon of the tribal society, and their use in recorded in almost all tribals as a stimulant. Tribal people do, of course, enjoy the sensation of being drunk. They express that they feel like king when they drink liquor fully. No social and religious festivals are performed practically without liquor. During dance the dancers of both sexes having a belief in getting more energy, spirit and rhythm drink indiscriminately and openly without any shyness, the distilled and undistilled liquor.

The Kinnaur liquor (Ghanti) is prepared in every household for their consumption. Every household are granted licence which can be acquired by paying Rs 25/- per annum to the state. The liquor can be prepared both by distillation and fermentation method. Arak or Rakh,or Phasur, Onguri are prepared by means of distillation process while Ningo, Chhang or Shudung and Bonga are prepared by means of fermentation process.

Arak is distilled from tag or Nanga Jo (Barley). The barley is heated over flame after putting sufficient water. After is it done, it is mixed with Phup powder (special ingredient of fermenting agent brought from places like Chango or Rampur).This is then kept in a closed bag for fermentation. After fermentation distillation is done by heating with water in an aircraft vessel. The steam water is then collected in a pot till the desired quantity is procured.

Dakhang Pashur (Anguri) is prepared out of grapes (Dukhang) by distillation process. Ningo, Chhang and Boja are prepared from Barley.

To produce Ningo, fermented barley is kept in a Jama pot with a hole at the bottom. A little water about 2 bottles are poured into the jama. And then this is kept covered for 25 days after which, the lower hole is opened and water percolate out of the pot. Then when the amount of water present is removed, the hole is closed again and then the remaining barley is taken in as Ningo.

For the making of Chhang, prepared ningo is removed from the pot but leaving behind some. Then water is added into it till about 25 mm above the grain. After 12 hours, the water is taken out of the pot in the same fashion as Ningo. This water is called Chhang. The remaining grain is given to the cattle as food.

Boja is also prepared from Barley. The fermented barley is kept in copper pot where some amount of water is added to it. Then it is rubbed with hands so that a thin paste is formed. Further it is filtered by means of sieve and then the liquid is served as Boja.

Brass and Bronze utensils are in common use in the area. Aluminium utensils are beginning to find favour with the people especially during the travel. Earthen pots are also used besides some wooden articles. Vessels made of the metals are procured from the plains.

Dress

Kinnauri wear woollen cloths throughout the year because of cold climate of the area. Men and women have a common head dress which a woollen cap is locally called *Thepang*. It is round woollen flat felt hat, with brim up turned. In winter, the upturned brim or ban is pulled down to protect ears from cold winds and snow. Pyjama, the dress of the men includes woollen shirts (*chamu-kurti*), long coat (*chhubha*), woollen pyjams (*chamu suttan*) (mostly of grey colour), Ghoya (woollen cloak), suluka (waist coat), sutan (woollen trouser), shew (cap) and kera (woollen rope tied around waists). *Chamu kurti* is an ordinary shirt worn by men and during winters *Chamu Kurti* is put on under *Chhuba*. Some people put on *tapru–suttan* embroided in various design and colours below the knees during the marriage and festival season.

The dress of the women includes woollen sari (*dhori*) full-sleeved blouse (*choli*), *chhanli* or shawl, Ra Ghoya (ordinary shirt), golak or ghoya or Yangluk or ligche (woollen shawl). Kera, sutan and shew. It is wrapped around the shoulders and its two ends are fasted together near the breast by means of a sliver hook called *digra*.

In Puh district, the dress pattern is different from the dress pattern of other part of district. In Hangrang district, people prefer to wear dark red or dark brown colour.

In Kinnaur district, *Chutan* akin to baby shawl is the dress of children. Rag is an ordinary shirt for the men.

Artificial Chambaka Flowers made by every lady in Kinnaur. Thin paper like material are cut into smaller pieces which is again cut into semi-circle shape. Such pieces are tied together to make with a straw to make a bunch of chambala which is used on cap.

Ornaments

Ornaments are the most valuable possession of Kinnauri women. They are normally made of silver and comprises of *mool, trimani kantaie, kandoch chalim, shanglavang, digra, tomach, lung, dhaglo* etc.

Men generally do not use ornaments. except *Murki* (ear ring) for the ear and rings for the finger. In older days, man used to wear daglo (steel bangles), *botone* se *Shanglya* (silver chain with buttons) and silver buttons. In Puh region, the men and shepherds carry *Chakmak* (a piece of steel for striking fire), a knife, a hatchet, a smoking pipe and goat hair rope around waist.

Metalware and Ornaments are done by the Domangs. They do all the job of solid and hollow casting, engraving, metal ware, ornament, musical instruments. It is said that these crafts have been imported from Tibet from the Tibetan craftsmen. They produce ornaments like Murki (Earring, *Balu* (nose- ring), *Chander sen har* (Necklace), *Long* (Nose ornament).The local people also do famous painting which is of Tibetan style made on paper made of ZIKO bark.

Communication

National Highways-22 is the most convenient communication network to Kinnaur. Buses ply regularly, except in winter and early spring when road conditions often impose long breaks. Himachal Pradesh Transport Corporation runs a number of buses to different parts of Kinnaur. There is one direct service from Shimla to Kaza plying on the Sutluj valley route, which takes 24 hours. In Kinnaur, at least four bus routes connect Reckong Peo with Shimla every day and other services go to Nichar, Sangla and Puh. Day services to travel to places within Kinnaur are also available but with less reliability than the long distance routes.

(i) By Air

There is no direct flight to Kinnaur and the nearest Airport is Shimla.

(ii) By Rail

There is no Rail service to Kinnaur and the nearest Railway Station at Shimla is connected by a narrow gauge line from Kalka (96Km.).

Education

The inspiration for a formal education was provided by the Moravian Mission particularly and the British rule in India. Perhaps the first regular primary school was started at Chini (Kalpa) in 1980 and another at Puh in 1899. Rev. J.T. Bruske and Mrs Bruske of the Moravian Mission started a school at Puh in 1899, but they left Puh in early 1900. In 1900, they started a school at Chini (Kalpa) for boys and girls which are regularly attended by twelve boys. The school at Chini was upgraded to lower middle in 1920, to middle school in 1944 and to High School in 1952. Next to it a primary school was started at Kalpa in 1914. The enlightened Raja of Bashahr Padam Singh who assumed powers in 1917 took a keen interest in education and established many schools. During that time, the problem of getting trained teachers was major as no one was interested to teach in isolated schools. To overcome this problem, several teachers were sent to training colleges and normal schools at state expenses. The Raja of Bashahr provided stipends and scholarships to motivate poor and deserving students for pursuit of studies.

The period after independence saw rapid expansion in the field of education. The Education Department, the Welfare Department, the Block Development agency and some voluntary social organisations like Parvatiya Adim Jati Sewak Sangh took initiative in opening educational institutions. (Chib, 1984).

At the time of formation of the district in 1960, there were two high schools, four middle schools and sixty–two primary schools and over a hundred primary schools in 1979. In addition to this, the state Social Welfare Board is running various Balwadi and Craft Training centres. According to census 2001, the total number of primary/junior schools, Middle/Senior Schools, Higher and Senior Secondary Schools and collage are 189, 33, 37 and 1 respectively. The detail of educational institutions is given in Table 2.6

Table 2.6: Educational Institutions in District

Year/ Tehsil	*Primary/Junior Basic School*	*Middle/Senior Basic School*	*Higher and Senior Secondary School*	*College*
1987-88	151	21	20	-
1988-89	153	30	22	-
1989-90	159	28	23	-
1990-91	162	27	24	-
1991-92	165	27	24	-
1992-93	165	27	24	-
1993-94	165	29	25	-
1994-95	163	28	26	-
1995-96	174	28	27	1
1996-97	181	26	28	1
1997-98	188	28	34	1
1998-99	189	29	36	1
1999-2000	189	33	37	1
2000-01	189	33	37	1
2001-02	189	33	37	1
Sangla & Kalpa Tehsil	51	11	11	1
Nichar Tehsil	74	12	11	-
Morang & Puh Tehsil	64	10	15	-

Source: District Kinnaur at Glance-2002.

The main reason for backwardness in the region is the lack of education in it. Education, an essential pre-requisite for the all-round development of any community is far from satisfactory levels. The causes of this low level of literacy are not difficult to ascertain. The difficult accessibility of schools, wrong time schedules and the indifferent attitude of population towards formal education and especially female education are some of the main factors responsible for low literacy level. Apart from this low quantity, the quality of education provided to the students is also defective. The education which is being given is devoid of any regional touch and is far away from the needs and requirement of lures them to the conveniences of urban life. The result is that in the absence of sufficient job opportunities in the region, the educated young men go to the plains in search of jobs.

Natural Calamities in Last Few Years

(i) Flood in Sutluj

Though many events occurred the decade but worst of all the events which can not be forgotten by the Kinnauri during the years to come is devasting flood in river Sutluj on the night intervening 31st July and 1st August, 2000 at 11.00 p.m. The flood started its play of the havoc at Khab and ended its play at Chaura in this district. As many as 56 people were washed away in the river Sutluj. 13 Bridges over river Sutluj completely washed away disrupting the normal life for about six months altogether.

Effects of Flood

• **38 Hectares Area washed away/damaged**	• **13 Bridges Washed Away**
• 80000 population affected	• 30 Kmtrs. NH22 washedAway/ Damaged
• 56 human lives lost	• Movement within District Restricted
• 58 houses washed away	• Shortage of foodgrains and essential commodities
• 10 houses partially damaged	• Failure of Power Supply & Communication
• 175 cattle heads lost	• Export of Horticulture Produce
• Property worth Rs. 22.26 Cr. destroyed	• Panic among the public

(ii) Flood in Panvin Khad

Besides above another devasting flood occurred in the hilly torrent known 'Pan Ghar', the tributary of the river Sutluj on its left bank at Wangtu due to cloud burst on the peaks of Panvin and Rohru on the intervening night of 11th and 12th August, 1997. As many as 19 persons, 4 bridges including RCC Bridge popularly known as 'Zangchham' were washed away in the flood. More than 103 houses were either washed away or damaged by the flood in the district.

(iii) Flood in Brua Khad

Flood in Brua Khad, the hilly torrent, the tributary of Baspa Khad occurred on 7th June, 2000 at 12.30 p.m. The flood washed away as many as 20 houses and Govt. Buildings of High/Primary School, Ayurvedic Hospital, Panchayat Ghar, Club Bhawan and Mahilamandal Bhawan. Affected houseless families had to be shifted in tents. Apart from above, 69 houses were rendered unsafe due to land slides caused by the flood.

(iv) Flood in Dulling Khad

On the intervening night of 4th/5th September, 1995 flood occurred in Dulling Khad, the hilly torrent, the tributary of Sutluj on its left bank at Tapri. Lower portion of Tapri Bazar consisting of 23 residential houses, HRTC workshop, Poultry farm, Official residence of SDO electricity and three hundred meters of Nation Highway washed away in the flood. As a result of which HRTC workshop and its employees were shifted from Tapri to Reckong Peo.

THE PEOPLE

According to the Surveyor General of India, the total area of Himachal Pradesh is 55,673 sq. km which is divided into 12 administrative districts. The total population of the Himachal Pradesh according to 2001 Census was 6,077,248 and a density of population in Himachal Pradesh is 109 persons per square km. According to Census 1991, the total scheduled caste population of the Himachal Pradesh is 13,10,296 and scheduled tribe population of the Himachal Pradesh which has its concentration in districts of Kinnaur and Lahul Spiti, parts of Chamba district and scattered in other districts is 2,18,398. About 69 per cent of the state tribal population fall under Tribal Sub-plan area.

Kinnar (1956, 1957a, 1957b) has published number of articles of Kinnauri folklore. Even in one of his article his beautifully described the formation of Kinnaur district and its people. Kirkpatruck (1878) as transcript of his travelogue highlighted the existence of polyandry in the Punjab state (now separate state, Himachal Pradesh).

Kaushal (1988) in his book explained the ancient tribes of Himachal Pradesh where he had made an attempt to give historical background of Kinnaur tribe in Himachal Pradesh.

Earlier, Kinnaur was the part of the princely state of Bushahr. The Bushahr merged with the territory of Himachal Pradesh after Independence and the Kinnaur region formed a Tehsil of Mahasu district. It came into being as an independent district in 1960 along with some part of the former Rampur Tehsil of Mahasu district.

Kinnaur District, the north eastern frontier district of Himachal Pradesh and border district of India lies in the western Himalayas on both banks of river Sutlej. It is situated between 31 degree 05′ 50″ and 32-degree 05′ 15″ north latitude and between 77 degree 45′ and 79-degree 00′ 35″ east longitude. It is about 80 kilometres in length and nearly 65 kilometres in breadth. The altitude in the Pradesh, a holy mountainous region in the lap of Himalayas range from 350 ft. to 6975 mts above mean sea level.

Historical Perspective

The Kinnauri has strong association with Hindu mythology. Most of the ancient literature have referred to these people as 'Super Being' higher and different than the common man the Homo Sapiens: and nearer to the order of God along with Gandharva, Yaksha, Kirat etc. According to the Raj Tarangi Kinnauri are said to have born from the shadow of Lord Brahma (Singh, 1982).

Kalidas, the great Sanskrit scholar, has composed verses in praise of Kinnaur and has given them importance in his famous Megadoota. According to Chakravarti (1971), Kinnauri might have their origin from Sumerians, branch of whom become famous by the name Simbiar; and the consecutive distribution of Simbeari in the Himalayan area give rise to the people called Kinnauri (Kaushal, 1965). According to the Shashi (1971), the word Kinnar means half giant and on this basis they are also known as Ashvamukha. (District Hand Book, 1961).

The Kinnaur have been declared as scheduled Area under the Fifth Schedule of the Constitution by the President of India as per the Schedule Areas (HP) Order 1975 dated 21st November 1975.

Bajpai (1991) focusing on the Kinnaur region, has stated that "it appears that princes being adventurous owing to territorial greed first went to these high hills and established themselves on varying territories in accordance with their powers. Particularly the area between Sutlej, its tributaries and Baspa up to Mansarover was under the rule of Thakkars from very early times. They were known by the place they had occupied like that of Chini Thakur and Kamru Thakur, under the overall sovereignty of Maurya and Gupta Kings later on. It was the Thakur of Kamru who proved strongest of all the other chiefs of the area and annexed their territories by force some time after the fall of the Kanauj Empire and laid the foundation of the state of Bushahr, to which the region of Kinnaur belonged till the dissolution of the state very recently".

Regarding the Bushahr state and its rulers, Bajpai (1991) mentions that "in the medieval period, though some of the hill states such as Kangra, Chamba and Sirmaur were attacked and made tributary to the Mughal emperor of Delhi, Bushahr state could not be reached by any adventurer of that period. The consolidating and addition of territories to Bushahr state continued during that period also. The Thakoorais of Dulaitoo, Kurungoloo and Kuaitro were annexed about Samwat 1611. Perhaps he was Raja Chatar Singh, who brought the whole area of the erstwhile Bushahr state under his control. He was considered as most virtuous ruler during his reign. Nothing particular is known about his successor Kalyan Singh. The Successor of Kalyan Singh according to genealogy was Raja Kehri Singh. He is described as the highest skill warrior of that time. Kehri Singh's successors were not of the same mettle. Besides mention in genealogy of Bushahr states nothing is known about Vijay Singh and Udai Singh. It is said that one Raja Ram Singh made Rampur his capital, in place of Sarahan and Kamru. During his reign a series of contests began with the Raja of Kulu and Bushahr had lost the territory of Seraj. It seems that territories which were annexed by Raja Kehri Singh became free during the weak rule of Raja Rudra Singh. But his successor Ugar Singh took them over by force of arms".

According to Punjab States Gazetteer—Shimla Hill States (1904), from 1803 to 1815 the erstwhile state of Bushahr faced the menace of Gurkha invasions. Immediately after the death of Raja Kesari Singh, the Gurkhas made a massive attach on Bushar. The minor ruler and his mother who could not withstand the attack fled away to Kamru leaving behind a rich treasury at Sarahan. The

Gurkhas looted the treasury and completely destroyed the records of the state. Regarding the Gurkha occupation, Bajpai (1991) has stated that the Gurkhas of Nepal had extended their dominions greatly during the end of the eighteenth Century and the beginning of nineteenth century. Amar Singh Thapa, the Gurkha leader, went up to Kangra valley. He was driven from the valley by the superior forces of Ranjit Singh, and those of Raja Sansar Chand of Kangra. The tract between Sutlej and Jamuna came under British protection by the Treaty of 1809 between Ranjit Singh and British Government. Thus the British Government took positive steps to expel the Gurkhas and after a long and desperate struggle, ochterlony completely defeated Amar Singh Thapa on 15th April 1815. On the conclusion of Gurkha war, Raja Mahendra Singh was granted a *sanad* on 6th November, 1815. It gave Khanetian Delath Thakurais to Buhahr, and a part of Rawain, which was a district of the state, was transferred to Keonthal; Kumharsain was constituted a separate Thakurai (ibid).

Dutta (1997) explained about the British contested over Shimla till states and trading practices of people rendering in these areas. According to him, "The British made administrative arrangements to control effectively the newly acquired territories. Sir David Ochterlong, Superintendent of Political Affairs and Agent to the Governor-General in the territories of protected Sikh and Hill Chiefs, was vested with combined political and military authority. Two assistants based at Sabathu and Nahan helped him in the administration. The supreme Government desired that these officers should pay attention to the survey of the states within their superintendence and collect and communicate their scientific information. The whole of the upper Sutlej valley was surveyed by J.A. Hodgson, Surveyor-General of India and his Deputy, J.D. Herbert who was also a distinguished astronomer. In his work, Hodgson and Herbert were helped by the three Gerard brothers (Alexander Gerard, J.G. Gerard and P. Gerard), who were all officers in the Bengal Army and distinguished explorers and geographers.

Francke (1913-26) of the Moravian Mission Board was requested by the Archaeological Survey of India to undertake an archaeological survey of the districts which once famed the kingdom of western Tibet and which had never been explored by any scholar with a thorough knowledge of Tibetan, local history and antiquities. He travelled up the Sutlej valley to Spiti and then to Ladakh. He ended up his journey by travelling from Ladakh to Kashmir.

From the foregoing account it would appear that during the princely days Kinnaur valley acted as a bulwark to the Bushahr state. However, with the lapse of paramountancy, the Kinnaur, them known as Chini Tehsil, was merged to form a part of then Mehasu District. The pargana Atharah/ Bish comprised of villages Nichar, Sungra, Kangos, Paunda, Bari, Taranda and Chauhra village with patwar circle at Paunda, the paragana Pandrah/ Bish constituted the revenue estates of Nathpa, Kandar, Barakamba, Chhotakamba, Garshu and Rupi with patwar circle at Rupi were in Rampur tahsil.

In fact, then Chini tahsil covered the entire Kinnaur valley beyond Wangtu which was created in 1891 by the then rulers Tikka Raghunath Singh. Thus, since 1891 onwards Chini tahsil continued to be in existence with its vast area beyond Wangtu up till 1960. Since 1847, it was a tahsil of the then Mahasu district. By 1960, the importance of reorganising border areas was realised and consequently in view of ethnic and cultural considerations, the areas which were partly in Rampur tahsil and partly in Chini tahsil were reorganised into a separate district forming the present Kinnaur district.

According to Punjab District Gazetter—Simla District (1904), the wilder parts of Bashahr beyond the Sutlej are thus described by Sir H. Davies: "Immediately to the south of Spiti and Lahul and the district of Kanawar, which forms the largest sub-division of the Bashahr principality, and consists of a series of rocky and precipitous ravines, descending rapidly to the bed of the Sutlej. The district is about 70 miles long by 40 and 20 broad at its northern and southern extremities respectively. In middle Kanawar the cultivated spots have an average elevation of 7,000 feet. The climate is genial, being beyond the influence of the periodical rains of India; and the winters are comparatively mild. Upper Kanawar more resembles the Alpine region of Tibet. Grain and fuel are produced abundantly; the poppy also flourishes. The Kanawaris are probably of Indian race, though in manners and religion they partially assimilate to the Tibetans. The people of the north are active traders, proceeding to Leh for charas, and to Gardokh for shawl-wool, giving in exchange money, clothes and spices. The mountain paths are scarcely practicable for laden mules, and merchandize is carried chiefly on the backs of sheep and goats."

Whatever may be their origin, many Hindu mythologies and literatures have considered them as belonging to a divine race. Along with other Himalayan god and goddesses like the Apsaras, the Yakshas, the Gandhavas, the Khasas, the Guhyakas, the Siddhas and others, the Kinnauris are also considered as having a divine origin living in the Himalayas. Along with those races they have also been mentioned in various ancient Indian literatures like Puranas, Upanishads, Mahabharta, Ramayana, Manusamhita etc. (Mahato and Raha, 1975). In Raghuvansam they have been described as a polyandrous tribe in the Tibeto-Himalayan region and in Nepal. Alain (1964) has found that in some ancient Indian literatures these people described as fabulous beings half human half bird with bird's legs and wings', are called Kinnaras or Kimpurusas meaning what kind (kim) of human being (narah, nurusa). They have also been mentioned as Aswamukha having human head with horse's body or horse's head with human body (ibid). The etymological meaning of the term Kinnaura is 'an ugly man' (Kim kutsita narah). Further these names (Kimpurasa and Aswamukha) might have been given to the defeated and enslaved Dasas (1947). Whatever their physical features may be, that they were good singers, have been mentioned in many books. In Meghdutam, mumarasambhavam, Raghuvansam, Mahabharata etc., they have been mentioned as celestial singers, expert in music and dance.

According to the Sankrityayan (1957) in the prehistoric times Kirata inhabited the region of present Kangra and almost the whole of the western Himalayas. The people residing in Lahul and Spiti, Kinnaur, Malana (Kulu) and some area of Uttaranchal belongs to Kirati stock. In ancient times Kirata were also called 'Mon', and it is interesting to note that the old capital of Kinnaur bore the name of 'Mone', possible because of the fact that it is related to the people of the Mon Tribe.

According to Hilton (1938), "the Raja had three wazirs who managed the affairs of their respective territories. These three wazirs were equal in rank and their offices were hereditary. Below them were several inferior officers also called wazirs, whose offices were not hereditary, but who were appointed by the Raja annually.

The district erstwhile known as Chini tehsil of the former Mahasu district, came into being as an independent district w.e.f. 1st May, 1960. Prior to merger of State on the eve of independence, Kinnaur valley was a part of erstwhile Bushahr State which had its headquarters at Rampur.(Ahluwalia, 1988; Agnihotri, 1959).

According government records, the history of Kinnaur is broadly divided into three parts:

(i) Early History

In the absence of authentic historical record the early history of Kinnaur region is obscure and the reference of the Kinnaura or Kannaura and there land is by the large confined to legends and mythological accounts. It would be worthwhile to look at the region of Kinnaur along with general conditions of northern India particularly the hilly regions of Himalayas during the period from 6th century B.C. India was divided in to sixteen great janpadas and several smaller ones. Among them Gandhara, Kamboja, Kuru, Koshal, Mull, Vajji, Panchal, Sakya were either in the southern Himalayas ranges or had territories extended up to Himalayan ranges. Among the states that were flourishing in the six century B.C. The kingdom of Magdha was the first to make a successful bid for supremacy under Bimbisara. Its emperor belonging to Sunga, Nanda and Maurya dynasties carried their banners upto the inhabited parts of inner Himalayan region. Chandragupta Maurya brought about its political unification under one scepter, negotiated an alliance with Parvataka (Himalayan King) before empire building.

With the help of several frontier tribes such as Kiratas, Kambojas, Panasikas and Valhika, he built up the great Mauryan Empire. The empire of Ashoka extended upto natural boundaries of India and beyond that in the west. After the collapse of the Mauryan empire the Kushanas established an extensive empire within and beyond India in the northwest. Emperor Kanishkas hegemony spread over Kashmir and the Central Asian regions of Kashgar, Yarkand and Khotan. His hold extended upto the territory of the Inner Himalayas and Kinnaur must have been the part of this empire. In the meantime northern India was divided into a number of small kingdoms and autonomous tribal states. Under such a divided country the Gupta Empire grew. Samudragupta's empire included the territories of Rohilkhand, Kumaon, Garhwal, Nepal and Assam. Its northern boundary was along the high Himalayas. Kinnaur must have been included in it too. Early in the seventh century A.D., Harsha came to power at Thaneshwar in A.D. 606. During the course of four decades he had established a most powerful empire in India. All the existing kingdoms of Kapisa, Kashmir, Kuluta, Satadru, Mon-li-pa-lo (Ladakh) and Suwarnagotra (in the high Himalayas) were incorporated in his empire. After the death of Harsha in A.D. 647 the country was once again divided into old principalities of the sixth century B.C.

It appears that the princes being adventurous owing to territorial greed first went to these high hills and established themselves on varying territories in accordance with their powers. Particularly the area between Sutluj, its tributaries and Baspa upto Mansarover was under the rule of Thakkers from very early times. They were known by the place they had occupied like that of Chini Thakur and Kamru Thakur, under the overall suzerainty of Mauryan and Gupta kings later on. It was the Thakur of Kamru who proved strongest of all the other chiefs of the area and annexed their territories by force sometime after the fall of the Kanauj Empire and laid the foundation of the state of Bushahr, to which the region of Kinnaur belonged till the dissolution of the state very recently.

(ii) Medieval Period History

By the beginning of the fourteenth century the entire area of Kinnaur was divided in seven parts, locally called Sat Khund. There was further splitting up of and the area came to be covered with much small hegemony, which were constantly warring against, or allying with, each other as conditions required. The neighbouring Bhots also found time to jump into the fray and did not desist from creating trouble. There are various forts like Labrang, Morang, and Kamru forts telling the story of that age.

In the medieval period, though some of the hill states such as Kangra, Chamba and Sirmaur were attacked and made tributary to the Mughal emperor at Delhi, Bushahr state could not be reached by any adventurer of that time. The consolidation and addition of territories of the Bushahr state continued during the period also. The Thakoorais of Dulaitoo, Kurungoloo and Kuaitro were annexed about Samwat 1611. Raja Chatar Singh who brought the whole area of the erswhile Bushahr State under his control. He was considered most virtuous ruler during his reign. Nothing particular known about his successor Kalyan Singh. According to genealogy, The successor of Kalyan Singh was Raja Kehri Singh. He was the highest skilled warrior of the time. Kehri Singh's successor was not of the same mettle. Besides mention in genealogy of Bushahr State, nothing is known about Vijay Singh and Udai Singh. It is said that one Raja Ram Singh made Rampur his capital in place of Sarahan and Kamru. During his reign a series of contests began with the Raja of Kulu and Bushahr had lost the territory of Seraj. It seems that the territories which were annexed by Raja Kehri Singh became free during the weak rule of Raja Rudra Singh. But his successor Ugar Singh took them over by force of arms.

(iii) Recent History

According to Punjab state Gazetteer—Shimla hill States from 1803 to 1815 the earstwhile states of Bushahr faced the menace of Gurkha invasions. Immediate after the death of Raja Kehri Singh, The Gurkhas made massive attack on Bushahr. The minor ruler and his mother who could not withstand the attack fled away to Namru leaving behind a rich treasury at Sarahan. The Gurkhas looted the treasury and completely destroyed the records of the state. Keeping Gurkhas of Nepal had extended their dominions greatly during the end of the eighteen century. Amar Singh Thapa, the Gurkha leader went up to Kangra valley. He was drawn from the valley by the superior forces of Ranjit Singh and those of raja Sansar Chand of Kangra. The tract between the Sutluj and Jamuna came under British protection by the treaty of 1809 between Ranjit Singh and the British Govt. Thus the British Government took positive step to expel the Gurkhas and after a long and desperate struggle, completely defeated Amar Singh Thapa on 15th April, 1815. On the conclusion of the Gurkha war Raja Mahendra Singh was granted a sanad on 6th November, 1815. It gave Khaneti and Delath thakurais to Bushahr and a part of Rawin, which was a district of the state, was transferred to Keonthal, Kumharsain was constituted a separate Thakurai.

From the foregoing account it would appear that during the princely days Kinnaur valley acted as a bulwark to the Bushahr state. However with the lapse of paramountacy, the Kinnaur then known as Chini Tehsil was merged to form a part of then Mahasu district. The pargana Atharahs Bish comprised of village Nichar, Sungra, Kangos, Ponda, Baro, Bari, Tranda, Chaura village with patwar circle at Ponda. The Paragana Bish consisted of the revenue estates of Nathpa, Kandhar, Barakamba, Chhotakamba, Garshu and Rupi with patwar circle at Rupi were in Rampur Tehsil.

In fact then Chini tehsil covered the entire Kinnaur valley beyond Wangtu which was created in 1891 by the then ruler Tika Raghunath Singh. Thus 1891 onwards Chinni Tehsil continued to be in existence with its vast area beyond Wangtu uptill 1960. Since 1947 it was a Tehsil of the then Mahasu district. By 1960 the importance of reorganising border area was realised and consequently in view of ethnic and cultural considerations the areas which were partly in Rampur Tehsil were reorganised into a separate District forming the present Kinnaur district.

Various studies has been done on Kinnaur district like The ethnographic profile of Kinnaurs has been written by various scholars and travellers like Atkinson (1882-86), Cunningham (1844), Gerard (1841), Herne (1876), Jacquement (1833), Pall, Murray-Aynshlay (1882), Sanskritayan (1957), Thomas (1852). Horne (1876) gives his travel account to villages visited in Kumaon, Garhwal and on the Satlej. Ahir (1971) explained the existence of temples in Kunum village (Kinnaur District) and considered it as pride of Himachal Pradesh. Ahluwalia (1998) in his book gave socio-cultural and economic aspect of people of Kinnaur district. Khullar (1972-73) described his expedition to Peo/Kinnaur Kriles from Kalpa I.T.B.P. Headquarters. In his article, he explained about the expedition approach and his field encounter with rough terrain of Kinnaur district. Haresh Kapadia (1985-86, 1993, 1999) described the expedition trip to Kinnaur district and narrates his encountered with hostile terrain of this Himalayan region. Handa (1987, 1999) in his writing describes about the structural and functional aspects of monasteries in life of Kinnaurs. He provides a detail account of spread of Buddhism in the Himalayan region. Dubey (1966) narrated his experience of expedition to Kinner–Kailash. Chakrabarti (1977) paper outlined the history of archaeological research in the western Himalayas since the middle of the 19th century. It briefly describes the archaeological work done in Himachal Pradesh region. Sandhu (1979) narrated his details of expedition to Baspa valley in Kinnaur District of Himachal Pradesh.

Jerath (1995) explain the about the living conditions and social aspects of Kinnaur. He gives a brief ethnographic profile of people inhabiting Kinnaur district.

Ethnic Composition of Kinnaur District

The inhabitant of the Kinnaur, i.e. Kinnauri has been referred at several places in Sanskrit literature. There are many variants of Kinnauri such as Koonauri, Kannaura, Koonawri, Kinnaura, Kinnaur, Kinnauries, Kinauri, and Kinnauri.

Hedin (1913) while writing about Chini commented, "the village is said to contain 500 inhabitants, who all belong to one tribe called Kanauri and the divided into three clans, each with its own language atleast a dialect widely different from the other two. He was mistaken in calling these strata as clans."

Rahul (1970) stated that the Kanets or Rajput of southern Kinnaur, the dominant group, are immigrants from the plains. The Chamangs including the Domangs are perhaps the indigenous people.

Shashi (1971) described, "According to the local folk tales, the Kolis and Badhis were brought from outside to work as artisans. They may not belong to the Khash tribe but outsiders also call them Kinners".

Raha (1974) gives a brief sketch about some of the ethnic groups of western and central Himalayas—the Ladakese of Ladak in Kashmir, the Lahulese and the Spitians of Lahul and Spiti district, the Kanets (Rajput) of Kinnaur district of Himachal Pradesh and the Bhotias of Pithoragarh, Chamoli and Uttrarkashi districts of Uttaranchal (erst while Uttar Pradesh).

According to Bhadaur (1978), "After the formation of the present Himachal Pradesh, the district of Kinnaur was declared as a scheduled area and the people considered as scheduled tribes. Broadly the Kinnaurs are divided into two strates, the upper caste being the land owners, and the remaining leather workers, weavers and carpenters and farm labourers.

Charak (1978) provide descriptive socio-cultural historical aspect of Kinnauris. According to him, the Kinnauri society is over-ridden by the Hindu caste system. The Kanets there are said to be divided into three grades, each comprising a number of septs, about seventy in all, peculiar to Lahul only. Besides Kanets the only two castes in Kinnaur are the Chamang who makes shoes and weave, and the Domang (Dums) who are blacksmiths and carpenters. Water or food which had been touched by the lower castes was not used by Kanets, nor were people at these castes allowed to enter a Kanet's house. If any Kanet happened to eat such food inadvertently be used to apply to his Raja who bade him make expiation-prayas chitta and pay some *nazrana*. This custom was called *Sajeran* or *Sacheran*.

Raha (1978) paper made an attempt to focus on the structural stratification and religious duality in Kinnaur culture. According to him, "with the exception of the Khasia, none of these ethnic groups in Kinnaur are divided into social divisions like clans or lineages. They are simply divided into a number of households. Only recently has some rudimentary form of occupational division been stated among the Koli, of lower Kinnaur, for example, *bonu* and *sui*. Those who weave as their main occupation are called *bonu* (weaving) or those who have suffered over territory are called *sui* (stitching).

Sen (1979) observed, "while the Kosia or the Rajput consider themselves as high caste, the other castes are grouped into low caste group".

According to Dhir (1990), "In Kinnaur, all are known as Kinnars which is a scheduled tribe. At least one third of the populations consists of scheduled castes Kohlis, Lohars and Badhi, but for purposes of outside worlds all residents of Kinnaur are Kinnars and hence tribes. These castes are in comparison backward to the tribes in Kinnaur".

Balokhra (1999) in book describes the geographical, people, history and administrative history, art and culture of Himachal Pradesh. In section on Kinnaur, he writes, "The Kinnaura belongs to different ethnic groups. Among them the Rajputs (also known as Kanet, Khas, Khoria or Khara) are the dominant group. Next comes Koli who are weavers and artisans, leather workers and ploughman, followed by the Lohar, Badhi and Nagalu, who are iron smiths and silver smiths, carpenters and masons and basket makers respectively".

Ibetson and Maclagan during first quarter of this 19th century found the Kanets (Rajputs or Khosia) of Kinnaur also divided into three grades (Khel). Each of these grades of the Kanet had a number of septs (*Khandan*) whose names did not appear among the Kanets of Bashahr proper. In case of first grade Kanet (Rajput), septs were different in different administrative units or parganas like Rajgoon,, Inner Tupka, outer Tupka, Pandrabis and Atharabis. Similarly the second grade Kanet (Rajput) also had different septs in different parganas. Ibetson mentioned about these paraganas, Inner Tupka, Shuwa and Rajgoon, where the septs were different. The third grade Kanets who asked as potters, also had a set of septs (Rose 1911).

According to Raha (1987), "The Khosia (Kanet) society is divided into three distinct states groups called the *Khel*. These three khels are the Orang, Maorang (also known as Orang-Mech) and the *Waza*. The three different khels of the Khosia society are further sub-divided into a number of smaller exogamous social units called the *Khandan*. Some of these *Khandans* in Kinnaur are further divided into the sub-*Khandan*. But the sub-khandans are definitely not found in case of all *Khandans*. Then come the smallest social unit of the Khosia society, the *Kim* or the household. A number of household comes under a *Khandan* or sub-*Khandan*. It is thus seen that the pattern of stratification of the Kinnaur society depends on the cultural and ecological designs of the areas".

The Kinnauri are divided into five ethnic groups i.e. Kanet, Koli, Badhi, Lohar and Nagloo. Mainly there are two divisions one of the Kanet and the other includes rest of the ethnic groups. The population of the district comprises of four distinct classes of Kanets (Rajputs), Chamangs {(Kolis, preparing shoes, weaving, tailoring and music drum beating)} and Domangs { Lohar (Blacksmith) and Badhi (carpentry), Nagloo (Basket makers)}. The Kanet constitute the majority of Kinnaur population. They are declared as Scheduled tribe people. The other four ethnic groups i.e Koli, Badhi, Lohar and Nagloo constitute the scheduled caste population of district.

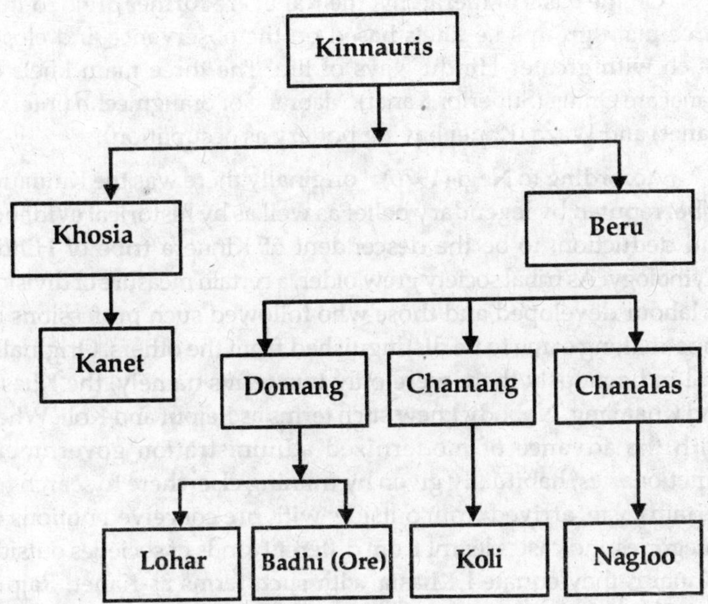

Fig. 2.1: Ethnic Composition of Kinnauris

The Kanet have treated them as their serfs having no status. These people (i.e Koli, Badhi, Lohar and Nagloo) mainly comprises of the artisan class and are considered untouchables whether they be Koli having the occupation of preparing shoes, weaving, tailoring and music drum beating, the Badhi having the carpentry and mason, Lohar-Blacksmith and Nagaloo -Basket making. (Chander, 1992; Chawla, 1981; Chib 1984; Mahato and Raha. 1975)

The language and the history or tradition of these people is not different and there is no marked difference in their religious and social customs and practices. They appear to be on continuum in this regard with the Kanet with some minor degree of variation in few cases. (Sharma, 1988; Bailey, 1975).

The Kanet are considered to be superior of the high class people (on the basis of social hierarchy and ritual purity) other groups are considered to be socially of low class and looked down upon by the Kanet. The Kanet do not have any occupation specialisation associated.

On the basis of hierarchy, the Kanet are further divided into three main groups i.e khels based on the observance and closer touch with greater Hindu ways of life. The three main khels of Kanet are Orang (Superior Kanet), Maorang or orangmeehh (inferior Kanet) and Waza (Kanet having pottery as occupation).

According to Negi (1976), "originally there was the Kinnaura tribe, reputed by legendary belief as well as by historical evidence and deduction, to be the descendent of Kinnera tribe of Hindu mythology. As tribal society grew older, a certain measure of division of labour developed and those who followed such professions as shoe-making came to be distinguished from the others. Originally and indigenously there were only two terms namely, the Khasia and Chamang. Nobody knew such terms as Rajput and Koli. When with the advance of modernized administration government functionaries, habitually given by training elsewhere to search for certain caste, arrived from outside, with pre conceived notions of categories and castes learnt from different kinds of societies outside Kinnaur, they equated Khasia with such terms as Kanet, Rajput and Chamang as Koli, whatever the terminology. The fundamental and more important feature is that Khasia or Chamang, Rajput and Koli both are Kinners/Kinnaurs/Kannaras and both have a common overall similarity and identity in social patterns, in dress and other mode of living in customs, and in faiths, beliefs, superstitions, worship, folklore, traditions etc. and this common bound is so strong that both fit completely into the same tribal picture entitled Kinner/ Kinnaura/ Kinaura".

The generic term Kinnauri includes the schedule tribe and schedule caste population of Kinnaur. (Khanna and Kapoor, 2000) The Kinnauri along the Tibetan border possess the Mongoloid and Tatar features and are of fair complexion and good looking with Persian features.

Kapoor (1990) highlighted the difference in usage of scheduled tribes and scheduled caste terminology among the Kinnauris. According to him, "of the four ethnic groups under the *Beru* category, the numerically most dominant are the Kolis. The rest there are very few in number and as stated earlier, only one or two households can be located in a village, and that too not in all the villages. These four ethnic groups are considered as the scheduled castes, but as Kinnaurs they are also treated as the Scheduled tribes like the Rajputs. These sometimes cause confusion".

TRIBE-CASTE IN KINNAUR

Scheduled Tribe in Kinnaur District

The Hindi equivalent for the English word "Tribe", as used in the Hindi version of the Constitution, is "Adim Jati". Thus, though not providing any specific definition, the Constitution does, by implication; give at least one clue-that of autochthony.

The scheduled tribe and scheduled castes are those tribe and caste which have been notified as such by the Presidential Order in pursuance of the Article 341 and 342 of the constitution. The Article 342 of the Constitution authorises the President to specify, by public notification, tribes or tribal communities, "which shall for the purposes of this constitution be deemed to be Scheduled Tribes", and also empowers Parliament to, by law, "include in or exclude from the list of Scheduled Tribes" specified in the Presidential notification.

From a study of the available clues, views and definitions, having a bearing on the subject and of the pragmatic selections of tribes actually made so far for the Presidential orders and the Parliamentary legislation, the following criteria may be taken to emerge for identifying a tribe to be scheduled:

1. Autochthony;
2. Groupism or a very strong community-fellowship, if not descent from a common ancestor or loyalty to a common headman or chief;
3. A principal, if not an exclusive, territorial habitat;
4. A distinctive way of life, primitive or backward by modern standards, and apart and aside from the main current of culture;
5. Economic, political and social backwardness.

The list of Scheduled Castes and Scheduled Tribes were notified for the first time under the constitution (Scheduled Caste)

Order 1950 and the constitution (Scheduled Tribes Order 1950 These list have been or modified or amended or supplemented from time to time. On reorganisation of the states the Schedule Castes and Schedule Tribes (modification) order came into force from October 1956. The first one affecting the Himachal Pradesh has been the Scheduled Castes and Scheduled tribes Order (amendment). The main objective of this Act was to remove the area restriction in respect of Scheduled castes and most of the Scheduled tribes. With the result most of the Scheduled castes and Scheduled tribes are not notified as such throughout the state, unlike in the 1956 order these were notified in relation to different regions the state registering area restrictions.

The following Presidential Orders, and the Act noted below, have, between them, scheduled the tribes shown alongside as the Scheduled Tribes of the territories now comprised in the State of Himachal Pradesh as given in Table 2.7.

Table 2.7: Presidential Orders and Tribe

A. Orders	*Tribes*
1. The Constitution (Scheduled Tribes) order 1950.	1. Gaddi
2. The Constitution (Scheduled Tribes) (Part 'C' States) Order, 1951.	2. Kanaura or Kinnara
3. The Constitution (Scheduled Tribes) (Union Territories) Order, 1951.	3. Gujjar
4. The Constitution (Scheduled Castes and Scheduled Tribes Lists-Modification) Order, 1956.	4. Jad
B. Act	*Tribes*
1. Central Act 31 of 1966 ; Section 28 (2) and Schedule xi.	1. Lamba
	2. Khampa
	3. Bhot or Bodh
	4. Pangwala
	5. Lahaula
	6. Swangla

In 1965, the Government of India appointed an "advisory committee on the Revision of the Lists of Scheduled Castes and Scheduled Tribes" to review the lists of Scheduled Castes and Tribes, preparatory to a revision of the lists by Parliament. (Mohan, 1998). According to the recommendations of the Committee, the revised list for what is now the State of Himachal Pradesh is given in Table 2.8.

Table 2.8: Revised List

S. No	*Tribe*	*Synonym*
1.	Banjara	
2.	Jad	
3.	Kanaura	Kinnaura
4.	Lahaula	
5.	Pangwala	
6.	Beta	
7.	Bodh	
8.	Chan	
9.	Domba	Gara
10.	Kanet (of Chhota and Bara Banghal)	Seok
11.	Swangla	

A bill, called the Scheduled Castes and Scheduled Tribes Order (Amendment) Bill 1967, was introduced in Parliament, in the light of the recommendations of this Advisory Committee. Parliament referred the bill to a Joint Select Committee. When the bill emerged from the melting pot of the Joint Select Committee, it provided for the tribes mentioned in Table 2.9 as the Scheduled Tribes of the territories now comprised in the state of Himachal Pradesh. (Negi, 1998).

The bill lapsed because the mid-term Lok Sabha general elections intervened. The new Parliament has not yet been seized of this issue of revising the lists of Scheduled Castes and Scheduled Tribes.

Table 2.9: Scheduled Tribes of Himachal Pradesh

1.	Beta; Beda.
2.	Bodh
3.	Chan
4.	Domba, Gara, Zeba
5.	Jad, Khampa, Lamba
6.	Kanaura (including Chamang and Damang)
7.	Kanet or Seok (in Chhota and Bara Banghal)
8.	Kolta
9.	Lahaula
10.	Pangwala
11.	Swangla

The increase in 1961, compared to 1951, and in 1971, as against 1961, is attributable to:

1. The addition of tribes to the Schedule (mainly);
2. The addition of the tribal population of the new areas integrated with Himachal Pradesh after the re-organization of the Punjabi; and
3. The normal increase of human population.

The Lokur committee and the Joint Select Committee of Parliament have proposed dropping Kinnara and retaining Kanaura only in pursuance of the general policy, advocated by them, of retaining only one of the synonymous tribal names where there happen to be more names than one of the same tribe. The resultant bill not yet having become an act, Kinnara and Kanaura both continue, alternatively, to be in official use.

Literally speaking, *Kanaura* means the inhabitant of Kinnaur. The name of the principal territorial habitat of the tribe has been variously spelt and pronounced as *Kanauring, Kanaur, Kanawar, Kunawar, Koonawar* etc. Kinnawar, and Kinnaur. The present official name is Kinnaur, though Kanaur is more in use in common parlance. Kinnaur is a separate district and a separate assembly constituency of Himachal Pradesh, originally and presumably. (Gupta, 1998).

The territory derived its name from the Tribe. Kinnaur (Kinnara Desa) was the abode of the Kinnara Tribe. The original name,

"Kinnaur", got corrupted into the various pronunciations and spellings given above. Thus, from the original Tribal name "Kinnara" to the original territorial name, "Kinnaur", i. e. the abode of "Kinnara," later, as shown above, corruptly pronounced and spelt, among else, as "Kanaur," and from this new and common territorial name "Kanaur" to the new and common Tribal name "Kanaura," i.e. the inhabitant of "Kanaur", seems to be genesis and the metamorphosis of these names. Alternatively, it may be that the new common Tribal name, "Kanaura", is a straight corruption of the original "Kinnara". (Bajpai, 1981; Bajpai , 1991; Banerjee and Banerjee, 1997).

Most of the Kinnaur population enumerated in the other districts, at the time of the census, consisted of temporary residents, though there do exist some permanent Kinnaur settlements and/or stray cases of domiciled Kanaura families outside Kinnaur. Kulu, the district with the largest domiciled population of this tribe, has not at all figured in the above census table because, owing to a bureaucratic time-lag in procedural adjustments following the re-organisation of Punjab, the Kinnauris population in Kulu (estimated at about 1000) has not been classed as tribal. The Kinnauris population of the Simla District (estimated to be about 200 souls) has also gone unreflected, in the tribal demographic statistics, by default for the same reason. This would bring the total to about 36320. Roughly, another nearly seven thousand might be added denoting under recording owing to misunderstandings. (Chandra, 1992; Chandra, 1974). Bisotra (1998) has underlined the existing management strategies, the problems and constraints in tribal development administration. Further he examined management issues centering around dispersed and isolate environment, stress, public relationing, personal and human resource management and management of resources and tries to work out new strategies in tribal administration.

Scheduled Caste of Kinnaur District

In the 1961 census the inhabitants of Kinnaur numbered 40,980 persons in a total area of 6,520 sq km; of these, 36,800 are members of the scheduled tribes and scheduled castes, 70 per cent of whom are the Khasia or Kanet and are classified as a scheduled tribe. The scheduled castes include the Chamang and Domang, who form the remaining 30 per cent of the scheduled population of the state.

The present day, Kannaura do not comprise a homogenous group and display significant territorial and ethnic diversity. For a better understanding of ethnic and cultural distribution of the Kinnaur district, it may be classified into three territorial units. These are, namely, Lower, Middle and Upper Kinnaur. Lower Kinnaur comprises the area between Chaura at the boundary of Kinnaur district with Rampur Buhahr and Kalpa including Nichar and Sangla valleys. The people of lower Kinnaur are primarily of the Mediterranean physical type. It is difficult to distinguish them from the people residing in the adjoining Shimla district with whom they have some affinity. The people of lower Kinnaur are mostly Hindus, though the ethno-historic factors have resulted in some Buddhist influence. The middle Kinnaur is the area between Kalpa and beyond Kanam, including Morang tahsil. The people of middle Kinnaur are of mixed racial strain. Some have marked Mongoloid and other marked Mediterranean features. In some cases there is an admixture of the above two in varying degrees. The inhabitants are Buddhists as well as Hindus. Many people have faith in both the religions. The remaining north-eastern part of the district i.e. the area between Puh and Hangrang valley, extending up to international border with Tibet, comprises upper Kinnaur. The predominant physical type of upper Kinnaur in the Mongoloid though a few persons with Mediterranean features are also seen in the area around Puh. Some persons show in blending of Mediterranean and Mongoloid elements in varying degrees. However, the people in Hangrang valley are almost universally Mongoloids. They mostly follow Mahayana Buddhist religion. (Nag, 1981).

According to Negi (1995), "The Kinnaur society is divided into two broad occupational groups—the peasants and the artisans possibly of diverse ethnic origin. These groups are represented by Kanets (Rajputs) and Scheduled Castes. The Kanet comprise the main cultivating community of the area and use honorific surname". Presenting an account of different castes/communities in Kinnaur district Bajpai (1970) states that there are two principal castes in the district, namely , Kanets or Rajputs and Scheduled Castes. "among the Kanets there are three grades. In the first grade Kanets there are as many as fifty sub-castes in the second grade there are seventeen sub-castes and in the third grade who work as potters have three sub-castes. Waza Kanets belong to the third category and are considered inferior among Kanets.

The Scheduled Castes include Chamangs and Domangs. Chamangs make and mend shoes and weave cloth. The Domangs are primarily blacksmiths. There is a third caste called Ores. The main profession of the Ores is carpentry. In the social status Ores are equal to the Domangs. Among Scheduled Castes, blacksmiths and carpenters, i.e. , Domangs and Ores consider themselves superior to Kolis or Chamangs(ibid,).

Though the basic racial strain among the Chamangs and Domangs is Mediterranean, somewhat coarser features are discernible among them as compared to the Rajputs. Some scholars identify them and their counterparts in the cis-Himalayas with darker people of mixed descent from Proto-Australoid and Mediterranean physical types (Nag, 1981).

According to Papiha et al (1980), "The largest group inhabiting Kinnaur is Khasia, the people being known as Kanets or Rajputs. The southern Khashia are the representatives of Hindu migrants from the northern plains of India during the Muslim invasions in the 12th and 13th centuries. The northern Khasia (Kanet) who are investigated here follow the Tibetan form of Buddhism and present more mongoloid facial features, and perhaps are the descendants from the admixture of the local population with immigrant tribes from Tibet during Tibet's expansion towards the south-west during the eighth or ninth century. The other group of the district is Beru, the scheduled castes, and consists of Chamang and Domang (commonly known as Koli and Lohar, nearest to the indigenous population. To the category of "scheduled," retained in the 1961 census, were assigned those tribes and castes with more primitive ways of life and therefore particularly threatened on contact with other peoples of India; the category provides them with some protection". Of all the castes, the Lohar, the Badhi and the Koli are found throughout Kinnaur. Of course it is not that all these castes are found in all the villages.

Occupation Affiliation of Ethnic Groups and Interaction

Negi (1990) paper highlights the nature and extent of Jajmani relations that have been developed among the Kinnauri tribe. He further explained how caste oriented socio-economic behaviour pattern has influenced the tribal groups and a definite pattern of Jajmani relation has been associated with the life of tribal group.

In Kinnaur district, Kanets are basically land owners and agriculturists whereas other ethnic groups are mostly artisans and agricultural service, landless labourers. There is master (*Dhonas*) and client (*Binans*) relationship (*Binanang*) exist between these ethnic groups. This relationship is traditional oral contract between these parties based on conventional practices and traditions. Under this relationship, the Koli, Badhi and Lohar provide regular services and goods to their patrons. This relationship not only revolves around economic aspects but also in socio-religion aspects also.

Kinnaur Society of today is a mixture of various ethnic groups with some basic difference in religions and social life. The element of unity among these ethnic groups can be classified as geographically defined territory, common name 'Kinnauris' and overall similarity in basic identity. According to Negi (1976), " immigrants there might have come to Kinnaur from all four quarters, but each and every one got absorbed into that certain something which has been the Kinnaura society. This society has no doubt been influence and changed in some degree, by these immigrations and by the changes of time, but its overall form exists, its soul lives and its genius subsist".

According to Joshi (1912), "Besides the Kanets, the only two castes in Kinnaur are the Chamang and Domang. Thus, the Chamang i.e. Koli who makes shoes and weave and the Domang i.e. Badhi and Lohar who are carpenters and blacksmith along with the Kanets, in fact, constitute the Kinnaura tribal society".

The craftsmanship is known only to middle and lower sections of highland society. The high castes may find it as social degradation to go for handicraft as alternative economy. It is a well known fact that all the time and the rear round, the population is fully engaged in subsistence economy. Only a few in a highland society has time and people to spare to be used in additional economic pursuits.

Lohars of Kinnaur is, as his name implies, a blacksmith, pure and simple. He is one of the village menials, receiving customary dues in the form of a share of the produce, in return for which he makes and mends all the iron implements of agriculture, the materials being provided by the husbandman. *Sunar* is the goldsmith and silversmith and the jeweller of the village. He is also to a large extent a money lender, taking ornaments in pawns and making cash advances upon them. The *Chamar* is the tanner and leather worker of the hills. The name *Chamar* is derived from the Sanskrit word *Charmakara* or worker in hides. He is also a labourer and a field worker in the village. They serve their master during all kinds of jobs, such as cutting grass, carrying wood and above all, they do an immense deal of hard work in the filed. All this they do as village menials, receiving fixed customary dues in the form of a share of the produce of the fields.

In Kinnaur different caste groups have their own specialised occupation.

- Lohar are blacksmiths and also silversmiths.
- Ores or Badhi are carpenters and mason.
- Koli are weavers and spinners and sometimes work on leather,.
- Nangalu are the basket makers.
- Kanets are the land owners and agriculturists.

Earlier, these artisan castes used to work for the Rajputs exclusively. At that time, in exchange of their services these artisan caste groups used to get cereals from the Rajput (Kanet). In some places of Kinnaur, the same system still exists. This type of symbolic co-operation among various ethnic groups is known as Shingmo or Halas when it is in the agricultural field and Binana or Zothit when it is in non-agricultural field.

Under the Shingmo or Halas relationship, a Koli cultivates the land of one or more Rajputs. It is his responsibility to performs all agricultural activities and also helps his client in irrigating the latter's land by means of the Chasma or Kul. In return for his services, he gets some amount (depending upon his labour) of cereals. It is an obligation in the part of the Rajput and the Koli, who come under this halas bond to invite the other when some social or religious ceremony is performed in house of either of them.

In the non-agricultural activities, *Binana* or *Zothit* is the mutual co-operation between the Rajput in one hand, and all other artisan groups on the other. These artisan castes supply articles produced by their specialised works to the Rajputs, and in exchange get traditionally fixed amount of cereals. *Bonu binana* (bonu =to weave) exist between Rajputs and Kolis (who weave woollen clothing for the Rajputs). The *Sui binana* (sui =to stitch) exists between the Rajput and Koli (who tailor the garments of the Rajputs). The auras ores binana exists between the Rajput and the Badhi (who do carpentry works for the Rajputs and sometimes also do house building job). Domang *binana* exist between the Rajput and the Lohar (they serve by preparing iron implements mostly for the agricultural and horticultural purposes and also silver ornaments). The *Chanalas binana* is exist between Rajput and the Nangalu (who prepares baskets for the Rajputs) only prevalent in lower Kinnaur.

The Lohar/Badhi prepare implements for farming and household use. The Lohar prepare different kinds of traditional ornaments both in gold and silver whereas Kanets are never associated with metal works. The technique used for making implements and ornaments still continues to be traditional. The workshop is located on the ground floor or in front of the house. The skin bellows (*Saph-khul*) are fixed on the ground called *Anning*. The whole process of purifying and melting is done to produce a finished product supplies the plough (*Tal-Col*) spinning sticks (*Pankt and Corn*) and variety of handles of various implements.

In social domain, in case of death in Kanet family, the Bandhi, make the wooden pier for carrying the corpse to the cremation ground, whereas *Chamang*/*Halas*/Kolis belong to the lowest strata of the society and so are their traditional occupations. They tills the Kanet's land with plough and pair of bull. While tilling the land, he is not assisted by other except his own caste members because of the belief that plough carriers some polluting substance. It has been reported that traditional service of the *Chamang* includes shoe-making, leather tanning and disposal of dead cattle, weaving and scavenging. The service of tanning and disposal of carcase are performed by only *Chamangs* because of the belief that they carry polluting substance. The yarn for home spun, woolen cloth is first spun at home with the help of spindles and then it is given to the *Chamang* for weaving. He begins his work by spreading it on the specially made ground pass it though pegs fixed at particular distances. On the occasions of marriages, festivals feasts particularly at the time of village *Devota's* visit, the *Chamang* cleans the premises of his master's house.

The inter ethnic relationship in schedule caste between *Chamang, Damang* and *ores* is not similar to their relationship with Kanet (Scheduled Caste). They largely exchange goods and services on reciprocal basis like in exchange of service like shoe-making or weaving, the *Chamang* get services of the *Domang* for make agriculture implements. The exchange of goods and services in scheduled caste are considered weak possibly because here both the parties are not economically strong enough to sustain it for a longer time.

It has been observed that apart from the services provided to scheduled tale (Kanet). The scheduled caste group have a special obligations towards village *devta*. In the *Devta* system, *Domang* and *Ores* are given a responsibility of playing a *Devta's* musical instruments. Since, these instruments do not carry any polluting substance, therefore while playing these instruments no purity and pollution in observed between them and *Devta*. The hornpipe (*Rousheek*) is only instrument exclusively played by the Kanet. Whereas big Kettle drum (*Nagara*) and pairs of small dreams carry polluting substance are played with great deal of precautions and kept at distance from the *Devta* as well other instruments. The performers of these drums come from the *Chamang* group. In lieu of their services to the village *devta,* they get fixed quality of grains annually. The *Chamang* services to *Devta* are based on hereditary whereas *Domang* and *Ores* are chosen by the *Devta* itself on performance basis.

In case of non availability of any particular caste group like Badhi caste in village or other neighbouring villages, then a few of Lohar performs the functions of the Badhi. They will work as carpenter and also as mason. These system is called *zothit* system.

In Kinnaur district, two more systems (*bhoara* and *mokomo kheyasimig*) of work exchange are present. In *bhoara* system, the members of the same caste (specially among the Rajputs), the related person come forward to help their relatives in economic activities and socio-religious festivities. These helpers are usually supplied with food and drink. In return the help seeker is oblige to help those helpers when they are in need. In *mokomo kheyasimig* system exists between a Rajput and his neighbours (*padesan*) or co-villagers (*doshang*). The conditions, in this case, are similar to those of *bhoara* system. (Chandra, 1987;Bajpai 1987.)

With the change in time, the occupational characteristic of these ethnic groups has changed considerably. Now the so to called upper and lower caste have started, accepting menial jobs in government services, working as daily wage labour in various government sponsored development. They are gradually giving up their traditional occupation as taking up occupation of other groups. Still prohibition on inter-dining is observed in rituals and public ceremonies.

On the basis of stratification in among the Kinnauris, the four ethnic groups i.e Koli, Badhi, Lohar and Nagloo were classified as scheduled caste population of district. Under such a dual classification (i.e Kinnaur as Scheduled Tribe area and Koli, Badhi, Lohar and Nagloo of Kinnaur district as Scheduled Caste) of habitants of Kinnaur district, the government officials faces lots of problem in administration and allocation of development budget for the people.

DOMESTIC UNIT—HOME/STRUCTURE

Kinnauri are polyandrous people. All brothers usually share one common wife that makes the polyandry of fraternal type. In the polyandrous family the relationship between son and fathers is not difficult to establish. Normally the practice is that the all the husbands are recognized as the fathers of each child. The eldest among the husbands is called *Teg Boda* (elder father) and others as *Jigich Boba* or *Goto* (Young Fathers). For day to day use the eldest among the living brothers is spoken of as the father of all the children born to a polyandrous wife. These days the institution of polyandry is gradually on its way out.

The joint family system prevails in Kinnaur. Under one roof not only father, sons, brothers but also uncles, nephews of the same descent live and sometimes even own property in common. The system is under serious strain due to the advent of modern civilization in the area.

ECONOMY

The economic structure of people is primarily agrarian and essentially rural in character with 68.7 per cent of the total working population engaged in agriculture. The size of land holding is exceedingly small and the methods of cultivation are by and large crude as well as primitive. They used to cultivate their land situated on the slopes of the mountain and also on the valleys in the form of terraces. "The spaces of arable land are few and the cultivation is commonly done in narrow strips along the brows of the mountains" (Gerard, 1841). According to Frazer (1820) they used to cultivate, twice a year, in the comparatively lower altitude land but only once in the higher altitude land. Usually they used to get a little yield of some inferior type of millets only by means of a crude method. As a result they had to depend on the different other places for cereals.

In the earliest time, the Kinnauri have primarily been an agro-pastoral tribe. This occupation had been followed by the people since time immemorial. Ecology is accountable for the practice of this occupation. The secondary sources of livelihood are: Horticulture, Agriculture, Forestry, Cottage Industry, Trade and Commerce etc occupation are equally important to them .This fact was noted by earliest British traveler James Balliec Frazer in 1814 but even after 24 years, in 1841, when Captain Alexander Gerard visited them, the situation/condition was still the same.

In olden times, the economy of the Kinnauris was "closed economy". The circulation of the economy was related which gave rise to the barter economy. Monetized market was altogether non-existent. With the passage of time, Periodic fairs were organized dedicated primarily to the trade and commercial activities. This ultimately led to the introduction of money economy. During the settlement of 1928 and even during the time of the visit of Rahul Sanskrityayan 1948, the condition of the people remained the same except the money lending system had become known.

Due to natural environment condition congenial to farming and non sufficient cereals, people have to devise alternative means of livelihood. During heavy snowfall, livestock breeding and animal husbandry were probably the nearest alternative and supplement to agriculture. The sheep, goat, ponies, horses have played an important role in their economy. These animals have provided meat, milk, wool, hairs, milk products, hides, manure and skin for various purposes. Before the entry of cash economy, the animal, animal products and fruits were the articles of exchange as the batter system as against salt and wool from Tibet. The Kinnauris were never completely nomadic yet many of them practice semi transhumance-nomadic pattern. They used to act as traders between Tibet and Rampur Bushahr. During Lavi Fair, they used to sale articles (like salt, borax, pashmina, sheep, goats and wool) from Tibet and purchase the articles (like cereals, fruits, cloth, sugar, jaggery) required in Tibet. With the restriction of movement between India and Tibet, the trade gradually declined.

However, no single economy is sufficient enough to fulfil the requirements of the household because of the peculiar geo-ecological conditions of the area. Over a period of time due to necessity, Kinnauris had became livestock breeders, semi transhuman-nomads, traders, spinner, weavers, blacksmither, silversmither, carpet, basket maker etc.

Agriculture

Agriculture is the largest source of livelihood of the people. According to Gerard (1841) 'Space of arable land is few and the cultivation is commonly in narrow stripes along the brows of the mountains. The crops for the most part are poor great ant of grain pervades the whole country'.

In earlier time, there was always deficient in food grains. It was supplemented by dried apricot and herbal roots collected from the wild forest and grounded in the form of flour and then used it for making chapattis.

Now, for majority of Kinnauris, agriculture is main source of livelihood. The soil, rainfall and temperature conditions are quite divergent in different areas. These varied natural conditions result in varying agro-economic practices and cropping pattern in the district.

There are two main crops in this region, Rabi and Kharif. Rabi crops such as barley, wheat, lentils, coriander seeds, and Field peas are sown from the middle of September to the middle of December. And it is harvested in April at the lower region and July in the higher region.

Kharif crops such as lesser millet, potato and maize are sown from March to the middle of July and harvested in from September to the end of November.

Barley (tag, cha, Tung, Nan, tak-swa) are cultivated as this crop is well adapted to all kinds of climatic region and all types of soil. This crop is used extensively as a bread cereal and for processing it into a kind of drinks. Wheat (Rajat, Jod/Jad, Oja, Tho/Do) are grown at the area where irrigation facilities are available or under rain-fed conditions. It is grown very early before severe winter. It is used for making bread or chapatti.

Coriander seeds (Bugo) are sown at the end of September to the middle of November. By April in lowland and May in highland, they are ready to be collected.

Kitchen garden is known to them for a very long time. Initially they use to grow peas, beans, green, turnips and potatoes too. In the present days, people grow more variety of plants and vegetables. Such as Cabbage, Turnips, Peas, Pumpkin, Tomatoes, Chillies, Radishes, Greens and Garlic. Tobacco and other drug yielding plant are also gradually cultivated.

Horticulture

Forty years ago, there was no systematic fruit cultivation, yet a good amount of fruits was available such as Apple, Apricot, Peach, Chestnut and Walnut. (Nadda, 1987). The land is endowed by nature with excellent climate and soil for the production of a variety of fruits. The scanty rainfall in this area is ideal for the growth of dry and temperate fruits. In the present situation with the setting up of research station, demonstration orchards and nurseries the yield in fruits has increased tremendously and even temperate vegetable seeds are planted till beyond Tapei. The Research Station in collaborate with Indian Council of Agricultural Research which is established at Kalpa with the sub-division at Ribba is aimed to conduct research on the different aspects of seed crops. This has immensely guided and encourages the cultivators.

State government is trying to commercialize the dry fruits like almonds, *chilgoza*, raisin and to popularize it among the people. The government has set up almonds farm at Spito, raisin farm at Sharbo and *Chilgoza* farm at Bocktu near Pangi.

Animal Husbandry

Animal Husbandry is the pillar of Kinnauris economy. The livestock consist of Goat, Sheep, Horse, Mule, Mare, Donkey, Yak, Cow, Ox, Dog and Cat. But Sheep and goat constitute alone more than half of the livestock. The manifold benefits derived from the sheep and goats are: milk, meat, wool, skin and hair. Both the animals are used as "pack-animals". At present times, both are considered as the most destructive animals from the standpoint of soil conservation. Besides goat, a special variety called Chigu is used extensively as it has been more improved and multiplied due to its valuable fleece called Pasham. Horses, Ponies, Mules and Donkeys are used as Beast of Burden as communication is extremely difficult due to the mountainous terrain and tract. Yak serves as a very important source of livelihood especially for those who stay at a higher region as the yield of milk is higher than that of cow. Male yak are called Yak/Yag and female as Breema. The cross breeding of cow and yak are mainly used for transportation. (Male-Churu/ Zo/Zofa and Female-Zomo).

One of the pre–dominant traditional occupation of Kinnauris is rearing of livestock. The Kinnauris used to keep different types of livestock, is proved from the writings of Frazer (1820) in which he has mentioned that the Kinnauris used to deep sheep, cattle and in some areas Saorajee or Yak in good numbers. Ponnies together with asses and mujles likewise bred and kept. Captain Gerard (1841) also mentioned the sheep and goad wealth of the Kinnauris. They used to take these animals before the advent of the winter to the forests in the foot hills and Terial areas for pasture and used to return when winter was over.

The Kinnauris had the flourishing trade with western Tibet and Ladakh until the beginning of 60s particularly in wool, pashmina, goats and sheep through the barter system. With the spread of education and increasing easy alternative avenues of life and sealing of Tibet border, the traditional occupation has been increasingly on the decline for last two decades.

Fodder for the animals was not known to the people before. The grass serves as the source of Fodder. The herds of cattle and flocks of sheep and goat are let loose to graze in the extensive pestures. During summer months; the main areas for grazing are in the Alpine pastures. But with the start of winter the herds/flocks are brought down to the village. Then they subsequently migrated to the lower parts of the hills even as far to the plains of Punjab and Uttar Pradesh.

One thing that should be clear is that, because of the regular migration of livestock, it does not indicate that people were living a nomadic life. The people have settled population. But the migration is only with the member of the family who is in charge of the livestock.

According to 1992 Livestock census, Kinnaur had as much as 1,15,141 livestock. The maximum number of livestock is in Nichar Tehsil and minimum no. of livestock is found in Hangrang Sub Tehsil, followed by Puh Tehsil. There are 58 Veterinary hospitals and dispensaries in the district. The total live stock in the district according to Livestock Census (1992) given in Table 2.10.

Table 2.10: Total Live Stock in District (in Number) According Livestock Census 1992

District/Tehsil/ Sub Tehsil	*Cow*	*Buffalo*	*Sheep*	*Goat*	*Other*	*Total*	*Poultry*
Nichar	9982	3	22207	10765	1478	44435	3340
Sangla	4469		5677	1499	624	12269	924
Kalpa	2996		9663	5064	832	18555	1051
Morang	2281		10077	5546	1194	19098	282
Puh	1687		5905	2677	1333	11602	170
Hangrang	841		4191	3071	1079	9182	28
Kinnaur	**22256**	**3**	**57720**	**28622**	**6540**	**115141**	**5795**

Yak is included in Cow category

Source: District Kinnaur at Glance-2002.

Fishing

This practice was known to people from very early time. There were no rules and regulation for fishing. During the reign of Raja, nominal fees were charged from the people for fishing. The sources that provide fishing are Baspa and Spiti River. Besides this many streams like Barnng, Tangling, Kozang /Kashang, Lippa or Kirrang, Shyaso, Leo, Hango, Chuling and Tirung are the fishing points.

The Fisheries Department was established in 1950-51. The rules and regulations were also laid down. Several measures have been taken towards the development of fisheries such as the construction Trout Farm at Sangla.

Land Utilisation Pattern

According to Tikka Raghunath Singh's settlement of 1897 Act, lands are classified into: Kiar, Bakhal, Karali, Newal, Kandas.

(i) Kiar

Kiar is a first class land. The yield is twice a year, a Rabi (rice) and Kharif (Maize) crop. This is mostly available by the riverside and is irrigated from the hill stream.

(ii) Bakhal

This is the land which is situated at the surrounding of the villages. The yield is twice a year with the use of manure. The fields are irrigated from the water carried by a channel to the field but as rule it is dependent on the rain. This type of field is only available in the lower Kinnaur.

(iii) Karali

This is mostly reclaimed land, which is situated at a distance form the village. It is neither manure nor irrigated. So it is entirely dependent on rain. The yield is only once in a year.

(iv) Newal

This land is considered a superior land and is mostly available at the bank of river Sutlej.It is a warm low lying land. The products can be twice or more in a year. It is ideal for the plantation of Apricot and Orchards.

(v) Kandas

This type of land is situated quite far from the dwelling areas. And it is considered productive.

The agriculture land under various agriculture produce is given in Table 2.11.

Table 2.11: Agriculture Land (in Hect) Under Various Agriculture Produce

Crops	*1999-2000*		*2000-2001*		*2001-2002*	
	Total Area	*Irrigated Area*	*Total Area*	*Irrigated Area*	*Total Area*	*Irrigated Area*
Rice	30	–	23	–	23	-
Maize	404	245	383	240	383	240
Wheat	371	92	379	125	436	125
Jo	1420	1049	1495	1168	1495	1168
Pulses	1340	340	1358	1325	1033	839
Tilhan	4	2	1	1	25	23
Potato	244	82	214	86	213	86

Source: District Kinnaur at Glance-2002.

The soils found in this region are sand, clay, stony and gravelly type. In Hang rang valley the soil contents are sandy loam, stony and gravelly type of soil. The soil of Nichar sub-division is mostly sandy and stony. This is suitable for growing crops like wheat and Barley and fruit trees. Sandy loam soil is suitable for lesser millets and for vineyards. Stony gravelly and sandy soils are not suitable for crop cultivation. Instead it can be put under fruit cultivation profitable.

Irrigation

Due to scarcity of rain water and typical topography of the region, people are compelled to devise this method i.e. Khul system. In the system, the , water from natural sources like spring (dhara)are carried to the desired place by channels made out of wood/ tins/ pipes. Agriculture field is irrigated by this narrow canals originating from springs. In some places, snow also serves as a good source of water for irrigation. When the snow melts, the water are collected and diverted through small check dams and taken to the field. The government has taken some initiative by providing pucca khuls with pipelines with an objective to reduce the expenditure on the repairs of khuls.

Handicraft/Cottage Industry and Natural Resources

The Kinnauris are absolutely secluded from the outside world. The villages are self-sufficient and their necessities are met among themselves through co-operation. During dry and cold winter, the people are forced to take up subsidiary occupation such as weaving woollen cloth for home consumption and for sale. Shoe (*Pono*) making, Carpentry, Blacksmithy, Basketry, Oil extraction, distilling and Bee-keeping were practised for use by every householder engaged in agriculture trade or keeping of livestock. (Gupta, 1998).

Weaving is generally done by the artisans, which is found in every village. Weaving is done only in winter as weavers are engaged in agricultural work during the other season. Towards the end of 19th century, Mr Schreve, Morian Missionary introduces the handloom from Europe which targets the people to make blanket. (Gupta, 1998).

The main articles manufactured are: Pattus (Shawls), Dohris (Woolen garment like saree), Chhanli (Blanket), Pattis (tweeds), Mufflers etc. These are made for household use or made against order for which the materials are provided by the customer and are paid only the labour charge.

Pono, the local crude footwear where the lower portion of this shoe is the leather and the upper portion is the wool, are made by the Chamang for the use by the local people. The stitching is done by leather thread. They are either embroidered or plain. Embroidered shoes are known as Tapru-se-Balzam-pono. In the present times, the Chamang have started manufacturing leather shoes.

The agricultural implements are made by the carpenters who belong to the ORES community. They are also engaged for the construction of houses and temples specially related to the wood works.

A carpenter is needed in every village for making wooden bowls and dishes for the use of the local people. Blacksmith are comprised by the member of the Domang community. They make all kinds of agricultural implements and musical implements too. The main articles made by them are :Bham (A large kettle drum), Nagari-cho (Pairs of kettle drums), Ronsheeng (hornpipe), Lumko (A vessel used for the distillation of local wine). The workshop is usually located on the ground floor. In case of a single storey house, the workplace is located in front of the house in the open space with having three walls only.

The skin Bellows are fixed to the hearth. The whole process of melting, purifying and welding are done to produce a finished product.

Basket weaving are done by the Chamang at places where raw materials like Rajial, Tako Katyang and Teg-shing are available. They make Koting or Kilta (Conical Basket), Chhatoch (Basket with handle), Changhor (basket).

Oil Extraction is done is almost every household. The oil is extracted from the kernel of Chul, Reg and Ker by traditional method. Firstly the Kernel of Apricot, Peaches and Walnuts are taken out and dried for 2/3 days. They are then put in a Kaniry (Stone Mortar) and pounded repeatedly with the pestle to make a thick paste. The paste is then boiled with high heat in a Cauldrn and later small cakes are made out of it. Each cake is placed on the edge of the mortar and pressed hard. In this process the oil oozes out of the cake and ran into the mortar. This oil is then kept in a metal or earthen pot, locally known as Gagri and Pathu. The oil cakes are used as cattle food.

Trade

Frazer (1814) remarked, "Indeed they are almost exclusively the commercial couriers between Hindustan and Tartary, as also between Tartary and Kashmir, frequently the routes from Leh to Ladakh, to Lhasa and Degrucha and Nepal on trading speculation".

After the Gurkha war (1815), British East India Company made enquiries about the trade potential of the hill regions, particularly of Bashahr state.

"The Kinnaurese, in the past, were very good traders and they were almost exclusively the commercial carrier between Hindustan and Tartary, and also between Tartary and Cashmere, frequenting routes from Leo in Ladhak, Lhassa and Degurcha and Nepal on trading speculation.... The Kunawar merchants used to carry on a trade not only to plains and neighbouring Chinese provinces but also the chief carriers between Garha, Ladakh and Cashmere and even push their commercial enterprises as far as Lhassa on the South-East exchanging commodities between that Degrucha and Nepals and between the Chinese towns in Little Tibet and in the north–west trading between Cashmere, Yarkhand, Kashgar, Garha and the various cities and people of these quarters " (Frazer, 1820).

In the past, trade was not only limited to Tibet and Ladakh. But it was extended to as far as Central Tibet, Arkand, Kashmir, Lahore, Delhi, Calcutta, Bhutan, Sikkim and Nepal. Captain R Rose, assistant agent to Governor General deputed Munshi Karimuddi to Garo and Ladakh for the purpose. He did submit a remarkable report to the government detailing routes, distance, commodities of trade.

The Kinnauris visit Tibet in May and June and return back in June and July. The Tibetans on the other hand visit the Kinnaur land in October and November and return back to their place in November and December. The traders usually travelled in batches of 20/25 carrying provisions with them as the Garo in Tibet had no permanent habitation. The traded items are salt, Sohaga, Goat's wool, Nerbussee (Adeng), shawl wool producing goats and tea. The Tibetans carry red blanket in their visit.

Salt is bartered for drug (a sort of raisin) and wheat at equal quantities which is finally sold at Rampur. For 1 seer of this is exchanged for 4 seers of rice or 8 seers of Mundooa attah or 16-20 seers for a Rupee.

Sohaga is bartered at a rate of 5 seers for 4 seers of drug and was sold at Rampur for 7/8 seers per rupee. Nerbussee or Nurbees (A drug) is brought form Nepal and sold to Kinnauris at rate of 1 rupee per 3 pieces. Goats producing shawl wool are purchased at a rate of 2 rupees each and sold for 3 rupees at Kinnaur. Box of tea of 10 seers cost 10 rupees at Garo. Saffron are purchased at 10 rupees for a seer and sold at 2 ruppes per tola.

According to M.K. Raha (1974), "Trade particularly with Tibet, definitely played a vital role in the life of the people of western and central Himalayas from the remote past till the closers of the border with Tibet. The people of these areas used to trade with Tibet. As the border was closed following the Chinese aggression over our country, the trade with the part of the world ceased to continue. At that time, they used to bring various trade goods from Tibet such as wool, Pashmina, salt, borax, butter and animals which they used to sell to the local people and also in the neighboring areas. Again while they used to go to Tibet for trade they used to carry with them cereals, sugar, molasses, spices, tea, metal utensils and implements, cotton, cloth etc. which had heavy demands in Tibet".

According to Sen (1978), "The important trade markets in Tibet visited by the Kinnaur traders were located at Dongbra, Gartok, Chang-Thang and Chumurti, at a distance of 60-80 kms from the border. The important trade market in the Indian side was located at Rampur at a distance of about 120 kms from the district Head quarter and on foot it was 14-16 days journey. Trade and animal husbandry are becoming less popular gradually because the educated section of population is not interested in trade and animal husbandry".

The chief article of export are Heeng (Asafoetida), Zeera (Cummin), Bhojeputtra, Opium, Charas, Mohru (a poisonous drug), iron, white puckers woolen stuff), suckland (Woolen stuff).

There are 3 fairs in a year which is organized in June, October and December at Rampur. At Lavi fair, tea from China and bars of Silver, stamped by authority from Yarkhand, toys from Russia, are occasionally held at Rampur fair. Mandarin chotick, cups and saucers are seen in Kinnaur.

According to Devies (1862), "From Rampur (at present in Simla District), which is still famous for the Lavi trade fair, and other places of the lower hills they used to buy rice, wheat, molasses, cotton clothing, metallic utensils and implements to sell partly in their own country and mostly in different trade centers. In exchange, they used to bring wool, pashmina, borax, rock salt, brick tea, ponies, yak tails and other items, which they used to sell partly in their villages and mostly in Rampur and other neighbouring states.

The trade has stopped with the closure of the border with Tibet due to Chinese aggression over our country. The artisan castes are still continuing with their traditional occupations. The traditional occupation of the Lohar is blacksmith, the same for the Badhi are carpentry and masonry works, for the Koli are agriculture, weaving, spinning and leather works and for the Nangalu is basket making.

In present day, the Kinnauris economy is based on agriculture and animal husbandry. With the change of time and introduction of new hill friendly technology, horticulture is getting more and more importance day by day. Besides, service, labour business and other occupations are making its place in the Kinnauri's economic system. Although the culturally affined artisan castes are still engaged with their traditional occupations and the same time they have adopted other economic activities like sheep and goad rearing, agriculture, labour etc.

Money and Banking

In the olden times, Money was not much in circulation. In early 19th century, Gerard noted, "Coins are rupees, both the common one of the Hindustan, the Muhamad Shahee, five or six per cent better and only half the value of the former. In course of time, British system of coinage was introduced. But now the decimal currency is in use.

Gradually Zamindar Sahukar/Moneylender appeared. They lend money on the security of land or jewellery. Earlier priest of the local deity temple serve as the person for getting loan. This way the temple under the priest served as the bank. The guarantee was only a solemn act before the deity and they never did cheat the deity.

In course of time, demand increased for various reasons such as construction of house, marriage expense, for investment in trade and commerce etc. This increase in the demand had also increased the rural indebtedness of the people and these burdens are carried on for generation to generation. In 1960, new border district of Kinnaur came into existence. And along with this came the credit facilities came into existence in the form of Co-operative credit society, Agricultural multipurpose society, Co-operative bank, Land mortgage bank etc.

In the meantime, India-China trade agreement of 1954 led to the closure of the trade with Tibet. Not only did the traders become idle but livestock was also denied grazing in Tibet. This problem was solved by sharing the existing grazing land in Himachal Pradesh. The petty traders who either do not own animals or do not have enough land or other means of livelihood, to make up the lost trade with Tibet are taking up alternative business such as shop, owning of improved breed of sheep etc. In the wake of Chinese aggression (1962), construction of roads on a large scale has provided employment to several people.

Human population are tackled by stacking a co-operative marketing and supply federation through which traders obtained loans for their business etc.

RELIGION

Kinnaur always consider themselves the inheritors of mythical Kinners. With this aura around them they have been great worshippers. It has been reported in various articles and books that long time back Kinnaurs knew no religion other than their own village deity. According to Negi (1976), "in Kinnaur the original religion, within the living memories of the people, centered around the principal village deity who commanded great respect. Animism has been the matrix". Village deity till recent times has been taken as granted the doctor, an astrologer, an arbitrator, a Judge and a chief executive etc.

In the earlier time, Kinnauri people used to worship gods like Indra, Krishna and Kubera. The traditional religion of Kinnaur is centered entirely on the principal village deity, which is divinity, a doctor, magistrate, a judge, the chief executive, an astrologer, the village hero etc. Hinduism has introduced Rama, Krishna, the Ramayana, the Mahabharata, the Gita, the Sukh Sagar, the Prem Sagar, some of the Puranas and some of the relevant systems of worship over a period of time. Later monastic Buddhism came to be established in the regions bordering Tibet. This led to the original form of animistic faith and worship is losing its traditional hold. The present situation is that village god and their retinues are still very strong and side by side there exist Laministic Buddhism and local version of Hinduism. The religious rites of birth, marriage and death are performed in accordance with the local version of Hinduism as interpreted by the local priests who generally are the Lamas particularly in the upper and central part of region. (Khosla, 1979).

In the mid nineteenth century, the religious situation described by Edward Thornton (1854) was ". the religion of Koonawur is Brahminism in the south; in the north, Lamaic Buddhism; in the middle, a mixture of the two systems. There prevails a regularly graduated transition from one to the other. Thus, Brahmins are not met with beyond Saharan, near the southern boundary, where they officiate at the shrine of the sanguinary female divinity Bhima Kali, to whom, at no remote period, they offered human sacrifices. At Kanum, about half way between the northern and southern frontiers, the sacred books are in Tibet and lamas are there first, met with, but

Kine of castes; thus partially amalgamating the two creeds. At Hungrang, on the northern frontier, the religion is pure Lamaic Budhhism…"

In earlier time, according to the Frazer (1820) and Thortan (1859) also highlighted the existence of Lama priest in Kinnaur district and confirmed non-existence of Brahmin beyond Sarahan. In the present time, the animistic traits of Hindus and Buddhist religion of people is not changed even after the spread of Hinduism in the adjoining areas throughout the centuries.

Generally the boys and girls of Kanet join Monasteries. The girls, who do not marry, devote their time studying Tibetan scriptures are called as *Zomos* and the boys who learn Tibetan scriptures or Buddhist doctrine are called lamas. Lamas are either gyolang like *brahmachari* or *durpu* who marry but never shave the head. Chandra (1982) observed, "until recently the Kinners had yet another way of resolving the problem of the so called surplus women. A part of the Kinners society is Buddhist. Lamaism, the local form of Buddhism, plays a dominant role in the life-style of the Kinners. The Buddhist monasteries shelter not only the lamas (monks) but also the *jomos* (nuns). Women having little prospects of getting married are initiated as *jomos* to serve the lamas and to earn merits for themselves for better future and salvation. This institution of *jumos* has certainly helped the Kinner society in solving the problem of surplus women".

Village are generally self-sustained units. Every village has a temple where people congregate for common worship. The village gods are carried in palanquins, on number of occasions, to places of religious interest or village fairs. When in trouble, the people go to the deities to seek their guidance and help. The village god is suspected to watch over the destiny of the village. He protects, rewards, threats, threatens and punishes the people, while they in turn, worship him/ her by singing and dancing. According to Chib (1977), "Like all other tribals Kanauras are extremely superstitions and believe in so many gods and goddess. Every village has its own *Devta* who in cases of annoyance is said to shower wrath in so many terms. The permission of *Devta* through a pujari is obtained for many purposes".

Now a day, the village gods, lamaistic Buddhism and a local version of Hinduism exist side by side. Most of the ceremonies are performed by lamas both in Hindu and Buddhist families. It is believed that the Bashahr Raja was originally from the plains. His religious functionaries were Brahmins, but his entire staff was from Kinnaur region. Over a period of time, the people had started imitating the Brahminical tradition.

According to Atkinson (1882), "the people of upper Knaor are of Tibetan origin and Buddhist in religion. Buddhism extends down the valley of the Sutlej as far as Sarahan between which and Pangi is a debatable ground common to Hindus and Buddhists. Further even Hedin (1913) writes, "In Chini *'Om Mani Padmehum'* has not all its own way with the soul of men". According to Thomas (1812), "The gradual transition, in ascending the Sutlej from Hinduism to Buddhism is very remarkable".

Despite early Christian mission at Puh and other parts of the region, Christianity has not been able to spread very much in the district. The missionary influence can be seen in the habits of cleanliness and the use of refined form of knitting was acquired by the people from the missionaries. (Negi, 1998).

Vaidya (1977) observes, "The monasteries in the Himalayans are great repository of the lamaistics religion and culture whether the monastery is flat-roofed as is found in the snow-bound regions or pagoda like tier-roofed as in the lower altitudes, it has some common characteristics. Unlike most of the Hindu temples, the Buddhist gompas are seldom dedicated to one god or goddess, rather they have the entire pantheon. Some of the gods find place on the altar in the form of metal and stucco images and others on walls and hangings. All the deities and divinities are placed in their own order of importance. The wall and hangings carry pictures of the legends connected with the various deities and divinities.

The village gods and goddesses here are not statues but are expected to behave like common human beings with sentiments and a sense of pride. Sometimes they arrange fights against counterparts, win or lose battle, do things unbecoming of godly beings and even share a joint family system like their devotees; they cut jokes with the villagers, are offended by trifles and threaten to harm the evil doers, and dance with the villagers while being carried in wooden or silver palanquins. These palanquins are made after prescribed intervals and a specific type of wood is used in accordance with the direction of the gods. The wood used to make a palanquin should be either from the forest of the deity or from any other place having links with the origin of the god or goddess.

The institution of village gods is not a matter of chance. There has been a legend behind each one of the gods and this practices has its roots on the mists of the hoary past. Right from the origin or manifestation of a particular god or goddess, the *gur*, the oracle who is the interpreter of the divinity, relates the whole story of the powers, miracles and curses inflicted by him from time to time. Filled with the power (*shakti*) of the deity, the *gur* shakes his body and goes into a trance before the audience and moving his head to and fro, he starts uttering, something in an unusual manner, sometimes in an unknown language, as if the deity itself is speaking. This is attended to carefully by the devotees as the prophecy or order of the benevolent spirit who is the village deity.

The village god control all the villagers and direct social customs. When this custom of village deities started is not known for certain, but the villagers know only that their activities and destinies are governed by these gods and that they cannot afford to disobey them at any cost. Thus it can be safely said that this institution is the major dictator of their activities, hopes and despairs, virtues and vices, natural and created misfortunes in a village society. The village god is the symbol of village culture. Even marriages and deaths are guided by the village god and it is the deity who directs the followers to allow or disallow any guest or stranger to the village. With this cultural phenomenon, the village society has grown like a family and the village fold are helpful to one another.

The culture of the area is woven around the customs involving village gods. In Kinnaur the village gods have eighteen masks in each palanquin, the heads being around and covered with black yak hair (*jata*). These palanquins are decorated at festivals and the devotees make the village gods dance while taking them on their shoulders. The dancing deities are asked for solutions to the problems of the devotees and the palanquins stops dancing as if the god were attending to the questioner. The queries are made in such a way that the reply from the deity is 'yes' or 'no'. After a brief pause, the palanquins again starts going upward, sometimes causing pain and discomfort to the bearers, then abruptly stops and moves forward or backward. If the palanquin moves toward to the questioner and bends forward, this is taken to signify an affirmative; if it bends backwards, it is considered to indicate a negative reply. The oracle replies to the questions on behalf of the deity and in order to establish the truth of what was said, he sometimes gives some mustered seeds or gains to those present. The seeds are counted and if the number is odd, it is considered that whatever was said by the oracle is correct; in the event of an even number, the oracle is told about the doubt over his prophecy. Though this is very simple method, people have faith in what is said by the oracle, and village disputes can be settled without being taken to the courts of law. For purposes of rejuvenation some deities are taken to their birthplace, which are located somewhere near the village or in some adjoining settlement.

In Kinnaur, ritual (deity going to Indrapuri (Heaven) during the winter and bring prosperity to their devotees) are performed to mark the departure, and when the deities are ceremonially welcomed back after several days, the villagers gather in an open space and make the oracle relate the achievements of the village god in Indrapuri. During the absence of the village god, no auspicious work is undertaken by the villagers and the whole settlement seems deserted and silent. The people like to stay indoors, specially after sunset, as it feared that evil spirits haunt the village and can cause harm to the natives. It is only after the return of the village god that festivities are arranged to celebrate the event. (Atkinson, 1976)

The temples of the village gods are managed by the temple committee constituted by the deity. The *bhandari* or manager is responsible for the property of the temple including the land belonging to the god. In earlier times it was often the case that all the village land belonged to the god, but now, under the existing; and reforms rules, the gods have been left with a small area. The *kayath* is the cashier of the celestial being and he maintains the accounts on behalf of the god. The *mali* is the oracle of the deity who conveys the wishes of the god after receiving message from him in a trance and is thus the medium of conversation with the god. The *pujari* or worshipper, though illiterate in many cases, is the person who conduct puja with incense. There is another attendant known as *shu-charas*, a sort of manager for the god who convenes meetings in the temples and makes other arrangement for the deity.

In Kinnaur, it is observed that some pieces of wood in the temple which were cut into two parts midway with some notches on them. These notches denoted the loan of grain taken by a particular villager. Whenever there is shortage of food grain with a particular family, the head of the family approaches the village god, who orders the *bhandari* to allow the family supplies at usual rate of interest. The interest is not paid in cash but in kind. These pieces of wood were called *rekhangs*, and the borrower was required to bring him half piece of wood with him to match up with the portion retained in the temple, before they were destroyed on the premises.(Ahir, 1971)

Bernier (1989) in his paper on Himalayan Towers Temples and Palaces of Himachal Pradesh describes the architectural models of temples in Himachal Pradesh. According to him, the remote town of Sarahan with its Bhima Koli temple complex is perhaps the most thrilling as well as the most neglected site in the history of Himachal Pradesh art".

The temple musicians play various folk tunes on their instrument on festive occasion in the morning and evening. They are called *bajantris*, and generally belong to a scheduled caste. The village gods inspect the band of musicians on various occasions to maintain the tradition of expertise in folk music. It is maintained that eighteen tunes on an identical number of musical instruments is the ideal standard repertoire of such a band, but it is difficult to find this number of instruments and musicians these days.

It is believed that the village gods are benevolent spirits who never fed displeased with their devotees. However, these spirits are accompanied by attendants known as *ganas*, who inflict curse on deviants. These curses are called *dosh* (displeasure) and the evildoer has to correct himself in line with the wishes of the god.

According to Chaudhary (1981), "Oath taken in the name of some supernatural force is known as *Darohi* and a *Darohi* has a tremendous impact. There is a peculiar institution in this district. Almost every important deity, Buddhist or Hindu or Animist has a man attached known as *Grokech* (mouth-piece). He is supposed to know by heart the legend about the origin of the deity. He recites the legends and get into a trance and his services are often taken for appeasing or avoiding a distress. This recital is known as *chironing*".

According to Chakravarti (1998), "Although Kinnaur is not a very big area, yet it has an age-old tradition. Despite being a Hindu region, Buddhist doctrine plays an important role in the life of the people of Kinnaur, like religion, music and dance have influenced their life from childhood to graveyard. It needs a vast and deep study to explore the nuances, the repertoire and the rich cultural heritage of the area".

Chandra (1981) made an attempt in the direction where religion has been viewed in the context of ecology at the local level in north-western Himalayan. He observed that being mountainous and situated at a high altitude Kinnaur presents a typical ecology for its inhabitants—that too having some variation at the local level. In the given ecological set-up the Kinnaur have evolved some adjustments with nature through various sets of arrangements at social as well as religious levels to make their survival and perpetuity easy. The hilly economy, social arrangements and the world of religion all put together provide a well adjusted plane to the Kinnaur. Religion has played an important role in their life by providing a balanced approach to meet the challenge of ecology of the area.

DANCE, FAIRS AND FESTIVALS

Kinnaurs are famous for their musical renderings and dances. The tradition of music and dance has been an asset of the people and till today the rituals, religious festivals and social assemblies are not wondered to be complete without music and dance. Dancers dance to the tune of musical instruments played upon by particular instrumentalists belonging to a particular low status ethnic group. Drums, *nagara, shehnai, bugjal,* flute, *ran singha, karnal* etc are the local musical instruments used in the dances. According to Das (1954), "In modern Kanauris have retained many characteristics of their Kinnaur ancestors. Even today their sportive habits and foundness for singing and dancing cannot be favourably compared with those of any other tribes. They are jolly and gay, being always cheerful and happy. They always sing in melodious voice wherever they go. It is a usual practice with them to sing and dance every evening. They are also in the habit of dancing before the village deities reciting hymns and singing their glories".

Sharma (1976) has classified songs of Kinnaur into five major categories and further into certain sub categories.

Kyang is the most popular dance of district Kinnaur. It is conducted in more than six forms. The main form of Kyang is called Dabar Kyang in which dancers dance in a circle led by the male members followed by the females. The leader is known as Ghumeri. He leads the dancers with a tuft in his hand swaying on lateral sides. This tuft is known as Chamver of the god. First of all, hymns of gods are sung during the time the leader goes on swaying the Chamvar. The leader with his left hand picked up by the right hand of the third person in the row is the only dancer having right hand free holding the Chamvar.

The traditional dresses of the dancers are richly decorated and embroidered. The ladies and agents both wear round cap with twigs of flowers tugged in front of them.

Tribal Festival is being celebrated since 1994 from 30 October to 2nd November every year at District Headquarter Reckong Peo and this festival has been declared as State Level festival and has been celebrated since 1987 under different names like Janjatiya Utsav, Phulaich Utsav and also as Tribal Festival. This festival not only depicts the panorama of rich culture heritage of district but also provides an opportunity to the local people to sell/exhibit their horticulture/agriculture produce, handicraft and artifacts. Besides the Kinnauri culture groups, participants from other districts/states also present and perform culture programmes symbolising national integration and brotherhood.

(i) Sazo

This festival is observed in the month of January. On this day the people take their bath in the natural springs and few even go to Sutluj river for bathing if they happen to live near the river. Poltus, rice, pulses, vegetables, meat, halva, chilta and pug are the principal dishes prepared on this occasion. In the morning the family god is worshiped with the food except meat. The hearth is also worshiped near noon time, the deity is brought out and worshiped with wine and halwa. A folk dance is held. Thereafter the deity is believed to have gone to Kinner Kailash.

(ii) Phagul or Suskar

It is celebrated in the month of February/March. In this festival the sprit of Kanda (Peaks) called Kali is mainly worshiped, the festivals lasts about a fornight and is celebrated all over Kinnaur. Each day of the festival is called by different names and several peculiar functions are held each day. On the last day a feast is prepared and people worship Kali on the roof of the houses and then partake of the food. It is believed that after the function and festival are celebrated with full zeal Kali the spirit feels happy and blesses the villagers with prosperity and plenty in the coming years.

(iii) Baisakhi or Beesh

It is celebrated in the month of April. The villagers prepare food like Poltu, Halwa and Keyshid. The image of the goddess is brought out of the temple and a fair is held in the Santang. It is an occasion to get together and to dance and drink. This festival marks the end of winter season also. New woolen clothes are worn from the wool spun during the winter.

(iv) Dakhraini

This festival is celebrated in the month of the July. On this day a feast is served. The deity is brought out and the villagers dance before her. Zongor and loskar flowers are brought from the kand peak and their garlands are offered to the goddess. After this these flowers are distributed among the villagers. One or two members from the family where death might have occurred before this festival go to the peak of the hill and ofter some food and fruits to shepherd in memory of the departed soul. A white flag on which some Buddhist mantras are written is fixes there as a sort of prayer for the peace of departed soul.

(v) Flaich Ukhayang

It is a festival of flowers celebrated in the month of September. This festival is celebrated through the Kinnaur District on different dates. Generally people celebrate it on the hill peaks near their villages. The village deity led by band is carried to the place of in procession. One he-goat is sacrificed. A fair is held throughout the day. Flowers of shuloo which have been brought from peaks for this purpose are woven into garlands. At the end of the fair these garlands are offered to the deity. Immediately after that the people accompany the goddess to the village adorned with flowers. The people sing and dance on the return journey.

(vi) Losar

Losar is celebrated in the month of December to welcome the new year. On this day in the morning a special preparation of parched barley mixed with butter milk is taken by all the family members and they put on garlands of chilgoza visits to the neighbours and friends are reciprocated and greetings of losuma tashi meaning happy new year are exchanged. While the elderly person betow their losuma shalkid or blessings. Two or three days before the losar festival khepa is observed. On this day it is customary to fetch small branches of a throny bush and place it on the doors. It is meant to ward off evil spirits. On the next day these throny twigs are removed and thrown far away from the village and a feast follows in the night.

According to Banerjee and Banerjee (1997), "Kinnaur and Kinner today do not epitomize singing and dancing. The inhabitants are educated and modern in outlook. The district has large hydro-electricity generation potential. The influence of outsiders has changed the outlook of the locals and the various developmental works have affected the ecology of area".

KINSHIP SYSTEM

Unity of the Household

Polyandry has long been a popular subject for speculation and occasionally for research by anthropologists. It can be most simply defined as that form of marriage in which a woman has more than one husband at a time. In fraternal polyandry, which is most common kind in Kinnaur, a group of brothers, real or classificatory, are collectively the husbands of a woman (or women).

This kind of polyandry has been reported from many parts of the world, but its best-documented and most prevalent occurrence is in Tibet, described by Prince Peter (1955) as "the largest and most flourishing polyandrous community in the world today," and in India. Mandelbaum (1972) notes that "in South India polyandry is of especially frequent occurrence. Six polyandrous tribes have been reported for Cochin; the Nayars of Travancore and the Izava of British Malabar have this form of marriage; while the Todas are the classic example of a polyandrous people in the textbooks of anthropology." The Singhalese are known to practice polyandry to some extent. The most consistent practitioners of polyandry in India today are probably the residents of certain sub-Himalayan hill areas in Himachal Pradesh, the northern Punjab, and north-western Uttaranchal.

Sen (1966), " the Himalayas came in contact with the Tibetans which brought about a change in their ethnic religious and cultural life. The purpose of pastoralism and trade, especially for the latter, the Himalayas and the Tibetans used to visit each other's territory and spend a considerable period of time. In Tibet, the inheritance rules are based on primogeniture and the people are predominantly polyandrous. The Himalayas also follow the rules of inheritance. Both in Tibet and in the Himalayas there are a number of young males who join the monastery to become monks and remained unmarried".

Polyandry thereby also serves "to reduce potential hostility between sibling brothers." Without polyandry there would be a tendency for children of brothers to break up the joint family in order that each group of siblings might pursue its own economic interests.

A widely cited economic advantage of fraternal polyandry is that it keeps family property, especially lands, intact in a patrilineal, patrilocal group. It accomplishes this by restricting the number of heirs and by keeping them together around a common wife. This virtue of polyandry is cited for Ceylon (Peter, 1955), Tibet (Peter, 1963), and the Himalayan hill area (Stulpanagel 1878). Raha (ed) (1987), edited book on polyandry cover three different polyandrous zones of India namely North and North west India, south India and east and northest India. It includes articles on polyandry system in various societies and love special nature like polyandry as was prevalent in ancient India, differential sex nature in polyandrous areas, bow and arrow ceremony among Todas and status and roles of women in polyandrous societies etc.

Beeraman (1961, 1962) made an attempt to compare the polyandry system in Himalayan area of India and Tibet. According to him, "In social life, probably the best known and most distinctive features in Tibet is the customs of fraternal polyandry whereby brothers share a common wife. In various form this is found from Ladakh to eastern Tibet. The occurrence of a similar practice among Paharis has led to considerable speculation as to possible diffusion of this marriage form between the two cultures, but they may well be independent developments". According to Beeraman (1962), "There are important differences between Pahari social organization and that of the rest of North India. Pahari marriage, for example, involves a bride price, while high caste plain marriage does not". Beereman (1978) in his paper on ecology, demography and domestic strategies highlights the modest population increase among land-owning high castes as contrasted to net population decrease among exploited low castes, relatively low fertility rate, strategies of adjustment in family and household composition in order to optimize the ratio of people of land, polyandry system as an example of such adjustment. Further he discussed the implications of these phenomena for development in the region and for the future of these and other Himalayan peoples.

Punjab District Gazetteer Simla District (1910) mentions that polyandry was in vogue among the people (Khas) particularly among the Kanets, of the Himalayas. It gives some detail of polyandry practised in Kinnaur at least, in the beginning of this century. "Polyandry prevails in the greater part of Kanawar and in some places in Rohan Tehsil.."(1910). Andrew Wilson (1885) also found polyandry to exist commonly from Toranda in the Sutlej valley, a few marches from Shimla upto Chinese Tibet.

Parmar (1975) and Chandra (1981) studied the polyandrous nature of Kinnaur tribe. Joshi (1984) traced the reason for polyandrous nature of Kinnaur people. According to him, "factors which contributed towards the continuance of polyandry among Kinnaur tribal society is that the Khasas who at one time inhibited the Himalayan religion from Kashmir to Nepal, including the Kinnaur region, were polyandrous in nature".

The local people hold different opinion about polyandry. Some people defined it on the grounds that it has kept the family a closely knit entity. It has prevented over population and the sub division or fragmentation of small agricultural holdings. It has enabled a family to derive full benefit from several sources for a proper living by way of pooling of labour. At the same time, the legend of Draupadi who had five husbands may have indirectly encouraged the people to adopt the system of polyandry. Since the eldest brother used to be out of the house for earning the livelihood and others used to engaged in looking after flock and for another trade or other occupation. In practice, it is not so often that all brothers are at home at a time which has given rise to polyandrous family in Kinnaur.

Raha and Coomar (1987) observed, "Polyandry which was the traditional institution of Kinnaur has perpetuated to the contemporary Kinnaurese society. This cultural trait perpetuated because it got the encouragement from the erstwhile Bushahar state authority. The encouragement from the king definitely boosted the desire of the Kinnaurese to remain united under the polyandrous hold with a common wife for all brothers. This system continued functioning in the Kinnaurese society for a pretty long time as the Kinnaurese accrued definite gain out of it. It helped them not only to remain united in the polyandrous household with the common wife but also to save fragmentation of property, to yield better crop, to have a suitable division of labour, to cope in a better way with the barter system developed in the peculiar Himalayan ecology and so on. Over a period of time, the process of gradual degeneration of polyandry in Kinnare has been quickened because of the various forces in the post-independency are acted our these areas—the forced like modernization, execution of various development and constructional projects, modern education, modern land reform acts, modern communication, cash and market communication, cash and market economy etc have enticed the people of Kinnaur, like many other Himalayan polyandrous societies, to discard their unique institution. Thus the century old polyandry in their high Himalayan society has been left at death's door".

Until recently, monogamy was the exception rather than the rule; polyandry (several brothers shared one wife) was more prevalent in most parts of the district. In case a joint wife proved to be barren, her sister was brought in as a second wife. Sometimes the youngest brother would prefer to bring another wife for himself on the plea that the common wife was very old. Gradually, love affairs usually result in separation of one of the husbands from the family. In such cases, the joint property must be partitioned, unless the new wife consents to be shared by all brothers. In case of refusal, she and her husband have to live separately, but the husband retains his right to share the original joint wife. In practice the original wife refuses to have anything to do with him. With the exposure to outside worlds and improvement in education system, monogamy is no longer an exception.

Partition and Rules of Inheritance

Inheritance of the property is patrilineal. In a polyandrous family, after the death of elder brother, his share is inherited to the surviving brothers or co- husband and in case of death of all brothers, the sons inherits the property in equal share. In Kinnaur, the rule of primogeniture (*jethang*) is followed. According to this, the eldest son has the first right over the property than the youngest brother (*Kanchang*). It is believed that eldest one can work hard on the field and get the yield desired and settle himself in life properly whereas the youngest, being a beginner, should have al least shelter to begin with. Therefore, the field is given to the eldest brother and the ancestral field is given to the youngest. The rest of the property is divided in equal share. As the Hindu Succession Act 1956 is not applicable to Kinnaur, the girls do not have any legal right over the property.

In earlier time, the rulers used to impose penalties on the partition of properties. In the case of division of movable property, a one half share was appropriated by the state and in the case of immovable property, official recognition was denied. (Monga, 1998)

Chandra (1981), "Some of avoidance of fragmentation of land in the large interest of household units is felt by these people. With the fragmentation of land there is a danger of less of productivity of land due to shortage of man-power and than the land size becomes unviable because the division of land, other property, animals, equipments etc. are also divided. As a result, it becomes almost impossible for the concerned members to run their economy smoothly at the household level for the want of manpower, animals and tools etc."

Chandra (1987) "Environment and ecology are yet other dimensions considered very important from the view point of the prevalence of the practice of polyandry. It provides a fresh look at the whole problem since both environment and ecology play vital determinately role in this aspect".

Chandra (1987) explores the nature and extent of environmental adaptation in relation to a particular community in the perspective of a geo-physical situation. In his paper, he made an attempt to show how best the Kinnauris have adopted to the hostile environment for their prosperity in the given conditions after myriad countervails, both at individual land as well as at corporate group level. It highlights adaptation at intra-level for various factors be it religion, material culture or social-systems. It is only after the environmental factors that these people have evolved specific social systems—marriage, kinship and household organizations. The Kinnauris who took this region as their home since long time past have shown tremendous amount of patience, sharpness and mental ability in accepting these challenges and have molded themselves in terms of their physical, socio-cultural requirements in such a way that the adjustments seems to provide ready answers to any question that relates to their environmental conditions or factor.

BIRTH

The Kinnauri do not consider pregnancy as an abnormal period. Delivery usually takes place in a separate room. A midwife called *api* in the local dialect is respected by the baby like a mother throughout her lifetime. Mother is made to eat hot *ghee* immediately after the delivery. *Chuli khali* (hill fruit like apricot or peaches) is boiled in water and the mother and baby are bathed in it. During the first fortnight after the delivery, a daily massage with *chuli* oil is given to the mother and child followed by a hot water bath. During the fortnight after the delivery, the mother is usually served with a balanced diet consisting of the *ghee*, animal fat, *chuli* oil, barley or *ogla sattus*, wheat *chapaties* with butter *halva*, coconut and rice. Chilies and sour or tart ingredients are avoided.

The mother and her family observe *'begnang'* for the seven days in which they are not allowed to enter places of worship as they are considered unclean. The musical instruments and other vessels are kept in the temple and also not to be touched by them. On the seventh day the whole house is cleaned and *gomutra* (cow's urine) and *gangajal* (Ganga water) sprinkled everywhere. A smearing (coating) of the cow dung is done in the Kitchen. The mother is then brought out with the child in verandah. Seven pebbles are placed over which mixture of *gomutra* and *gangajal* is sprinkled. The mother then goes round these pebbles seven times and on the completion of each circle she throws away one of the pebbles. In certain cases a Lama also performs the *hawan* on seventh day. During this occasion close relatives and members of the clan (*biradari*) depute their representative to convey blessings to the newborn. They bring with them a few grams of *Ghee* or *chuli* oil and two kilograms of wheat or *ogla* flour or *kanwni* rice as the present for the mother. A feast is arranged for these guests. The offerings are received only from people of the same caste strictly on the basis of give and take.

In upper region of Kinnaur, various other ceremonies (like *Khaskis*, *Bedhai* and *bose*) are observed at the time of birth. In *Khaskis*, on the fifteenth day after the delivery, the entire clan is invited and a big feast is organised with great pomp and show. During the feast, the Chamangs and Domangs are also invited. They are not allowed to enter the house of Kanet and are entertained separately. Champak flowers brought from the lower hills are presented to the father or grandfather of the child. The *api* is given food daily during her attendance on the mother for eight days and thereafter she is given clothes and some cash. In *Bedhai*, Chamangs and Domangs offer congratulation in the first month of the year (*Chait Sankranti*). Musical instruments are worshiped by offering of wine and wheat or barely bread. After worship, the offerings are distributed among those present on the occasion. In *bose* ceremony, the local deity is invited to bestow his blessings to child between first and third years.

A *Krachogmig* {mundan (shaving of the head) ceremony} takes place when the child attains the age of 1 or 2 years. The day is fixed in consultation with Lama. This is done on the full moon day or during the bright half of the month. i.e. *shukla—paksha*. Each odd month is considered auspicious. The *Mundan* is performed by the maternal uncle or in his absence some other person whose mother and father are alive and hairs are buried.

In case a child happens to cut the upper tooth first then this is considered a bad omen for the mother's parents and brother. To ward off the evil effect, the mother or her husband would send a massage regarding this to the maternal uncle. Then the child's maternal uncle would bring a few cloths for the child. Before entering house, he would drop these clothes through *Dusrang* (outlet for the smoke). Mother would collect them and put them on the child immediately. The maternal uncle would then enter the house through the main door.

In polyandry family, the relation between son and father is not difficult to establish. Normally the practice is that all the husbands are recognised as the father of each child. The eldest among the husband is called *Teg-Boba* (elder father) and the others as a *Goto* or *Jigich–Boba* (younger father). The eldest among the living brothers is spoken of as the father of all the children born to a polyandrous wife. The wife names the father of her different children, only in case of division.

Chandra (1972), "The Kanet children, born at the polyandrous union call their mother's husbands with the same kinship term Baba without least distinction but before the term *Baba, Teg, Maghung* and *Zigchit* are prefixed for the eldest, intermediate and the youngest father respectively. However this kinship term *Baba* meaning father does not carry any distinctive connotation with it regarding the biological or social fatherhood".

Raha, Mahato and Negi (1976) has analyzed that Kinnauris kinship terms show certain intensity features. The three cultural divisions of Kinnaur based on religion, bear distinctiveness in some of their terminologies. While some of the terms are common although Kinnaur, some others are different in different cultural zones. The kinship terms all over Kinnaur are distinctly classificatory barring a few which are denotative in nature. Different Kinnauris kinship terminologies have been found to obey various criteria of distinction while some go against these rules. Another interesting feature is the impact of the polyandry. But that polyandry is perishing can also be revealed from some other terms which are at present distinct terms but should have been put under a common one.

MARRIAGE

Kapur (1993) in his book describes the marriage and divorce customs of schedule tribe of Himachal Pradesh in details. In one of his chapter on Kinnaur, he gives a historical ethnographical account of Kinnauris and explains the various types of marriages and divorce.

Marriage is popularly known in Kinnaur as *Rehja*. It is settled much earlier when boys and girls are very young. The standard and legal marriage is called as *Janetang* or *Janekang* a suitable bride is looked for by the parents of boy and then the negotiations are opened with the girl's parents generally through relatives. If the bride's parents seem agreeable to the proposal, two persons called *majomi* (go between) settles a marriage by offering a piece of cloth (*khatak*) and bottle of liquor. Acceptance of the proposal is followed by an offering to the family deity or the village deity and, in some places; a token item of ornaments is left behind by the middleman. After some time, the bridegroom father along with some other person goes to the prospective bride's house and agrees upon the day of marriage and decides how much money of bridegroom's side will give to the bride's family. For the purchase of cloths and ornament for the bride, money is paid before the date of the marriage. If the wife leaves her husband and goes with another man, this money has to be returned.

An offering to the family deity or village deity follows the acceptance of the proposal. A marriage date is then fixed in consultation with a Lama or the village priest. On the appointed date, the bridegroom goes to the bride's place and is accompanied by the small party of half a dozen or so along with the village band. The party stays there for two feasts held in the honour of the party at the bride's place.

The bridegroom returns home with bride after the merrymaking during the night. The ladies belonging to the bride's village but married in the groom's village go out to welcome the bride and take with them wine and other eatables for the guest. A *Puja* (worship) is performed to placate the evil spirits supposed to have come with the party than the party goes to the groom's house. The sister or mother of the groom welcomes the bride. The symbolic ritual called *oopagey* is performed. In this Groom's maternal uncle wraps a turban around the head of groom and his brothers. Garland of *chilgoza, chuli,* and walnuts are draped round their necks by the bride.

After the marriage is over, the groom's father sets asides some piece of land for the bride. This is confirmed in writing on a plain or stamped paper. The paper remains with the girl's father or maternal uncle in the presence of a Middleman. Similarly surety bond is given by the bride's father to the effect that said document should become inoperative in case of divorce. A list of weighed utensils is than prepared before they are given. In the event of divorce these utensils or cost thereof has to be reimbursed by the groom's father. This custom is called *'bandobust'* ceremony. In the end a ceremony called *'borshimik'* takes place where the pride garlands all the members of a marriage party with the *chuli* or *chilgoza* embracing them and weeping bitterly while bidding farewell.

According to Raha and Mahateo (1985), "The newly wed bride becomes the wife of all the brothers through Turban tying ceremony. In this ritual, which is considered as the most important ritual that symbolizes polyandry marriage, all brothers sit in a row. The bride sits before them. The grooms i.e. brothers and their bride are garlanded... The maternal uncle of the bridegrooms then takes a piece of white cloth, make *paga* (turban) on the head of each brother with that piece of cloth, with accompaniment of music. This indicates the marriage of all the brothers with the common wife. Gazetteer of Kinnaur also refers to a ceremony of putting on turban (*Pag Likshimu*) by each brother to get the status of husband (Mamgain, 1971).

In the study conducted by Ramesh Chandra (1973), it has been observed that at the ethnic group level, the polyandrous form in popular amongst all groups; but it is specially among Kanet and Koli. Also all the ethnic groups showed an aptitude for monogamous marriage. The transitional form of marriage lie poly-monogamous is found to be the highest among the Kanet and then the Koli and the Lohar follow in order. The rare polyandrous and polygynous form of marriage was confined among the Kanet, Koli and the Badhi in his sample study. Further, according to him, the change in marriage form is directly linked with changes in the occupational pattern. This is supported by the facts that the persons involved in the polyandrous form of marriage are engaged in a complex set of occupations viz. agriculture tending sheep herds, trade etc. while these involved in monogamous unions tend to have an economy based mainly on labour, service or other means where joint efforts of family lend are not necessarily needed. The people engaged in the intermediary forms of marriage tend to combine an economy based on individual endeavor with sharing the joint economy with their brothers.

Types of Marriage

Other forms of marriages are *Dumtangshis* (Love Marriage), *Dubdhub* (Marriage by capture) and *Har* (enticing away someone wife.)

(i) Dumtangshis or Bennang-Hachis or the Jushis (Love Marriage) and Dubdhub (Marriage by Capture)

In this form of marriage, the boy takes his beloved to his house secretly and in *Dubdhub*, the boy with the help of his friends forcibly carries away the girl of his choice to his house.

The boy's father deputes two middle men to the girl's parents requesting for their pardon for the outrage committed by his son. The girl's father, after some resentment, asks them to make amends in terms of money, called *izaat* for atonement for this graceful act. This money has to be deposited at once by the middlemen. Within a month, the girl is sent back to her house—on some auspicious date the boy accompanied by the middleman and two ladies escort the girl back to his house. On this occasion, she is given new clothes and a few ornaments. The boy offers *chilgoza* and *chuli* garlands as a mark of respect to his mother-in-law and other elderly women in the family. After that the groom returns with his men leaving behind the bride. On a suitable date, her father and brother take her to her husband. All the relatives collect at the girl's house on this occasion and absolve the girl of the disgrace of being forcibly carried away of elopement.

(ii) Har (Enticing Away Someone Wife)

In *Har* marriage, woman run away or elopes with a man other than her husband. In this case, the husband will send representative to the new husband and demand a refund of the expenditure incurred by him during the marriage. He will have to be repaid this amount, called *hari* and is mutually decided upon.

Marriage Gift

At the time of deciding the date of marriage, they also agree upon how much money the bridegroom's side will give to the bride's family. This money varies from rupees fifty to one thousand, depending upon the capacity and status of the family concerned. This money is paid before the date of the marriage and is used in the purchase of clothes and ornaments for the bride. This expenditure is accounted and taken note of. In the event of the wife leaving her husband and going with another man, this money has to be returned.

The dowry system has never existed in the Kinnaur district. The gifts and presents are customarily given to the bride by her parents and relatives and *udanang* are regarded as an exclusive property of the women. At the time of marriage, the bridegroom is given a turban, a *khakharis* a *nangeh* (a dish) or a *batich* (cup) and a *nang* (platter) and sometimes he is given some garments. These gifts can not be regarded as dowry.

Termination of Marriage Bond

Divorce is permissible among all the group of Kinnaur. If the woman wants a divorce she usually does it through her parent. In the divorce proceedings, the parents of the women and her husband or husbands gather at a place on an appointment day and settle the accounts. The utensils, ornaments and cash given to the women at the time of marriage by the parents have to be returned by the husband or the husbands. Finally a twig is placed between the couple and they break this twig; this symbolizes the breaking of the marriage.

DEATH

Earlier the Kinnauri used to keep the dead body on the hill top to be eaten by the animals and birds. But due to the impact of Hinduism in Kinnaur, the body is cremated. Infants up to age of the two years who happen to die are buried. In case of children above two years and below five years, the body is thrown into the river.

Panchratna (alloy of gold, silver, copper, brass and iron) is kept in the mouth before the death of the person. A Lama is often called to stand by and to recite some mantras and to pray for the eternal peace of the soul. The dead body is bathed in the warm water. Brass lamp is lit with *chuli* or mustered oil on the spot of the death. This continues to burn for seven-day continuously. A pot full of barley is also placed next to the lamp.

A band accompanies the funeral procession by plying sad tunes. The man leading the procession carries an urn with live coals. Barley is scattered at frequent intervals all along the last journey. One man from each household in the village accompanies the corpse for collecting wood and other material for the pyre. The son or nearest relative light the pyre from the direction in which head is placed. The ashes are carried mostly to the Ganga or Sutluj. Close relatives shave their head in the respect to the dead. During this period the family members do not eat meat, oil, fried articles of food, onion, garlic or turmeric, nor do they season with *tudka* or *chhuk* (spices fried in heated oil) for the eight days. On the eight-day the family members purify themselves then the house is cleaned both actually and ritually. In the evening they get together and a feast is held in which no wood taboos are observed (called *kolyashinmg*). On the fifteenth day a general feast is held for the villagers in the temple. In this feast *chiltas* are distributed and these are taken home. *Asting* is the last ritual after the death is known. It is performed on the thirteenth day following the day of death. A grand dinner is given to the relatives and villagers in whom *poltus* are distributed.

POLITICAL ORGANISATION

The political organization of the Kinnauri revolves around the village deity and panchayat along with the modern system of administration. The village deity who commands great respect among the people and his words are respected by the villagers in order to avoid future curse of God. There is a Panchayat consisting of two or three noblemen of the village. The main function of the village Panchayat is to settle dispute among the people of the village. Members of the Panchayat include old and respectable persons of the village community; they know the facts of the dispute. Personal hearing is given to opposite parties to put forward their claims and counter claims. After hearing the parties, the Panchayat decides the dispute then and there to the satisfaction of all concerned. The old system of administration of justice generally and in personal matters particularly, still enjoys the confidence of the majority of the people in the region though the modern system of administration also exist in the area.

In some of the village it has been observed that they still have their traditional council called *Char Bhai*, which consists of *Char*, who belong to Rajput (Kanet) community, *Halmandi* and *Tokiya* who belong to the Koli community. They dispense justice in matter arising out of dispute of economic, social or religious nature. The council is also presided over by a *Grokch* who is believed to the mouth piece of their deity and his verdict is presumed to be the will of the deity and is supreme. The offenders are punished by ex communication fines and punitive feasts, depending on the severity of the offence.

Apart from punishment by modern jurisdiction system, the village deity or fraternity (biradari) also plays an important role. In earlier days, the deity or fraternity (*biradari*) decision used be taken as final. According one of the oldest folklore, *Dhunglu*, named robber, raped a Rajput girl was sentenced to death by different panchayats around Chini for his misconduct. It is said that he was tied to the tree branch standing along Tannnangse precipice. The branch was cut and he was pushed in the deep rivine down where he was torn to pieces.

For offence like kidnapping, divorce, blame etc, the *Izzat* money as a fine is rendered to the aggrieved party or the village deity or Panchayat. Another punishment known as *Tichig Michig* is awarded for grabbing of somebody's immovable property, forceful marriage or violating devata's direction. *Chhetpa* (fine levied by the Biradari) is punishment awarded for not coming to deity's shrine or for coming late or somebody absents himself from the meeting called of all the villagers.

The punishment *Ti Darang Shishe*, exemplary punishment awarded to an incorrigible errant. The person is summoned to the temple where the oracle of deity throws a shower of cold water on him out of *krow* (bowl kept in the temple and used in many rituals) and prays for his downfall to all levels.

After formation of district in 1960, the civil and criminal courts were opened. All the division and sub divisions have sub-judge courts now. In the matter of criminal courts, besides the District Magistrate, there are several courts in every tahsil and sub tehsils under tehsildars and naib-tehsildars all over the district. After 25 January 1971 when the state was granted a full statehood the above set up at the state headquarters and the function of various agencies in district became the concern of general administration. Now the separate head of departments correspond directly with their respective district level officers. (Kapoor, 1998; Negi, 1987).

Suresh Kapur (1998) in his article emphasis on significant of customary law for the community of its origin which decide the disputes arising out of matrimony, succession, adoption, maintenance and the family matters. Further he advocates that tribal customs and beliefs are meant for tribal people only. The special measures should be adopted by the govt. in order to protect the tribal from social injustice and various other forms of exploitation.

KINNAURI WOMEN

Kinnauri women enjoy a unique position. She is not only the keeper of her hearth but contributes equally to the outside work. She equally participates in dances and folk songs. The women are exposed to laborious work like weeding, irrigation, carrying load and water, cooking food etc. Women play a crucial role in agriculture activities except ploughing. Chuhan (1998), "The Kinnauris have been practising polyandry as a system of marriage but with the pace of modernization and development, it is now fading away. The Kinnauri women are beautiful, modest and hard working as any other woman in the hills and spend most of the day working on the agricultural fields".

It has been observed that the Kinnauri, due to difficult geographical terrain, have lesser interaction with the outside world. Still the influence of the plain culture can be observed in their day to day life. The polyandrous system among the Kinnauri is also changing with the influence of the Hinduism. The migration of the Kinnauri to the nearby districts and states to bring about the transition in all spheres of life of these people. In the present administrative scenario of the country, one can easily observe the change in this ethnic group from schedule tribe to schedule caste. Sen (1996), "The people of Kinnar have accepted the recent innovations and changes brought out by the administration. The developmental activities were proved to be useful to the people and were accepted slowly at the initial stage but when people found the programme beneficial and useful, they started taking initiative to invent their sources and even did not depend much on the govt. departments." Kapoor (1990) emplaned on urgent need for the proper understand of categorization of communities in the compartmental definition of tribe and caste. Anthropologically, there is dire need of giving the clear-cut definition for defining these ethnic groups in the Kinnauri.

3

Methodology

The focus of chapter is on the various methods that were chosen for collecting data during the progress of this work. It deals with the different dilemmas that were faced by the researchers in the initial stages of the study. Once these barriers were effectively dealt with the progress was feasible in the study. The greatest disadvantage to overcome was the unfamiliarity with a new area among the alien people. Gradually the researchers were able to overcome this obstacle. This chapter further delineates the reasons for choosing Kinnaur as the locale for study. It talks about the selection of sample, how and where it was selected? The chapter describes the various methods that were chosen and the reasons why they were opted for? It also makes an attempt to portray the difficulties faced in interviewing the respondents, because of restricted time they had. Being totally engrossed in their family and work, they sparingly gave time to the researcher.

WHY KINNAUR ?—THE LOCALE FOR THE STUDY

As a student of anthropology, our interest have been always about the awesome and fascinating culture of the tribals of India which is very moved by the backwardness of the tribals and often thought of doing something about it.

Having kept this area of interest in our mind, our next step was to identify the area of study. Many works are being done on the tribals of India but we have seen that very few work has been done in those areas which have different ecological settings. As we went through literature, we observed that there is no research work pertaining to anthropological exploration to health dimensions of Kinnauri tribe of Himachal Pradesh. Our interest on this tribe was also enhanced more because of their polyandrous nature, being in an isolated and excluded area, the often heard landslides, the Tibetans as their very close neighbours, the hospitable behaviour and world famous trekking routes and valleys!

Keeping all these factors in mind, Kinnaur District was chosen to be the locale for the research work.

RESEARCH DESIGN, SAMPLE AND METHODOLOGY

Research design involves combining the essential elements of investigation into an effective problem-solving sequences. It is usually an idealized blueprint or road map that helps the ethnographer conceptualised how each step will follow the one before to build knowledge and understanding. A useful research design limits the scope of the endeavour, links theory to method and guides the ethnographer.

Field work is the most characteristic element of any ethnographic research design. It is exploratory in nature. The foremost step is to undergo an exploratory visit, so, a pilot study was conducted. The survey period helps to learn the basics, understanding the native language, the kinship ties, census information, historical data and the basic structure and functions of the culture under study for the months to come. The pilot study helps in giving clearer geographic and conceptual boundaries.

THE SAMPLE

For present study, in total of 15 villages and 743 Household were selected. The respondents from the 15 villages were further selected using the simple random sampling technique. The Head of the Household was asked if he was willing to be a part of the research sample. It would take a minimum two to three hours to interview and get detailed responses from the head of the Household. He would carry on with his work while responding to the questions. This was important, as the researchers did not want to disrupt the daily routine of the respondents. After the completion of the first interview the researchers moved further. As the interviews were time consuming, mentally exhausting, not to forget the physical exhaustion in the blazing sun of the afternoon, difficult geographical terrain and dispersed household pattern in the hilly region. Since focus of the study was to understand the health profile of four ethnic groups, therefore the entire sample is divided into four ethnic groups i.e. Kanet, Koli, Badhi and Lohar. In the selected villages attempts has been made to cover all four ethnic groups. Due to numeric limitation of Badhi and Lohar in the villages, the number of sample is lesser. The name of villages and ethnic groups covered is given in Table 3.1.

Table 3.1: Name of the Village and Ethnic Groups Covered

S. No	*Name of Village*	*Name of the Ethnic Groups*				*Total*
		Kanet	*Koli*	*Badhi*	*Lohar*	
1.	Roghi	17	31	–	2	**50**
2.	Kalpa	–	–	3	18	**21**
3.	Kothi	7	11	5	8	**31**
4.	Kamru	–	–	1	5	**6**
5.	Bharang	109	79	8	3	**199**
6.	Sangla	–	–	24	11	**35**
7.	Tangling	50	45	2	12	**109**
8.	Kashmir	1	–	15	–	**16**
9.	Kanei	39	70	2	15	**126**
10.	Shogthong	2	–	–	–	**2**
11.	Sapni	24	64	–	–	**88**
12.	Pangi	–	–	14	2	**16**
13.	Shayso	26	–	–	–	**26**
14.	Kilba	–	–	11	1	**12**
15.	Powari	–	–	–	6	**6**
	Total	**275**	**300**	**85**	**83**	**743**

RESEARCH METHODOLOGY

Various methods of data collection were brought into use while studying the sample. They were beneficial in eliciting different kinds of information. The several methods used were Survey, interviews, observations, group discussions and the home visits. Secondary sources of data collection were brought to use at different points of data collection. This helped in substantial data collection both qualitatively and quantitatively.

Secondary Data

In order to elucidate the existing health infrastructure and also information regarding the tribal population, data from secondary sources were collected. In the present study, data were collected from District headquarters, Sub-Divisional headquarters, Block headquarters and District Hospital, Primary Health Centres, Sub-Centres. During the course of study, the various other government departments like Economic and Statistical Department, Women and Child Development Department and NGOs were contacted. During the study, the secondary data was collected from various libraries in Delhi and Himachal Pradesh.

In the initial stages the pilot study was performed to test the accuracy of the interview schedule.

Village Profile Schedule

Village profile schedule was developed to collect the detailed information about the village. The village profile schedule was used to obtain description of the community household structure, sources of electricity, availability of important public health measure and educational facility. This information was collected from two key personnel in the villages.

Household Schedule

The household schedule was prepared to elucidate the information from each household. The information from the Households was obtained regarding:

- Family Size
- Economic
- Habitation Pattern
- Demography
- Morbidity
- Maternal and Child Health
- Family Planning
- Health Utilization and Awareness

Interview Schedule

The elaborate interview schedule was designed to elicit the information from the respondents. The schedule was always kept in hand while interviewing the respondents. It was the means of keeping a check that the interview was not going astray. A household Schedule was developed consisting of closed, open and multiple response questions. To confirm the responses in the schedule, the same question was asked in different ways two or three times during the interview (triangulation). The responses which were found to be consistent were considered reliable.

Interview schedule was the basic and the most important instrument used during the data collection. The major advantage of using an interview schedule during this study was its adaptability. It helped in meeting and talking to respondents, reading their facial expressions, probing responses, investigating motives and feelings, following up ideas, which would not have been possible in a questionnaire.

Questionnaires could have concealed a lot of information, as the responses would have to be taken at face value. The responses during the interviews were developed and clarified at the site itself. The shape of the informal interviews conducted was determined by the individual respondents themselves and these helped in yielding a rich and valuable data. Though the interviews were not completely rigid, they were also not unstructured. A guided framework had been established, which was always available to ensure that all the areas crucial to study were covered. Though the conversations about many topics were interesting and could have produced useful insights into many problems, it had to be remembered that this was an interview and not just an interesting conversation. Thus, the respondent was allowed a considerable degree of latitude within the established framework. During all the interviews the researcher's main aim was to be completely objective, and thus avoid any kind of bias.

Pilot Study

To ensure the authenticity of the household schedule it was essential to put it to a pretest, which helped in developing the continuity during collection of data from Household. It was advantageous in improving the efficiency of smooth flow from one topic to another, without any interruptions. A sample of 60 households was selected during the pilot study (from June 1999 to

August 1999). A few questions in the household schedule had to be modified. After testing the proficiency of the household schedule, the actual process of interviewing the respondents was activated.

Rapport Establishment

In any study the foremost step in commencing an interview requires rapport formation. The reasons for easy and comfortable rapport formation were:

- Help from the local friends
- Doctors/ANM-help in establishing rapport
- The Women/Head of household felt that this was a chance, which would provide them with an excuse to talk to the researchers and ask few questions in return.
- Their inherent culture: Himachali on the whole are friendly people. They greet guests and others with a friendly smile on their face. Majority of the times they are seen to exhibit respect for the outsider.

Interview the Respondents

The whole process of data collection was not very easy. The main reason for this was the busy schedules of the respondents. For most of the respondents, their houses were their workplaces. It was difficult to interview these respondents at home as majority of the respondents were involved in their economic activity, for hours together. Most of them felt that during their work they could manage to take out free time to answer. They complained about their hectic schedules at home, which offered limited chances of being able to sit and take out time to reply.

The interviews were conducted in the mornings, afternoons and in the evenings. This helped in gathering a variety of data, as the respondents were involved in different kinds of activities at different hours of day.

Majority of the respondents were eager to know of how these interviews would be useful to them. They inquired if they would receive any monetary benefits or benefits in some other form. It was important to let the respondents know that the information's being collected were for the research and which would highlight their problems to the outside world. This study would also help in bringing in light the health conditions of the people. Many of the respondents were in such poor conditions that one would feel very depressed towards the end of the interview.

Focus Group Discussion (FGD)

Focus Group Discussions are primarily group discussion, unguided or guided to a limited extent, with generally ordered guidelines. A focus group is a congregation of individuals as far as possible homogenous, who are most likely to be benefited and affected by or related to the issues of the research project. Besides ensuring homogeneity among the participants in terms of age, sex, educational level and other such attributes, the criteria of related to the issues of social structure, local customs and group dynamic need consideration. The group was limited to 6-8 members to avoid interpersonal discussion and confusion. Care was taken that the focus group was not dominated by anyone member of the group. The focus group discussed a particular topic under the direction of moderator (researcher) who promoted interaction and ensured that the discussion remained on the topic of interest. A funnel approach i.e. general to specific question was followed.

Focused group discussions were common at the households/ PHC. Towards the end of the interview a general group discussion would initiate, where they would talk about their common problems both at health and development. Efforts were made by the researchers to bring focus in these group discussions, so as to gain insights in the areas of research. Majority of respondents were, keen to know about alternate ways of improving their health and status in life. These focused group discussions helped in enriching the study with various insights, which were not possible to achieve through interviews with individual respondents. The reason being the group discussions led to various contradictions and cross questionings, which only the respondents could raise among them.

Observations

Observations were brought in use at various stages of data collection. Detailed observations were made during the home visits conducted during the study. Home visits were conducted especially to observe the physical and social environment of the respondent. A minimum of about three or four hours was spent in a household.

The respondents were amiable and extended warm hospitality to the researcher. Observations were made regarding the basic infrastructure facilities, health, hygiene and sanitation maintained in home and outside the houses. Further observations were made to gain knowledge regarding their interaction level with patients and Health Service Provider.

Occasionally, a few days were spent at the respondent's house. Thus the researchers got the chance to observe the respondents at home, involved in their family lives. This was an excellent opportunity to observe the interaction between the family members. These visits proved to be fruitful in eliciting rich information regarding the family lives of the respondents.

TIME DURATION

A lengthy period was required for data collection. The main reason being the researchers have to get acquainted with a tribal area, the people and their language. Long periods, were spent with the respondents, as they were enthusiastic to talk to the researchers. Many a times they had to be interrupted gently as they used to get carried away with their talks, case studies required longer duration, as in-depth information had to be collected and along with them home visits had to be made. The intensive field work was conducted from July 2000 to August 2000, January 2001 to July 2001, December 2001-March 2002 and final validation of data and information from March 2003-July 2003.

ANALYSING THE SAMPLE

Any amount of information collected through questionnaires, interviews, observations is of little value unless it is analysed correctly. Both the descriptive and the inferential statistical methods were used in analysing the data. At various stages, pictures in detail in form of charts, tables, percentages, averages etc have been presented and implications have been drawn. These methods have assisted in not only providing useful analysis to the study but at the same time making the analysis illustrative and interesting to read.

DEMOGRAPHIC ANALYSIS

Population and its current and prospective rates of growth affect socio-economic development and associated welfare values substantially. Such demographic challenges to development have heightened the significance of in depth study on population of an area.

The term 'Demography' is used as synonymous with "demographic analysis", which is primarily concerned with quantitative relations among the demographic phenomena in abstraction from their association with other phenomena. Demography may also be conceived in a broad sense to include, in addition to the quantitative study of population, the study of interrelationships between population and socio-economic, cultural and other variable. Demographic variables which have been utilized in the present study are as follows:

Sex Ratio

The Sex–ratio is the number of females per thousand male.

Masculinity Proportion (MP)

Masculinity is the measurer of sex-composition. The masculinity proportions has been calculated as per the standard formula.

Dependency Ratio (DR)

Another measure to study the structure of the population is the dependency ratio. This measure indicates the number of dependents per 1000 workers and may be computed on the basis of three broad age groups, which are below 15, between 15-59 and 60 and above. The population in the age group 15-59 is considered to be the working population; that below 15 as the young dependents. The dependency ratio is computed as per the standard formula.

Index of Aging has been Calculated as per the Standard Formula

Measures of Economic Characteristics (Work Force Participation)

The crude activity rate is economically active population to total population, whereas general activity rate is economically active population to population aged 15-49 years and calculated as per the standard formula.

Child Woman Ratio (CWR)

The ratio of children under 5 years old to women of child bearing age, referred to as the child women ratio and it has been calculated as per the standard formula.

Crude Birth Rate (CBR)

The crude birth rate is the ratio of the total registered live birth in some specified year in a particular area to the total mid year population of that area multiplied by 1,000.

General Fertility Rate (GFR)

The general fertility rate is defined as the number of births per 1000 women of child bearing age. The general fertility rate relates birth (at a specified time) with population at risk of childbearing i.e to the females at reproductive ages of 15-49 years.

Age Specific Fertility Rate (ASFR)

It is defined as the number of live births to a mother of a specified age group per thousand mid-year female population of the same age group.

Total Fertility Rate (TFR)

It is defined as the average number of children that would be born alive to a woman during her reproductive span (15-49 years) of life confirming to the age specific fertility rate in a given year and it has been calculated as per the standard formula.

Gross Reproductive Rate (GRR)

The gross reproductive fertility rate is defined as the number of female children a cohort of women would have passing through the reproductive ages and bearing children according to fixed schedule of fertility, if they all survive to the end of the child bearing period calculated as per the standard formula.

General Marital Fertility Rate (GMFR)

The General Marital Fertility rate is expressed as the number of live births per thousand married women in the reproductive age group 15-49 years.

Age Specific Marital Fertility Rate (ASMFR)

The Age Specific Marital Fertility Rate (ASMFR) describes the fertility experience of married females by age at a specified time and have been calculated as per the standard formula.

Total Marital Fertility Rate (TMFR)

Total Marital Fertility Rate summaries the pattern of marital fertility exhibited by the ASMFRs and presents a single index of total marital fertility. In other words, total marital fertility rate is cumulative value of age specific marital fertilities at the end of the reproductive period.

Age Specific Mortality Rate (ASMR)

The Age Specific Mortality Rate (ASMR) describes the number of deaths in particular age group in mid year population of same age group.

Crude Death Rate (CDR)

The Crude Death Rate is the most simple and the most commonly used measures of mortality, which can quickly calculated and at the same time, easily understood. It is a ratio of the total registered deaths of a specified year to the total mid-year population, multiplied by 1,000.

Still Birth Rate

Still birth is a term which is accounted in reproductive wastage. Still birth is sometimes referred as foetal death. It represents the death which is occurred before the complete expulsion or extraction from the womb of mother. From 28 weeks of conception to the exact child bearing period, the embryonic loss is accounted as still birth or foetal death.

Neo Natal Mortality Rate (NNMR)

The Neo Natal Mortality rate is described as the number of infant deaths upto 28 days of life to the number of live births during the year.

Post Neonatal Mortality Rate (PNMR)

Post Neonatal Mortality is described as the number of infant deaths above 28 days to less than 1 year of life to the number of live births during the year.

Infant Mortality Rate (IMR)

'Infants are defined in demography as an exact age group, namely, age Zero or those children in the first year of life, who have not yet reached age one'. The Infant Mortality Rate is generally computed as a ratio of infant deaths (deaths of children under one year of age) registered in calendar year to the total number of live births registered in the same year.

Under Five Mortality Rate (U5MR)

The under five mortality rate is the number of deaths of children under five years of age per thousand live births.

BARRIERS IN THE RESEARCH

Research in a Himalayan region requires certain intensive efforts. Conducting a research among unfamiliar people and hostile environmental conditions needed prior groundwork. In order to avoid the danger of bias creeping in, the continuous, intentional efforts had to be made, to be with the people and learn as much as possible about them.

Language

One of the most critical means of gaining acceptance into a culture is mastering the language of that culture. Although, the second language of the Kinnauries is Hindi. To conduct a successful research, rapport formation with the interviewees was very essential. An unknown language, Kinnauri language was the first important obstacle that had to be overcome. Understanding the language required intensive and patient efforts.

Review of Literature

Till no such anthropological research on health trends and its dimension is conducted on the Kinnaur tribe, therefore, not much data is available. It was beneficial to review the other general information available on the Kinnauri tribe. An insight into general published documents/books/articles was of great help to the researchers, in finalizing the objectives and in conduction of this research.

Visits

Familiarization with the people requires other efforts, like travelling and moving around in the tribal area. There was a chance to visit and see personally the different ecological conditions in Kinnaur. Apart from these, the researchers also planned a few other visits, which included the NGOs, Department of Health and Family Welfare, Collectorate Office etc.

Bias

The danger of bias creeping in during the interviews is always there, mainly because the persons interviewing are human beings and not robots. The interviews would invariably result in the researchers getting emotionally involved during the discussions. Yet, the researchers tried to be mainly objective and consistent in all the interviews. The awareness to avoid bias during the interviews was a constant self-control on the researchers.

LIMITATIONS OF THE STUDY

- Non uniformity of sample size among ethnic groups. Due to lesser population of three ethnic groups i.e Koli, Lohar and Badhai in the village, the sample size of these groups is lesser than the sample size of Kanet ethnic group.
- Due to flood situation in August 2000, the fieldwork was stopped due to lack of proper transportation services to the district. The fieldwork was reinstated after few months. This natural calamities had unnecessary delayed the field work in the district.

4

Health Dimensions Among Four Ethnic Groups

This chapter discuse the various demographic indicators like fertility, mortality, maternal and child health care, health seeking behaviour and family planning system of four ethnic groups and pooled (Total). Further, an attempt has also been made to understand the Health Care System in the district Kinnaur of Himachal Pradesh in order to get insight view of the existence of traditional and other health systems in Kinnaur, health institutions, infrastructure availability, supply logistic system etc.

BACKGROUND CHARACTERISTICS

The state of Himachal Pradesh situated in the north western part of India is a land of lush green forests, deep river valleys, beautiful plateaus and snow-capped lofty mountains. It is bordered with Jammu and Kashmir in the North, Punjab in the West and South-West, Haryana in the South, Uttaranchal in the South-East and with China in the East. The area as a whole is hilly, mountainous and the major part is inaccessible, physiographically complex, snow covered and forest clad. The altitude varies from 450 to 6500 mts. above sea level in Himachal Pradesh.

For the present study, the Kinnaur district of Himachal Pradesh has been selected. The geographical location of Kinnaur is between 31-05'-20" and 32-05'-20" north altitude and between 77-45'-0' and 79-00'-50" east longitude. The entire district is spread over the Himalayan mountainous terrain. The Kinnaur valley is stretched over about 80 kilometers in length and 64 kilometers in breadth. The area lying within the district territory is nearly 6679 sq. kilometers. This is 11.7 per cent of the total area of Himachal Pradesh. The Kinnaur district is divided into three blocks, Kalpa, Nicher and Poo. The tehsil in Kalpa, Nichar and Poo blocks are Kalpa and Sangla, Nichar and Morang and Poo respectively. (Hangrang is sub teshil in Poo Block).

The inhabitants of Kinnaur district; i.e. Kinnauri has been spelt out with many variants such as Konawri, Koonauri, Kanauri, Kannauri, Kinnaura and Kinnauri (presently in use). The ethnic population of Kinnaur district, Kinnauri are divided into five groups i.e. Kanet, Koli, Badhi, Lohar and Nagloo. The entire population is classified as Scheduled Caste or Scheduled Tribe. The population of the district comprises of five distinct classes of Kanets (Rajputs), Chamangs {(Kolis, preparing shoes, weaving, tailoring and music drum beating)} and Domangs {Lohar (Blacksmith) and Badhi (carpentry), Nagloo (Basket makers)}. The Kanet constitute the majority of Kinnaur population. They are declared as Scheduled tribe people. The other four ethnic groups i.e Koli, Badhi, Lohar and Nagloo constitute the scheduled caste population of district. The Koli, Badhi, Lohar and Nagloo mainly comprises of the artisan class and are considered untouchables whether they be Koli having the occupation of preparing shoes, weaving, tailoring and music drum beating, the Badhi having the carpentry and mason, Lohar–Blacksmith and Nagaloo–Basket making.

Till date various social scientists and quite a few anthropologists have studied the social cultural dimensions of Kinnauris. But no attempt has been made to get the insight of demographic dimension and health system of the ethnic groups. The anthropological interpretation of health system and comparative analysis of health situation of various ethnic groups reflects knowledge, belief and practices related to health among the ethnic population of Kinnaur district. All the computation and analysis have been presented separately for each of population group and then are pooled together to work out for Kinnauris group.

Household Coverage and Total Population

The present study has been undertaken among the four ethnic groups namely Kanet (Scheduled Tribe), Koli (Scheduled Caste),

Badhi (Scheduled Caste) and Lohar (Scheduled Caste) which are collectively known as Kinnauris (broadly categorised as Scheduled Tribe of HP) constitute the major share of total population. In total 743 households were surveyed from 15 villages of Kinnaur district. The total population of the sampled villages is 5970 (3123 males and 2847 female). The maximum population covered in Koli group (41.9 per cent) followed by Kanet (36.1 per cent), Koli (11.7 per cent) and Lohar (10.3 per cent). Table 4.1 shows the total number of households and population in each ethnic group.

Table 4.1: Total Number of Households and Population in Each Ethnic Group

Ethnic Group	*Total Household*		*Population*					
			Male		*Female*		*Total*	
	No.	*%age*	*No.*	*%age*	*No.*	*%age*	*No.*	*%age*
Kanet	275	37.0	1123	36.0	1033	36.3	2156	36.1
Koli	300	40.4	1298	41.6	1206	42.4	2504	41.9
Badhi	85	11.4	377	12.1	324	11.4	701	11.7
Lohar	83	11.2	325	10.4	284	10.0	609	10.2
Total	**743**	**100.0**	**3123**	**100.0**	**2847**	**100.0**	**5970**	**100.0**

Under the study due to numerical dominance, the Koli and Kanet population have been recorded maximum in the sample. The number of Badhi and Lohar is less in total population studies due to their low population in the village as compare to other population.

Land Holding

The land holding or size of land owned by the households is indicator for estimation of overall economic condition and its potential role in determining the health status of the family. Due to hilly terrain in Kinnaur district, the main source of livelihood is land. Therefore the size of land owned is indicative of relative economic status and resources. The majority of Kanet (40.0 per cent) had land holding between 10–30 hectares where Koli (42.0 per cent), Badhi (40.0 per cent) and Lohar (47.0 per cent) have land holding less than 10 hectares. It has been observed that the maximum per cent of households having a land holding less than 10 hectares

(37.4 per cent) followed by 10-30 hectares (33.6 per cent), landless (15.6 per cent), 31-50 hectares (11.8 per cent) and 51 and above (1.5 per cent) as shown in Table 4.2.

Table 4.2: Land Holding Among the Kinnauris

Total Land (in Hect)	*Kanet*		*Koli*		*Badhi*		*Lohar*		*Total (Pooled)*	
	No.	*%age*	*No.*	*%age*	*No.*	*%age*	*No.*	*%age*	*No.*	*%age*
Landless	8	2.9	67	22.3	18	21.2	23	27.7	116	15.6
Less than 10	79	28.7	126	42.0	34	40.0	39	47.0	278	37.4
10-30	110	40.0	88	29.3	31	36.5	21	25.3	250	33.6
31-50	69	25.1	17	5.7	2	2.4		0.0	88	11.8
51 and above	9	3.3	2	0.7		0.0		0.0	11	1.5
Total	**275**	**100.0**	**300**	**100.0**	**85**	**100.0**	**83**	**100.0**	**743**	**100.0**

On comparing the land holding pattern of various ethnic groups in the pooled population, it is has been observed that maximum numbers of people without land are found out in Lohar followed by Koli and Badhi. The reason for this disparity could be attributed to engagement of Lohar in other economic activities, non availability of capital fund for land purchasing and poor economic conditions. The reason for Koli having is due to reason that since the majority of people of Koli group are engaged in government/ private jobs at Block HQ, therefore their land holding is less than 10 hectares.

House Ownership

In the present study, majority (80.5 per cent) of people in all four ethnic groups reside in their own house. It has been observed that 84.3 per cent of Koli owned their house followed by Kanet (82.5 per cent), Lohar (72.3 per cent) and Badhi (68.2 per cent).

Housing Condition

In the present study, majority of Kanet and Koli group stay in Pucca House whereas majority of Badhi and Lohar stay in Semi Pucca and Kachha houses respectively. It has been observed that 90.2 per cent of the Kanet population stay in Pucca house followed by Koli (89.7 per cent), Badhi (27.1 per cent) and Lohar

(22.9 per cent). It has been observed that majority of population reside in Pucca house (75.2 per cent) followed by Semi Pucca (13.9 per cent), Kachha (10.9 per cent). The reasons for high percentage of Kanet people residing in Pucca Houses are economically well being, availability of loan for house construction under schemes for scheduled tribe, high education status and concern over good living conditions. Whereas, the majority of Lohar people still reside in Kachha house due to poor economic condition and attachment to ancestral houses.

Housing Floors

In the present study, majority of people in all ethnic groups (except Lohar) have double storyed houses. It has been observed that 92.0 per cent of the Koli population stay in double story house followed by Kanet (84.0 per cent), Badhi (51.8 per cent) and Lohar (26.5 per cent). The majority of Koli and Kanet people prefer to have double storyed house due to large family size and high status in the village.

Number of Rooms

In the present study, majority of people (pooled) (39.6 per cent) reside in houses with 3-5 rooms. The maximum number of Kanet (44.7 per cent) and Koli (41.0 per cent) reside in house with 6-8 rooms whereas Badhi (49.4 per cent) and Lohar (48.2 per cent). The majority of Koli and Kanet people have houses with more number of rooms due to large family size and high status in the village whereas Badhi and Lohar still prefer to reside in houses with less number of rooms due to their poor economic conditions and less expenditure on expansion of houses.

Source of Water Supply

In present study, the main source of water is stored at common place and used later on through small water channels. It has been observed that majority of people Kanet (64.0 per cent) followed by Koli (63.3 per cent), Lohar (47.0 per cent) and Badhi (44.7 per cent) used store water as main source of water. In hilly region like Kinnaur, main source of water supply of people is stored water and water streams. It was reported that most of the people due to non-availability of water at high altitude faces some problem in getting water for their household activities. Mainly, water collection activity is confined to the women in the family. It is the responsibility of women to fetch water from nearest available water source for household activities.

Availability of Drainage Facilities

In present study, in majority drainage facilities is available in their respective houses as reported by Kanet (86.9 per cent) followed by Koli (77.3 per cent), Badhi (44.7 per cent) and Lohar (34.9 per cent).

System of Disposal of Human Waste

In present study, the system of disposal of human waste is septic toilets as reported by Kanet (62.9 per cent) followed by Koli (53.3 per cent), Badhi (20.0 per cent) and Lohar (10.8 per cent). Still most of Lohar (62.7 per cent) and Badhi (50.6 per cent) people did not have toilets in their house. Still Badhi and Lohar people prefer to defecate at open place due to their belief system that defecation in house is not good for house hygienic conditions and maintenance of toilets in house takes lots of time.

Housing of Animals

In present study, the housing of animals had been kept separately (at the corner of house) lower floor of the house as reported by Kanet (88.7 per cent) followed by Koli (92.0 per cent), Badhi (68.2 per cent) and Lohar (61.4 per cent). The reason for keeping housing of animals separately is to maintain the hygienic condition of house.

POPULATION COMPOSITION

The population composition is taken as the internal structure of a ethnic group with regards to one or more demographic traits such as age, sex, marital status and economic activities, education characteristics etc at a given period of time. This section deals with demographic profile such as age and sex composition, marital status, age at marriage, age at menopause, age at menarche etc.

Age Composition

The age composition or age structure is basic and important demographic characteristic of the population. The measure of age composition is the per cent distribution of a population at various age categories simultaneously cross classified by sex. The structure of a population is closely associated with the number of births, deaths and migration. The population pyramid is constructed for an overall comprehension of the age structure. The present data on age is collected in terms of competed whole or full years. The age data has been further categorised into five years groups instead of year numbers.

The age distributions of complied population (pooled data of four ethnic groups) have shown 33.4 per cent of population under 15 years and only 4.4 per cent in the age group of 60+. It shows that more or less young population composition with moderately high fertility which can be well compared with growing population structure (high proportion of young and childbearing population and less proportion of old people) of any developing region. On comparing the inter population variation, the child population in the age group 0- 14 years is highest in Badhi (36.7 per cent) followed by Lohar (35.0 per cent), Kanet (34.0 per cent) and lowest value among Koli (31.5 per cent). On comparing the population (36.1 per cent) of age group 0-14 years of India, it has been observed that percentage of population of Badhi between age group 0-14 years is higher than percentage of population between age group 0-14 years in India where percentage of population between age group 0-14 years of Kanet, Badhi, Lohar is lesser than percentage of population between age group 0-14 years in India.

The next broad age group of 15-49 years, comparing of population at fertile and peak productive ages, the highest per cent is found in Koli (57.9 per cent) followed by Lohar (54.0 per cent), Kanet (52.9 per cent) and lowest value among Badhi (51.6 per cent).

It has been observed that majority of population in all four ethnic groups is in childbearing age. The proportion of old age people i.e age group 60 + is lowest among all four ethnic groups. This indicates high mortality or low survival to old age group. The population distribution in five year groups is shown in Table 4.3.

Table 4.3: Demographic Profile of All Ethnic Groups

Age	Kanet				Koli			
	Male	Female	Total	Sex Ratio	Male	Female	Total	Sex Ratio
1-4	134	112	246	836	136	123	259	904
	4.3%	3.9%	4.1%		4.4%	4.3%	4.3%	
5-9	129	121	250	938	141	129	270	915
	4.1%	4.3%	4.2%		4.5%	4.5%	4.5%	
10-14	121	115	236	950	135	125	260	926
	3.9%	4.0%	4.0%		4.3%	4.4%	4.4%	
15-19	103	94	197	913	122	127	249	1041
	3.3%	3.3%	3.3%		3.9%	4.5%	4.2%	
20-24	110	103	213	936	129	124	253	961
	3.5%	3.6%	3.6%		4.1%	4.4%	4.2%	
25-29	92	89	181	967	127	119	246	937
	2.9%	3.1%	3.0%		4.1%	4.2%	4.1%	
30-34	90	89	179	989	128	116	244	906
	2.9%	3.1%	3.0%		4.1%	4.1%	4.1%	

(Table 4.3 Contd...)

Age	Kanet				Koli			
	Male	Female	Total	Sex Ratio	Male	Female	Total	Sex Ratio
35-39	79	72	151	911	97	90	187	928
	2.5%	2.5%	2.5%		3.1%	3.2%	3.1%	
40-44	70	65	135	929	82	75	157	915
	2.2%	2.3%	2.3%		2.6%	2.6%	2.6%	
45-49	44	40	84	909	60	53	113	883
	1.4%	1.4%	1.4%		1.9%	1.9%	1.9%	
50-54	53	50	103	943	59	55	114	932
	1.7%	1.8%	1.7%		1.9%	1.9%	1.9%	
55-59	34	31	65	912	33	28	61	848
	1.1%	1.1%	1.1%		1.1%	1.0%	1.0%	
60-64	35	31	66	886	34	30	64	882
	1.1%	1.1%	1.1%		1.1%	1.1%	1.1%	
65+	29	21	50	724	15	12	27	800
	0.9%	0.7%	0.8%		0.5%	0.4%	0.5%	
Numbers.	**1123**	**1033**	**2156**	**920**	**1298**	**1206**	**2504**	**929**

(Table 4.3 Contd...)

Age	Badhi				Lohar				Total (Pooled)			
	Male	Female	Total	Sex Ratio	Male	Female	Total	Sex Ratio	Male	Female	Total	Sex Ratio
1-4	35	29	64	829	31	27	58	871	336	291	627	866
	1.1%	1.0%	1.1%		1.0%	0.9%	1.0%		10.8%	10.2%	10.5%	
5-9	56	45	101	804	47	41	88	872	373	336	709	901
	1.8%	1.6%	1.7%		1.5%	1.4%	1.5%		11.9%	11.8%	11.9%	
10-14	49	43	92	878	36	31	67	861	341	314	655	921
	1.6%	1.5%	1.5%		1.2%	1.1%	1.1%		10.9%	11.0%	11.0%	
15-19	35	33	68	943	29	25	54	862	289	279	568	965
	1.1%	1.2%	1.1%		0.9%	0.9%	0.9%		9.3%	9.8%	9.5%	
20-24	36	33	69	917	34	31	65	912	309	291	600	942
	1.2%	1.2%	1.2%		1.1%	1.1%	1.1%		9.9%	10.2%	10.1%	
25-29	29	26	55	897	28	26	54	929	276	260	536	942
	0.9%	0.9%	0.9%		0.9%	0.9%	0.9%		8.8%	9.1%	9.0%	
30-34	30	26	56	867	28	25	53	893	276	256	532	928
	1.0%	0.9%	0.9%		0.9%	0.9%	0.9%		8.8%	9.0%	8.9%	

(Table 4.3 Contd...)

Age	Badhi				Lohar				Total (Pooled)			
	Male	Female	Total	Sex Ratio	Male	Female	Total	Sex Ratio	Male	Female	Total	Sex Ratio
35-39	25	22	47	880	23	20	43	870	224	204	428	911
	0.8%	0.8%	0.8%		0.7%	0.7%	0.7%		7.2%	7.2%	7.2%	
40-44	20	17	37	850	17	15	32	882	189	172	361	910
	0.6%	0.6%	0.6%		0.5%	0.5%	0.5%		6.1%	6.0%	6.0%	
45-49	16	14	30	875	15	13	28	867	135	120	255	889
	0.5%	0.5%	0.5%		0.5%	0.5%	0.5%		4.3%	4.2%	4.3%	
50-54	17	14	31	824	14	12	26	857	143	131	274	916
	0.5%	0.5%	0.5%		0.4%	0.4%	0.4%		4.6%	4.6%	4.6%	
55-59	11	9	20	818	9	7	16	778	87	75	162	862
	0.4%	0.3%	0.3%		0.3%	0.2%	0.3%		2.8%	2.6%	2.7%	
60-64	11	8	19	727	10	8	18	800	90	77	167	856
	0.4%	0.3%	0.3%		0.3%	0.3%	0.3%		2.9%	2.7%	2.8%	
65+	7	5	12	714	4	3	7	750	55	41	96	745
	0.2%	0.2%	0.2%		0.1%	0.1%	0.1%		1.8%	1.4%	1.6%	
Numbers.	**377**	**324**	**701**	**859**	**325**	**284**	**609**	**874**	**3123**	**2847**	**5970**	**912**

Sex Ratio

In most of developing countries, sex ratio is titled in favour of males, unlike the trend observed in developed countries. The sex ratio for four ethnic groups and total (pooled) group indicate excess of males over females i.e. sex ratio values are less than 1000. The Koli (929) showed the highest sex ratio closely followed by Kanet (920), Lohar (874) and Badhi (859) as shown in Table 4.4. The sex ratio estimate for the total (pooled) worked out to be 912 which indicate higher proportion of males than females—a true characteristic of developing region. This sex ratio is found to be lesser than sex ratio (970) of Himachal Pradesh state (2001) and higher than sex ratio of Kinnaur district (851) (District Statistical Handbook, 2002). The early marriage, frequent child bearing, low status of women strong male preference, low socio- economic status could be the factors for low sex ratio in Badhi and Lohar group. It has been reported that birth of female child in the family is not accepted with pleasure and preference for male child is quite high in all the four ethnic groups.

Table 4.4: Sex Ratio

	Kanet	*Koli*	*Badhi*	*Lohar*	*Total (Pooled)*
Sex Ratio	920	929	859	874	912

The sex ratio of all four ethnic groups is higher than reported sex ratio of district Kinnaur (851) (District Statistical Handbook, 2002). The sex ratio of total (pooled) is found to be lower than the sex ratio of Hamirpur (1102), Kangra (1027), Mandi (1014), Una (997), Bilaspur (992), Chamba (961), Kullu (928)and higher than the Sirmaur (901), Shimla (898), Solan (853), Lahul-Spiti (804) districts. The high sex ratio of Kanet and Koli ethnic groups has improvement the overall total sex ratio of total (pooled) population. Further in comparison with other Himalayan population, the high level of literacy, proper medical care, better socio-economic conditions of ethnic groups of district Kinnaur especially Koli and Kanet population is the main reason for high sex ratio.

It has been observed that sex ratio of total (pooled) is higher than sex ratio of Jaunsari (889); Gaddi (870), Brahmin (890), Rajput (870), Scheduled Caste (850) of Himachal Pradesh; Sherpas (744), Lepchas (849) of Sikkim and lower than Semsa (1005) of Assam;

Buddhist (921), Hindus (1051), Bhutias (996), Tamangs (984) of Sikkim; Johari Bhotia (1132), Rang Bhotia (1209), Marchha Bhotia (1168) of Uttaranchal (Kapoor 1996; Kshatriya et al. 1997; Patra 2001; and Ghosh and Limbu 2002).

Masculinity Proportion

Masculinity proportion is the other measure of sex composition depicting the balance of the sexes in a population, in terms of proportion or per cents at specified time. This measure shows the opposite trend noticed in case of the overall sex ratio. The highest masculinity proportion has been observed in Badhi followed by Lohar, Kanet and Koli. The masculinity proportion of total (pooled) was estimated as 52.3 (per cent) as shown in Table 4.5. The high masculinity proportion could be attributed to strong preference of male child, frequent child bearing and early marriage etc.

Table 4.5: Masculinity Proportion (in per cent)

	Kanet	*Koli*	*Badhi*	*Lohar*	*Total (Pooled)*
Masculinity Proportion	52.1	51.8	53.8	53.4	52.3

When compared with other Himalayan population, it was observed that masculinity proportion of the total (pooled) is found to be lower than Lepchas(54.09), Sherpas (57.35) of Sikkim; Gaddis (53.48), Brahmin (52.9), Scheduled caste (54.05) of Himachal Pradesh; Rajis (53.81) of Uttaranchal and higher than Buddhist (52.06), Hindus (48.74), Bhutias (50.1), Tamangs (50.41) of Sikkim; Bodhs (51.1), Baltis (51.0), Brokas (51.4) and Arghuns (51.9) of Jammu and Kashmir (Kapoor 1996; Kshatriya et al. 1997; Patra 2001; and Ghosh and Limbu 2002).

Marital Status

Marriage is considered a socially and legally recognised childbearing institution. The distribution of population in a region classified by marital status is another essential part of the health and development study. In Kinnaur district, marriage is considered nearly universal in all ethnic groups. The following tables exhibit age and sex wise distribution of marital status of four ethnic groups of Kinnaur and total (pooled).

Among Kanet, it has been observed that 52.3 per cent population (24.8 per cent of male and 27.6 per cent female), 43.4 per cent population (24.8 per cent of male and 18.6 per cent female), 4.3 per cent population (2.5 per cent of male and 1.8 per cent female) are under married, unmarried and widow/divorced category respectively as shown in Table 4.6.

Table 4.6: Marital Status of Kanet

Age	*Unmarried*						*Married*					
	Male		*Female*		*Total*		*Male*		*Female*		*Total*	
	No.	*%age*	*No.*	*%age*	*No.*	*%age*	*No.*	*%age*	*No.*	*%age*	*No.*	*%age*
1-4	134	25.0	112	27.9	246	26.3		0.0		0.0	0	0.0
5-9	129	24.1	121	30.2	250	26.7		0.0		0.0	0	0.0
10-14	121	22.6	115	28.7	236	25.2		0.0		0.0	0	0.0
15-19	80	15.0	31	7.7	111	11.9	23	4.3	63	10.6	86	7.6
20-24	36	6.7	11	2.7	47	5.0	65	12.2	84	14.1	149	13.2
25-29	14	2.6	6	1.5	20	2.1	73	13.7	81	13.6	154	13.7
30-34	11	2.1	2	0.5	13	1.4	76	14.2	83	14.0	159	14.1
35-39	5	0.9	1	0.2	6	0.6	70	13.1	71	12.0	141	12.5
40-44	2	0.4	1	0.2	3	0.3	64	12.0	55	9.3	119	10.5
45-49	1	0.2		0.0	1	0.1	36	6.7	37	6.2	73	6.5

(Table 4.6 Contd...)

Age	Unmarried						Married					
	Male		Female		Total		Male		Female		Total	
	No.	%age	No.	%age	No.	%age	No.	%age	No.	%age	No.	%age
50-54	1	0.2	1	0.2	2	0.2	49	9.2	47	7.9	96	8.5
55-59		0.0		0.0	0	0.0	26	4.9	28	4.7	54	4.8
60-64	1	0.2		0.0	1	0.1	29	5.4	26	4.4	55	4.9
65-+		0.0		0.0	0	0.0	23	4.3	19	3.2	42	3.7
Total	**535**	**100.0**	**401**	**100.0**	**936**	**100.0**	**534**	**100.0**	**594**	**100.0**	**1128**	**100.0**

(Table 4.6 Contd...)

Age	Widow/Divorced						Total					
	Male		Female		Total		Male		Female		Total	
	No.	%age	No.	%age	No.	%age	No.	%age	No.	%age	No.	%age
1-4		0.0		0.0	0	0.0	134	11.9	112	10.8	246	11.4
5-9		0.0		0.0	0	0.0	129	11.5	121	11.7	250	11.6
10-14		0.0		0.0	0	0.0	121	10.8	115	11.1	236	10.9
15-19		0.0		0.0	0	0.0	103	9.2	94	9.1	197	9.1
20-24	9	16.7	8	21.1	17	18.5	110	9.8	103	10.0	213	9.9
25-29	5	9.3	2	5.3	7	7.6	92	8.2	89	8.6	181	8.4
30-34	3	5.6	4	10.5	7	7.6	90	8.0	89	8.6	179	8.3
35-39	4	7.4		0.0	4	4.3	79	7.0	72	7.0	151	7.0
40-44	4	7.4	9	23.7	13	14.1	70	6.2	65	6.3	135	6.3
45-49	7	13.0	3	7.9	10	10.9	44	3.9	40	3.9	84	3.9
50-54	3	5.6	2	5.3	5	5.4	53	4.7	50	4.8	103	4.8
55-59	8	14.8	3	7.9	11	12.0	34	3.0	31	3.0	65	3.0
60-64	5	9.3	5	13.2	10	10.9	35	3.1	31	3.0	66	3.1
65+	6	11.1	2	5.3	8	8.7	29	2.6	21	2.0	50	2.3
Total	**54**	**100.0**	**38**	**100.0**	**92**	**100.0**	**1123**	**100.0**	**1033**	**100.0**	**2156**	**100.0**

Among Koli, it has been observed that 55.6 per cent population (27.8 per cent of male and 27.8 per cent female), 41.2 per cent population (22.6 per cent of male and 18.6 per cent female), 3.2 per cent population (1.4 per cent of male and 1.8 per cent female) are under married, unmarried and widow/divorced category respectively as shown in Table 4.7.

Table 4.7: Marital Status of Koli

Age	Unmarried						Married					
	Male		Female		Total		Male		Female		Total	
	No.	%age	No.	%age	No.	%age	No.	%age	No.	%age	No.	%age
1-4	136	24.0	123	26.5	259	25.1		0.0		0.0	0	0.0
5-9	141	24.9	129	27.7	270	26.2		0.0		0.0	0	0.0
10-14	135	23.8	125	26.9	260	25.2		0.0		0.0	0	0.0
15-19	83	14.6	45	9.7	128	12.4	39	5.6	82	11.8	121	8.7
20-24	34	6.0	24	5.2	58	5.6	92	13.2	96	13.8	188	13.5
25-29	18	3.2	15	3.2	33	3.2	104	15.0	100	14.3	204	14.7
30-34	11	1.9	3	0.6	14	1.4	113	16.3	107	15.4	220	15.8
35-39	4	0.7	1	0.2	5	0.5	89	12.8	83	11.9	172	12.4
40-44	3	0.5		0.0	3	0.3	74	10.6	71	10.2	145	10.4

(Table 4.7 Contd...)

Age	Unmarried						Married					
	Male		Female		Total		Male		Female		Total	
	No.	%age	No.	%age	No.	%age	No.	%age	No.	%age	No.	%age
45-49	1	0.2		0.0	1	0.1	56	8.1	51	7.3	107	7.7
50-54		0.0		0.0	0	0.0	55	7.9	49	7.0	104	7.5
55-59	1	0.2		0.0	1	0.1	30	4.3	24	3.4	54	3.9
60-64		0.0		0.0	0	0.0	32	4.6	27	3.9	59	4.2
65-+		0.0		0.0	0	0.0	11	1.6	7	1.0	18	1.3
Total	**567**	**100.0**	**465**	**100.0**	**1032**	**100.0**	**695**	**100.0**	**697**	**100.0**	**1392**	**100.0**

(Table 4.7 Contd...)

Age	Widow/Divorced						Total					
	Male		Female		Total		Male		Female		Total	
	No.	%age	No.	%age	No.	%age	No.	%age	No.	%age	No.	%age
1-4		0.0		0.0	0	0.0	136	10.5	123	10.2	259	10.3
5-9		0.0		0.0	0	0.0	141	10.9	129	10.7	270	10.8
10-14		0.0		0.0	0	0.0	135	10.4	125	10.4	260	10.4
15-19		0.0		0.0	0	0.0	122	9.4	127	10.5	249	9.9
20-24	3	8.3	4	9.1	7	8.8	129	9.9	124	10.3	253	10.1
25-29	5	13.9	4	9.1	9	11.3	127	9.8	119	9.9	246	9.8
30-34	4	11.1	6	13.6	10	12.5	128	9.9	116	9.6	244	9.7
35-39	4	11.1	6	13.6	10	12.5	97	7.5	90	7.5	187	7.5
40-44	5	13.9	4	9.1	9	11.3	82	6.3	75	6.2	157	6.3
45-49	3	8.3	2	4.5	5	6.3	60	4.6	53	4.4	113	4.5
50-54	4	11.1	6	13.6	10	12.5	59	4.5	55	4.6	114	4.6
55-59	2	5.6	4	9.1	6	7.5	33	2.5	28	2.3	61	2.4
60-64	2	5.6	3	6.8	5	6.3	34	2.6	30	2.5	64	2.6
65+	4	11.1	5	11.4	9	11.3	15	1.2	12	1.0	27	1.1
Total	**36**	**100.0**	**44**	**100.0**	**80**	**100.0**	**1298**	**100.0**	**1206**	**100.0**	**2504**	**100.0**

Among Badhi, it has been observed that 43.1 per cent population (21.5 per cent of male and 21.5 per cent female), 51.4 per cent population (39.8 per cent of male and 21.5 per cent female), 5.6 per cent population (2.4 per cent of male and 3.1 per cent female) are under married, unmarried and widow/divorced category respectively as shown in Table 4.8.

Table 4.8: Marital Status of Badhi

Age	*Unmarried*						*Married*					
	Male		*Female*		*Total*		*Male*		*Female*		*Total*	
	No.	*%age*	*No.*	*%age*	*No.*	*%age*	*No.*	*%age*	*No.*	*%age*	*No.*	*%age*
1-4	35	16.7	29	19.2	64	17.8		0.0		0.0	0	0.0
5-9	56	26.8	45	29.8	101	28.1		0.0		0.0	0	0.0
10-14	49	23.4	43	28.5	92	25.6		0.0		0.0	0	0.0
15-19	27	12.9	16	10.6	43	11.9	8	5.3	17	11.3	25	8.3
20-24	22	10.5	12	7.9	34	9.4	14	9.3	21	13.9	35	11.6
25-29	12	5.7	5	3.3	17	4.7	17	11.3	19	12.6	36	11.9
30-34	2	1.0	1	0.7	3	0.8	27	17.9	24	15.9	51	16.9
35-39	4	1.9		0.0	4	1.1	21	13.9	22	14.6	43	14.2
40-44	1	0.5		0.0	1	0.3	18	11.9	16	10.6	34	11.3
45-49		0.0		0.0	0	0.0	14	9.3	11	7.3	25	8.3

(Table 4.8 Contd...)

Age	Unmarried						Married					
	Male		Female		Total		Male		Female		Total	
	No.	%age	No.	%age	No.	%age	No.	%age	No.	%age	No.	%age
50-54		0.0		0.0	0	0.0	14	9.3	12	7.9	26	8.6
55-59	1	0.5		0.0	1	0.3	7	4.6	5	3.3	12	4.0
60-64		0.0		0.0	0	0.0	7	4.6	3	2.0	10	3.3
65+		0.0		0.0	0	0.0	4	2.6	1	0.7	5	1.7
Total	**209**	**100.0**	**151**	**100.0**	**360**	**100.0**	**151**	**100.0**	**151**	**100.0**	**302**	**100.0**
	29.8%		**21.5%**		**51.4%**		**21.5%**		**21.5%**		**43.1%**	

(Table 4.8 Contd...)

Age	Widow/Divorced						Total					
	Male		Female		Total		Male		Female		Total	
	No.	%age	No.	%age	No.	%age	No.	%age	No.	%age	No.	%age
1-4		0.0		0.0	0	0.0	35	9.3	29	9.0	64	9.1
5-9		0.0		0.0	0	0.0	56	14.9	45	13.9	101	14.4
10-14		0.0		0.0	0	0.0	49	13.0	43	13.3	92	13.1
15-19		0.0		0.0	0	0.0	35	9.3	33	10.2	68	9.7
20-24		0.0		0.0	0	0.0	36	9.5	33	10.2	69	9.8
25-29		0.0	2	9.1	2	5.1	29	7.7	26	8.0	55	7.8
30-34	1	5.9	1	4.5	2	5.1	30	8.0	26	8.0	56	8.0
35-39		0.0		0.0	0	0.0	25	6.6	22	6.8	47	6.7
40-44	1	5.9	1	4.5	2	5.1	20	5.3	17	5.2	37	5.3
45-49	2	11.8	3	13.6	5	12.8	16	4.2	14	4.3	30	4.3
50-54	3	17.6	2	9.1	5	12.8	17	4.5	14	4.3	31	4.4
55-59	3	17.6	4	18.2	7	17.9	11	2.9	9	2.8	20	2.9
60-64	4	23.5	5	22.7	9	23.1	11	2.9	8	2.5	19	2.7
65+	3	17.6	4	18.2	7	17.9	7	1.9	5	1.5	12	1.7
Total	**17**	**100.0**	**22**	**100.0**	**39**	**100.0**	**377**	**100.0**	**324**	**100.0**	**701**	**100.0**
	2.4%		**3.1%**		**5.6%**		**53.8%**		**46.2%**		**100.0%**	

Among Lohar, it has been observed that 46.6 per cent population (23.2 per cent of male and 23.5 per cent female), 48.6 per cent population (27.6 per cent of male and 21.0 per cent female), 4.8 per cent population (2.6 per cent of male and 2.1 per cent female) are under married, unmarried and widow/divorced category respectively as shown in Table 4.9.

Table 4.9: Marital Status of Lohar

Age	*Unmarried*						*Married*					
	Male		*Female*		*Total*		*Male*		*Female*		*Total*	
	No.	*%age*	*No.*	*%age*	*No.*	*%age*	*No.*	*%age*	*No.*	*%age*	*No.*	*%age*
1-4	31	18.5	27	21.1	58	19.6		0.0		0.0	0	0.0
5-9	47	28.0	41	32.0	88	29.7		0.0		0.0	0	0.0
10-14	36	21.4	31	24.2	67	22.6		0.0		0.0	0	0.0
15-19	20	11.9	14	10.9	34	11.5	9	6.4	11	7.7	20	7.0
20-24	19	11.3	11	8.6	30	10.1	15	10.6	20	14.0	35	12.3
25-29	9	5.4	4	3.1	13	4.4	18	12.8	20	14.0	38	13.4
30-34	4	2.4		0.0	4	1.4	24	17.0	25	17.5	49	17.3
35-39	1	0.6		0.0	1	0.3	21	14.9	20	14.0	41	14.4
40-44		0.0		0.0	0	0.0	17	12.1	14	9.8	31	10.9

(Table 4.9 Contd...)

Age	Unmarried						Married					
	Male		*Female*		*Total*		*Male*		*Female*		*Total*	
	No.	*%age*	*No.*	*%age*	*No.*	*%age*	*No.*	*%age*	*No.*	*%age*	*No.*	*%age*
45-49	1	0.6		0.0	1	0.3	13	9.2	11	7.7	24	8.5
50-54		0.0		0.0	0	0.0	9	6.4	8	5.6	17	6.0
55-59		0.0		0.0	0	0.0	7	5.0	6	4.2	13	4.6
60-64		0.0		0.0	0	0.0	6	4.3	5	3.5	11	3.9
65+		0.0		0.0	0	0.0	2	1.4	3	2.1	5	1.8
Total	**168**	**100.0**	**128**	**100.0**	**296**	**100.0**	**141**	**100.0**	**143**	**100.0**	**284**	**100.0**

(Table 4.9 Contd...)

Age	Widow/Divorced						Total					
	Male		Female		Total		Male		Female		Total	
	No.	%age	No.	%age	No.	%age	No.	%age	No.	%age	No.	%age
1-4		0.0		0.0	0	0.0	31	9.5	27	9.5	58	9.5
5-9		0.0		0.0	0	0.0	47	14.5	41	14.4	88	14.4
10-14		0.0		0.0	0	0.0	36	11.1	31	10.9	67	11.0
15-19		0.0		0.0	0	0.0	29	8.9	25	8.8	54	8.9
20-24		0.0		0.0	0	0.0	34	10.5	31	10.9	65	10.7
25-29	1	6.3	2	15.4	3	10.3	28	8.6	26	9.2	54	8.9
30-34		0.0		0.0	0	0.0	28	8.6	25	8.8	53	8.7
35-39	1	6.3		0.0	1	3.4	23	7.1	20	7.0	43	7.1
40-44		0.0	1	7.7	1	3.4	17	5.2	15	5.3	32	5.3
45-49	1	6.3	2	15.4	3	10.3	15	4.6	13	4.6	28	4.6
50-54	5	31.3	4	30.8	9	31.0	14	4.3	12	4.2	26	4.3
55-59	2	12.5	1	7.7	3	10.3	9	2.8	7	2.5	16	2.6
60-64	4	25.0	3	23.1	7	24.1	10	3.1	8	2.8	18	3.0
65+	2	12.5		0.0	2	6.9	4	1.2	3	1.1	7	1.1
Total	**16**	**100.0**	**13**	**100.0**	**29**	**100.0**	**325**	**100.0**	**284**	**100.0**	**609**	**100.0**

Among Total (Pooled), it has been observed that 52.0 per cent population (25.5 per cent of male and 26.5 per cent female), 44.0 per cent population (24.8 per cent of male and 19.2 per cent female), 4.0 per cent population (2.1 per cent of male and 2.0 per cent female) are under married, unmarried and widow/divorced category respectively as shown in Table 4.10.

Table 4.10: Marital Status of Total (Pooled)

Age	Unmarried						Married					
	Male		Female		Total		Male		Female		Total	
	No.	%age	No.	%age	No.	%age	No.	%age	No.	%age	No.	%age
1-4	336	22.7	291	11.1	627	23.9	0	0.0	0	0.0	0	0.0
5-9	373	25.2	336	12.8	709	27.0	0	0.0	0	0.0	0	0.0
10-14	341	23.1	314	12.0	655	25.0	0	0.0	0	0.0	0	0.0
15-19	210	14.2	106	4.0	316	12.0	79	5.2	173	10.9	252	8.1
20-24	111	7.5	58	2.2	169	6.4	186	12.2	221	13.9	407	13.1
25-29	53	3.6	30	1.1	83	3.2	212	13.9	220	13.9	432	13.9
30-34	28	1.9	6	0.2	34	1.3	240	15.8	239	15.1	479	15.4
35-39	14	0.9	2	0.1	16	0.6	201	13.2	196	12.4	397	12.8
40-44	6	0.4	1	0.0	7	0.3	173	11.4	156	9.8	329	10.6

(Table 4.10 Contd...)

Age	Unmarried						Married					
	Male		Female		Total		Male		Female		Total	
	No.	%age	No.	%age	No.	%age	No.	%age	No.	%age	No.	%age
45-49	3	0.2	0	0.0	3	0.1	119	7.8	110	6.9	229	7.4
50-54	1	0.1	1	0.0	2	0.1	127	8.3	116	7.3	243	7.8
55-59	2	0.1	0	0.0	2	0.1	70	4.6	63	4.0	133	4.3
60-64	1	0.1	0	0.0	1	0.0	74	4.9	61	3.8	135	4.3
65+	0	0.0	0	0.0	0	0.0	40	2.6	30	1.9	70	2.3
Total	**1479**	**100.0**	**1145**	**43.6**	**2624**	**100.0**	**1521**	**100.0**	**1585**	**100.0**	**3106**	**100.0**
	24.8%		**19.2%**		**44.0%**		**25.5%**		**26.5%**		**52.0%**	

(Table 4.10 Contd...)

Age	Widow/Divorced						Total					
	Male		Female		Total		Male		Female		Total	
	No.	%age	No.	%age	No.	%age	No.	%age	No.	%age	No.	%age
1-4	0	0.0	0	0.0	0	0.0	336	10.8	291	10.2	627	10.5
5-9	0	0.0	0	0.0	0	0.0	373	11.9	336	11.8	709	11.9
10-14	0	0.0	0	0.0	0	0.0	341	10.9	314	11.0	655	11.0
15-19	0	0.0	0	0.0	0	0.0	289	9.3	279	9.8	568	9.5
20-24	12	9.8	12	10.3	24	10.0	309	9.9	291	10.2	600	10.1
25-29	11	8.9	10	8.5	21	8.8	276	8.8	260	9.1	536	9.0
30-34	8	6.5	11	9.4	19	7.9	276	8.8	256	9.0	532	8.9
35-39	9	7.3	6	5.1	15	6.3	224	7.2	204	7.2	428	7.2
40-44	10	8.1	15	12.8	25	10.4	189	6.1	172	6.0	361	6.0
45-49	13	10.6	10	8.5	23	9.6	135	4.3	120	4.2	255	4.3
50-54	15	12.2	14	12.0	29	12.1	143	4.6	131	4.6	274	4.6
55-59	15	12.2	12	10.3	27	11.3	87	2.8	75	2.6	162	2.7
60-64	15	12.2	16	13.7	31	12.9	90	2.9	77	2.7	167	2.8
65+	15	12.2	11	9.4	26	10.8	55	1.8	41	1.4	96	1.6
Total	**123**	**100.0**	**117**	**100.0**	**240**	**100.0**	**3123**	**100.0**	**2847**	**100.0**	**5970**	**100.0**
	2.1%		**2.0%**		**4.0%**		**52.3%**		**47.7%**		**100.0%**	

In the all the four ethnic groups and total (pooled), it has been observed that no children under 14 years are married, the percentage of male unmarried population is high as compared to female, the percentage of unmarried males is higher than females, female married population in reproductive age group is higher and the percentage of widow/divorces is low.

Dependency Ratio

The dependency ratio measures the impact of age composition on the livelihood activity of a population. It is estimated in reference to the proportion of young (0 -14 years) and old (60+ years) in the total population, who are presumably economically inactive and proportion of population at productive ages (15- 59 years). The dependency ratio describes the burden of dependency or dependency load supported by potentially economically active population. The total dependency ratio is estimated as summation of young age dependency ratio and old age dependency ratio. The young age dependency ratio, old age dependency ratio and total dependency ratio is given in Table 4.11.

Table 4.11: Dependency Ratio

Ratio	*Kanet*	*Koli*	*Badhi*	*Lohar*	*Total (Pooled)*
Young Age Dependency Ratio (YADR) (0-14 years)	55.96	48.58	62.23	57.41	53.58
Old Age Dependency Ratio (OADR) (60+ years)	8.87	5.60	7.51	6.74	7.08
Total Dependency Ratio (TDR)	**64.83**	**54.19**	**69.73**	**64.15**	**60.66**

In study ethnic groups, the dependency load is due to the large young age dependency while the old age dependency comprises only a small part of the total dependency due to high fertility and young age composition.

Badhi and Lohar are found showing higher young age dependency ratio as compared to Kanet and Koli. The young age dependency ratio for total (pooled) worked out to be 53.58. The young age dependency ratio is found to be lower than scheduled caste of Himachal Pradesh (86.00), India (73.30) due to less population growth in age group 0- 14 years.

The old age dependency ratio seems to be quite low in the all study ethnic groups as compared to young age dependency ratio. The highest old age dependency ratio has been observed in Kanet followed by Badhi, Lohar and Koli. The old age dependency ratio for total (pooled) worked out to be 7.08. The old age dependency ratio is found to be lower than India (12.0) due to lack of old age care, harsh environmental conditions, low socio-economic condition, high mortality rate in old age population.

The total dependency ratio was found to be highest among Badhi (69.73) followed by Kanet (64.83), Lohar (64.15) and Koli (54.19). The total dependency ratio for total (pooled) worked out to be 60.66. The total dependency ratio is found to be lower than Brahmins (77.94), Rajputs (75.44) and Scheduled caste (91.16) of HP. (Bhasin and Bhasin, 1993). It is also lower than population groups {Baltis (85.63), Brokpas (80.39), Bodhas (72.78)} of Jammu and Kashmir and population groups {Lepchas (81.63), Bhutia (88.21), Sherpas (78.95), Tamangs (72.02) and Buddhistis (83.22) of Sikkim. It is higher than Hindus (65.59) of Sikkim. and Arghuns (55.92) of Jammu and Kashmir (Chauhan et al. 1991; Patra 2001; SRS 2002).

Index of Aging

Index of aging measures the relative number of old persons in a given population. It can be described as the ratio of old persons aged 60 years and above to the child population aged 0- 14 years. The highest index of aging estimate has been observed in Kanet (15.85) followed by Badhi (12.06), Lohar (11.74) and Koli (11.53) as shown in Table 4.12. It could be attributed to high proportion of population under 15 years and high mortality in aged.

Table 4.12: Index of Aging

Index	*Kanet*	*Koli*	*Badhi*	*Lohar*	*Total (Pooled)*
Index of Aging	15.85	11.53	12.06	11.74	13.21

The index of aging of total (pooled) found to be 13.21 which is higher than Rajputs (8.31), Gaddis (7.95), Scheduled caste (6.00) of HP; {Baltis (6.91), Brokpas (10.07), Arghuns (11.60)} of Jammu and Kashmir; {Lepchas (4.88), Bhutia (7.65), Tamangs (8.42), Buddhists (6.12) and Hindus (6.77) of Sikkim and lower than India (16.00) (Chauhan et al. 1991; Patra 2001; SRS 2002).

Number of Members in a Family

In study ethnic groups, it has been observed that the per cent of family comprises of two members among the Kanet, Koli, Badhi and Lohar is 4.4 per cent, 4.7 per cent, 2.4 per cent and 4.8 per cent respectively. In Kanet, Koli, Badhi and Lohar, maximum number of members in family are found to be 4, 6 5 and 7 respectively. The mean number of members in Kanet, Koli, Badhi and Lohar are observed to be 5.99, 6.09, 6.87 and 7.28 respectively. The mean number of members in total (pooled) group is 6.28.

Household Income

Household income is relative indicators of socio-economic status and material comfort, wealth resources is complex determinant of good health; as its potential role is often determined by various other socio-economic factors affecting fertility. The annual household income distribution i.e. total money income received by any earning member of a household has been studied. The lowest income has been observed Lohar (25.3 per cent) followed by Badhi (24.7 per cent), Kanet (18.2 per cent) and Koli (16.3 per cent) whereas the highest income grade was found out in Kanet (33.5 per cent) followed by Koli (28.7 per cent), Badhi (24.7 per cent) and Lohar (14.5 per cent) as shown in Table 4.13. It can be interpreted here that Kanet are economically better placed than other ethnic groups in the region due to high land holding and business.

Table 4.13: Household Income

Total Income (In Rs)	*Kanet*		*Koli*		*Badhi*		*Lohar*		*Total (Pooled)*	
	No.	*%age*	*No.*	*%age*	*No.*	*%age*	*No.*	*%age*	*No.*	*%age*
10,000 and less	50	18.2	49	16.3	21	24.7	21	25.3	141	19.0
10,001-30,000	64	23.3	94	31.3	23	27.1	31	37.3	212	28.5
30,001-50,000	69	25.1	71	23.7	20	23.5	19	22.9	179	24.1
50,001 and above	92	33.5	86	28.7	21	24.7	12	14.5	211	28.4
Total	**275**	**100.0**	**300**	**100.0**	**85**	**100.0**	**83**	**100.0**	**743**	**100.0**

Economic Characteristic of Women

The economic characteristics are widely acknowledged to expert dominant influence in a person's life and often act as relative indicators of the level of socio-economic status and health. It is believed that attitude towards work and occupation, working conditions and nature work always have influence on health condition of the individual. Since Kinnaur district is poor in resources due to limited cultivable area and relatively underdeveloped place, with adverse climatic condition. The household activities continue to the major vocation of most of the women in study ethnic group. The maximum economically inactive economic characteristic of women is in Lohar (92.4 per cent) followed by Badhi (90.2 per cent), Koli (85.4 per cent) and Kanet (81.6 per cent) as shown in Table 4.14. The women respondent has reported themselves as non-worker or housewife inspite of their active participation in various activities in the field as well as at home. When gainful activity was considered, the Kanet women seemed to be most economically active followed by Koli, Badhi and Lohar. The reasons for more economically activeness of Kanet women are economically well off, education and better position in Kinnaur society.

Table 4.14: Economic Characteristic of Women

Economic Characteristics	*Kanet*	*Koli*	*Badhi*	*Lohar*	*Total (Pooled)*
Economically Inactive	81.6	85.4	90.2	92.4	87.4
Economically Active (cultivation/ agricultural, unskilled labour, trade, commerce and service)	18.4	14.6	9.8	7.6	12.6

Economic Characteristic of Men

It has been observed that most of the men in total (pooled) are economically active. Only few has reported themselves as non-workers due to unemployedness and retired status.

The maximum economically active men are found out in Badhi (98.2 per cent) followed by Lohar (97.1 per cent), Koli (94.6 per cent) and Kanet (93.4 per cent) as shown in Table 4.15. Further it has been observed that most of the people are found to be engaged in cultivation/agriculture, unskilled labour like carpentry, weaving, servicing, repair work etc. The second largest economically active group constitute the people engaged in white collar jobs, professionals, technicians and skilled worker. These observations reflect the ongoing changes in the economic scenario in the Kinnaur district resulting in a large scale shift from traditional occupation as well as a tendency of upward occupational mobility among men.

Table 4.15: Economic Characteristic of Men

Economic Characteristics	*Kanet*	*Koli*	*Badhi*	*Lohar*	*Total (Pooled)*
Economically Inactive	6.6	5.4	1.8	2.9	4.2
Economically Active(cultivation/ agricultural, unskilled labour, trade, commerce and service)	93.4	94.6	98.2	97.1	95.8

Literacy Rate

Literacy rate express the number of literate persons who are able to read and wri'e with some understanding in a population at a specified time. The literacy rate as per cent of total population of particular population group is taken.

The literacy rate (as per cent of particular ethnic group) has been observed the highest among Kanet (51.3 per cent) followed by Koli (51.2 per cent), Badhi (50.6 per cent) and Lohar (50.2 per cent). The literacy rate of total (pooled) is found to be 51.1 per cent which is lesser than literacy rate (77.13 per cent) of Himachal Pradesh state, Chamba (63.73 per cent) district, Kangra (80.68 per cent), Lahul-Spiti (73.17 per cent), Kullu (73.36 per cent), Mandi (75.86 per cent), Hamirpur (83.16 per cent), Una (81.09 per cent), Bilaspur (78.80 per cent), Solan (77.16 per cent), Sirmaur (70.85 per cent) and Shimla (79.68 per cent) district. The low literacy rate can be attributed to far away distance of schools, socio- economic conditions etc.

There is large inter-population variation in the levels of male and female literacy. Among Kanet, male literacy rate (55.7 per cent) is found to be highest followed by Koli (53.7 per cent), Lohar (52.3 per cent) and Badhi (52.0 per cent). In Badhi, the female literacy rate (49.1 per cent) is highest followed by Koli (48.6 per cent), Lohar (47.9 per cent) and Kanet (46.5 per cent). In Total (Pooled), the literacy rate of total (pooled) group is 51.1 per cent with 54.1 per cent male literacy rate and 47.8 female literacy rate.

The future analysis of age specific literacy rate shows great variation in age patterns in all study ethnic group due to progress in education system over the years. At age 5- 9 years, when the formal schooling should have already started, the literacy rates for male registered large differentials, with highest recorded by Kanet followed Koli, Badhi and Lohar. It suggests that formal schooling of male children at proper ages is almost universal among Kanet and Koli.

FERTILITY REGULATION FACTORS

Fertility denoting the actual birth performance of women, is the major positive force in the vital processes, continuously replenishing the population; thereby counteracting continuous reduction by mortality. The fitness, growth of population, survival and good health condition depend on fertility pattern of the population. The fertility is an event that occurs over time. The knowledge of the current fertility levels, differential and trends (at a specified time/period) as well as cumulative fertility/family size estimates for the population are of vital importance.

Age at Menarche of Mothers

The menarche is considered as an important physiological episode. It represents the specific stage of first periodical regular flow of blood from womb in all the health normal females. The first menstruation start is taken into account as the age of puberty and maturity and from this stage, females are biologically capable to conceive. It has been observed that age at menarche in Total (Pooled) varies from 12 to 15 years. The maximum per cent of women in Total (Pooled) menstruated at the age of 14 years (43.6 per cent) and 13 years (38.0 per cent) whereas only 11.7 per cent and 6.7 per cent women achieved menstruation cycle at the age of 12 and 15 years respectively as shown in Table 4.16.

The mean age of menarche of women is found to similar (13.5 years) in four ethnic groups and Total (Pooled) except Lohar (13.4 years). The mean age of menarche is found be lesser than the scheduled caste of Himachal Pradesh (14.04 years). The low mean age of menarche in Lohar can be attributed to poor living conditions, malnutrition and higher family size.

Table 4.16: Age at Menarche of Mothers

Age at Menarche (in Years)	*Kanet*		*Koli*		*Badhi*		*Lohar*		*Total (Pooled)*	
	No.	*%age*	*No.*	*%age*	*No.*	*%age*	*No.*	*%age*	*No.*	*%age*
12	32	11.6	31	10.3	11	12.9	13	15.7	87	11.7
13	102	37.1	120	40.0	31	36.5	29	34.9	282	38.0
14	126	45.8	131	43.7	35	41.2	32	38.6	324	43.6
15	15	5.5	18	6.0	8	9.4	9	10.8	50	6.7
Total	**275**	**100.0**	**300**	**100.0**	**85**	**100.0**	**83**	**100.0**	**743**	**100.0**
Mean age at Menarche	**13.5**		**13.5**		**13.5**		**13.4**		**13.5**	

Age at Menarche of Daughters

It has been observed that in Total (Pooled), maximum monarchical age of daughter varies from the age of 13 years (43.7 per cent) to 14 years (40.4 per cent). The menstrual cycle of 11.4 per cent daughters is started at 12 years whereas only 4.5 per cent attained menstruation cycle at the age of 15 years as shown in Table 4.17.

Table 4.17: Age at Menarche of Daughters

Age at Menarche (in years)	*Kanet*		*Koli*		*Badhi*		*Lohar*		*Total (Pooled)*	
	No.	*%age*	*No.*	*%age*	*No.*	*%age*	*No.*	*%age*	*No.*	*%age*
12	9	10.1	11	11.0	4	13.8	4	14.8	28	11.4
13	41	46.1	45	45.0	11	37.9	10	37.0	107	43.7
14	35	39.3	40	40.0	12	41.4	12	44.4	99	40.4
15	4	4.5	4	4.0	2	6.9	1	3.7	11	4.5
Total	**89**	**100.0**	**100**	**100.0**	**29**	**100.0**	**27**	**100.0**	**245**	**100.0**
Mean age at Menarche	**13.4**		**13.4**		**13.4**		**13.4**		**13.4**	

The mean age of menarche of daughter is found to similar (13.5 years) in four ethnic groups and Total (Pooled). The mean age of menarche is found be lesser than the scheduled caste of Himachal Pradesh (14.04 years).

Age at Menopause

Menopause is considered as the end of reproductive life of women and from this stage, they are not capable for conception. The menopause is a critical biological phase in women's life, during which ovulation and menstruation are arrested and consequently the reproductive function ceases. It is very complex process of change which occurs in the biological structure of women, accompanied by a variety of physiological events, sometime resulting in extremely unpleasant symptoms including hot flushes, headaches, insomnia and itching.

It has been observed that in total (Pooled), maximum menopausal age varies from the age of 44 years (37.8 per cent) to 45 years (24.6 per cent). The menopausal age of 17.4 per cent women is started at 43 years whereas only 10.2 per cent attained menopause at the age of 47 years as shown in Table 4.18.

Table 4.18: Age at Menopause

Age at Menopause (In years)	*Kanet*		*Koli*		*Badhi*		*Lohar*		*Total (Pooled)*	
	No.	*%age*	*No.*	*%age*	*No.*	*%age*	*No.*	*%age*	*No.*	*%age*
43	30	17.0	33	17.8	10	18.5	9	18.8	82	17.7
44	67	38:1	71	38.4	19	35.2	18	37.5	175	37.8
45	45	25.6	47	25.4	12	22.2	10	20.8	114	24.6
46	18	10.2	16	8.6	6	11.1	5	10.4	45	9.7
47	16	9.1	18	9.7	7	13.0	6	12.5	47	10.2
Total	**176**	**100.0**	**185**	**100.0**	**54**	**100.0**	**48**	**100.0**	**463**	**100.0**
Mean	**44.6**		**44.5**		**44.6**		**44.6**		**44.6**	

The mean age of menopause is found to similar (44.6 years) in four ethnic groups and Total (Pooled) except 44.5 years in Koli group. The difference in age at menopause in various populations may be influenced by genetical and environmental factors. The mean age of menopause is found be higher than the scheduled caste (44.32 years), Gaddis (44.41) and Brahmin (44.1 years) of Himachal Pradesh and lower than the Rajput of Himachal Pradesh (44.81 years).

Age at Marriage (Women)

Age at marriage is a very important demographic and health characteristic. The mean age at marriage indicates average time of family formation. Age at marriage is important factor for influencing population growth and age at marriage of women helps in determining the reproductive span of women.

The maximum mean age at marriage of women is found in Kanet (21.3 years) followed by Koli (21.2 years), Badhi (20.9 years) and Lohar (20.1 years). The mean age at marriage of total (pooled) is found to be 21.1 years. The reason for high mean of marriage of women can be attributed to awareness of about proper age at marriage, literacy and high status of women in the society.

It has been observed that in total (pooled), maximum number of women married at the age of 19 years (11.5 per cent), 20 years (13.5 per cent) and 21 years (14.3 per cent) respectively. The women married at 18 years or earlier accounts for 18.6 per cent of total and women married at 25 years and after accounts for 13.5 per cent of total population. The maximum number of women in Kanet (14.4 per cent) , Koli (14.4 per cent) and Badhi (15.0 per cent) married at age of 21 years where as in Lohar, the maximum number (14.7 per cent) of women married at the age of 20 years.

The mean age at marriage of women is found to be higher than mean age at marriage of women of Himachal Pradesh (20.8 years) Brahmin (18.81 years), Rajput (18.32 years), Scheduled caste (18.42 years) and Gaddis (18.22 years) of Himachal Pradesh.

Age at Marriage (Men)

The maximum mean age at marriage of men is found in Koli (24.0 yrs.) followed by Kanet (23.8 yrs.), Badhi (22.7 yrs.) and Lohar (22.6 yrs.). The mean age at marriage of total (pooled) is found to be 23.6 yrs.. The reason for high mean of marriage of men can be attributed to awareness of about proper age at marriage and literacy.

It has been observed that in pooled, maximum number of men married at the age of 24 years (14.8 per cent), 25 years (13.8 per cent) and 26 years (12.4 per cent) respectively. The men married at 21 years or earlier accounts for 22.2 per cent of total and men married at 25 years and after accounts for 42.4 per cent of total population.

The maximum number of men in Kanet (15.7 per cent) , Koli (15.6 per cent) married at age of 24 years whereas Badhi (12.5 per cent) and Lohar (12.7 per cent) married at age of 25 years.

The mean age at marriage of men is found to be higher than mean age at marriage of men of Brahmin (22.4 yrs.), Rajput (22.65 yrs.), Scheduled caste (22.21 yrs.) and lower than Gaddis (24.42 yrs.) of Himachal Pradesh.

MEASURES OF FERTILITY

The measure of fertility ratios/ rates helps in understanding the relation between general conditions of the people and level of fertility in the population. In demographic studies, fertility occupies the central position. It is considered that population growth is depended on fertility. The term 'fertility' is to understand the actual reproductive performance of women. In simple word 'fertility' is considered as the capacity to produce both living and dead offspring. The ability of population to adapt to its environment depends on its fertility and population growth. Fertility is a genetic concept that encompasses a complex time-dependent process covering events ranging from exposure to intercourse and child bearing to the formation and dissolution of unions. It is complex multivariate system representing a set of different reproduction strategies. The fertility measures of four ethnic groups and total (pooled) are studied in detail in following pages.

Distribution of Reproductive Women and Number of Live Births

It has been observed that during last one year, a total of 1582 women produced 153 live births. The maximum per cent of children are born by women in age group of 25-29 years (39.2 per cent) followed by age group of 20-24 years (22.2 per cent) and age group of 30-34 years(20.9 per cent). No children are born by the women in age group 45-49 years whereas the minimum per cent of children are by the women in age group of 40-44 years (2.6 per cent) followed by age group of 15-19 (5.9 per cent` years and 35-39 years (9.2 per cent) as shown in Table 4.19.

Table 4.19: Age Group-wise Distribution of Reproductive Women and Live Births (in Last One Year)

Age Group in Years		Kanet		Koli		Badhi		Lohar		Total (Pooled)	
		No. of Women	No. of Live Birth	No. of Women	No. of Live Birth	No. of Women	No. of Live Birth	No. of Women	No. of Live Birth	No. of Women	No. of Live Birth
15-19	No.	94	3	127	4	33	1	25	1	279	9
	%	17.0	5.4	18.0	6.8	19.3	5.0	16.1	5.6	17.6	5.9
20-24	No.	103	12	124	12	33	5	31	5	291	34
	%	18.7	21.4	17.6	20.3	19.3	25.0	20.0	27.8	18.4	22.2
25-29	No.	89	22	119	24	26	7	26	7	260	60
	%	16.1	39.3	16.9	40.7	15.2	35.0	16.8	38.9	16.4	39.2
30-34	No.	89	12	116	11	26	5	25	4	256	32
	%	16.1	21.4	16.5	18.6	15.2	25.0	16.1	22.2	16.2	20.9
35-39	No.	72	5	90	6	22	2	20	1	204	14
	%	13.0	8.9	12.8	10.2	12.9	10.0	12.9	5.6	12.9	9.2

(Table 4.19 Contd...)

Age Group in Years		Kanet		Koli		Badhi		Lohar		Total (Pooled)	
		No. of Women	No. of Live Birth	No. of Women	No. of Live Birth	No. of Women	No. of Live Birth	No. of Women	No. of Live Birth	No. of Women	No. of Live Birth
40-44	No.	65	2	75	2	17	0	15	0	172	4
	%	11.8	3.6	10.7	3.4	9.9	0.0	9.7	0.0	10.9	2.6
45-49	No.	40	0	53	0	14	0	13	0	120	0
	%	7.2	0.0	7.5	0.0	8.2	0.0	8.4	0.0	7.6	0.0
Total	**No.**	**552**	**56**	**704**	**59**	**171**	**20**	**155**	**18**	**1582**	**153**
	%	**100.0**	**100.0**	**100.0**	**100.0**	**100.0**	**100.0**	**100.0**	**100.0**	**100.0**	**100.0**

Child Women Ratio

The age sex distribution of a population in a region is mainly the function of past fertility, mortality, migration. The child women ratio is considered as the effective measure of fertility as it reflect only the number of surviving children under 5 years (or 9 years) of age. In the present study, the following two have been considered (a) ratio of children aged 0-4 years to 1000 women aged 15-44 years (b) ratio of children aged 0-9 years to 1000 women aged 15-49 years. In first and second variant, the Kanet has shown the highest ratio followed by Lohar, Badhi and Koli as shown in Table 4.20.

Table 4.20: Child Women Ratio

Ethnic Group	*Child Women Ratio*	
	Children (0-4)/ Women (15-44)	*Children (0-9)/ Women (15-49)*
Kanet	480.47	898.55
Koli	397.85	751.42
Badhi	407.64	964.91
Lohar	408.45	941.94
Total (Pooled)	**428.86**	**844.50**

The child women ratio of total (pooled) is lower than Buddhist (553.4), Hindus (516.1) of Sikkim; Bodhs (510.1), Baltis (689.3) of Jammu and Kashmir; Marchha Bhotia (481.4), Dharchula Bhotia (662.6) of Uttaranchal and higher than Brokpas (296.2), Arghuns (360.1) of Jammu and Kashmir; Johar Bhotia (325.1) of Uttaranchal as given in Table 4.33.

Crude Birth Rate (CBR)

Crude Birth Rate is used to measure the fertility. It indicates the general magnitude of the fertility level of a population/region at a specified time. Crude birth rate plays an important role in fertility rate. It has been defined that crude birth rate is ratio of total registered live birth to the total population in some specific year.

In the present study, the crude birth rate highest among Lohar (29.56) followed by Badhi (28.53), Kanet (25.97) and Koli (23.56). The Crude Birth Rate of Koli is lower because more number of Koli are adapting to family planning as compared to the other groups as shown in Table 4.21. Whereas the reason for high crude birth rate of Lohar and Badhi can be attributed to male sex preference, paucity of family planning method adoption, illiteracy, low socio-economic status etc. The crude birth rate of total (pooled) is 25.63.

Table 4.21: Crude Birth Rate

Ethnic Group	*Crude Birth Rate*
Kanet	25.97
Koli	23.56
Badhi	28.53
Lohar	29.56
Total (Pooled)	**25.63**

The crude birth rate of total (pooled) is lower than Buddhist (21.8), Bhutias (22.2), Lepchas (20.8), Sherpas (19.6), Tamangs (24.3) of Sikkim; Bodhs (24.4), Baltis (23.5), Arghuns (14.2) of Jammu and Kashmir; Himachal Pradesh (28.2); Gaddis (28.40), Brahmin (25.8), Scheduled caste (48.60) of Himachal Pradesh; Johari Bhotia (48.21), Rang Bhotia (53.6), Marchha Bhotia (35.98), Rajis (46) of Uttaranchal and higher than Brokpas (27.1) of Jammu and Kashmir; Hindus (28.3) of Sikkim; Rajput (21.70) of Himachal Pradesh; India (24.8) as given in Table 4.30.

General Fertility Rate

The general fertility rate is defined as the number of births per 1000 women of child bearing age. The general marital fertility rate also does not indicate a definite pattern except that though the crude birth rate is moderate, general fertility rate is high. The general fertility rate relates birth (at a specified time) with population at risk of childbearing i.e to the females at reproductive ages of 15-49 years. It is affected by the distribution of female by age in the reproductive span.

The general fertility rate is found to be highest among Badhi (116.96) followed by Lohar (116.13), Kanet (101.45) and Koli (83.81) as shown in Table 4.22. The general fertility rate of total (pooled) is 96.71.

Table 4.22: General Fertility Rate

Ethnic Group	*General Fertility Rate*
Kanet	101.45
Koli	83.81
Badhi	116.96
Lohar	116.13
Total (Pooled)	**96.71**

The general fertility rate of total (pooled) is higher than Buddhist (92.60), Bhutias (93.2), Lepchas(92.1), Sherpas (90.9), Tamangs (92.5) of Sikkim; Mao (65.3) of Manipur; Bodhs (95.7), Arghuns (49.8) of Jammu and Kashmir; Brahmin (93.3) Rajput (83.9) of Himachal Pradesh; Johari Bhotia (175.2), Rang Bhotia (186.4), Marchha Bhotia (162.4), Rajis (213.2) of Uttaranchal and lower than Baltis (98.5), Brokpas (119) of Jammu and Kashmir; Hindu (108.6) of Sikkim; Gaddis (110.5), scheduled caste of HP (201.50) and India (116.60) as given in Table 4.30.

Age Specific Fertility Rate (ASFR)

Since the rate of childbearing is not uniform throughout all ages, the age specific fertility rate reveals the extent and pattern of age differentials. The Age specific fertility Rates are expressed as the number of live births at a specific time to women at a particular age. Age of mother is an important factor affecting the fertility level and the rate of child bearing is not uniform through out all ages. In fact, fertility is heavily concentrated between ages 20 to 29 years. The ASFRs are lower at younger fertile ages i.e 15-19 years. The age group showing peak fertility or the highest ASFR is found common to all i.e 25-29 years in all four ethnic groups and total (pooled). The ASFRs for two ethnic groups (Badhi and Lohar) in 25-29 age group is similar as shown in Table 4.23 *(See the next page)*.

Total Fertility Rate

The total fertility rate summaries the pattern of fertility exhibited by age specifc birth rates of women and presents a single index of total fertility. Total Fertility rate expressed per women refers to the number of children a hypothetical average women would have if, during her lifetime her child bearing behaviour remain same as that of a cross- section of women at the time of observation. It has been observed that total fertility rate has been found to be maximum in Badhi (3.67) followed by Lohar (3.40), Kanet (3.15) and Koli (2.59) as shown in Table 4.24. The reason for high total fertility rate of Badhi and Lohar can be attributed to low socio-economic status, high extent of illiteracy, male sex preference, less family planning adoption etc. The total fertility rate of total (pooled) is 2.98.

Table 4.23: Age Specific Fertility Rate

Age Group	*Kanet*	*Koli*	*Badhi*	*Lohar*	*Total (Pooled)*
15-19	31.91	31.50	30.30	40.00	32.26
20-24	116.50	96.77	151.52	161.29	116.84
25-29	247.19	201.68	269.23	269.23	230.77
30-34	134.83	94.83	192.31	160.00	125.00
35-39	69.44	66.67	90.91	50.00	68.63
40-44	30.77	26.67	0.00	0.00	23.26
45-49	0	0	0	0	0
Total	**101.45**	**83.81**	**116.96**	**116.13**	**96.71**

Table 4.24: Total Fertility Rate

	Kanet	*Koli*	*Badhi*	*Lohar*	*Total (Pooled)*
Total Fertility Rate Per Women	3.15	2.59	3.67	3.40	2.98

The total fertility rate of total (pooled) is lower than Buddhist (3.10), Hindus (3.20), Bhutias (3.1), Lepchas(3), Sherpas (3), Tamangs (3.1) of Sikkim; Mao (4.97) of Manipur; Baltis (3.1), Brokpas (3.0) of Jammu and Kashmir; Gaddi (3.5), Brahmin (3.2), Bods (3.6), Rajputs (3.2), scheduled caste (5.91) of HP; Johari Bhotia (6.2), Rang Bhotia (5.8), Marchha Bhotia (6.3), Rajis (7.4) of Uttaranchal and lower than Bodhs (2.7), Arghuns (1.6) of Jammu and Kashmir; Himachal Pradesh (2.14) and India (2.85) as given in Table 4.30.

Gross Reproductive Fertility Rate

The gross reproductive fertility rate is defined as the number of female children a cohort of women would have passing through the reproductive ages and bearing children according to fixed scheduled of fertility, if they all survive to the end of the child bearing period. It has been observed that Gross Reproductive Fertility Rate has been found to be maximum in Kanet (2.15) followed by Lohar (1.50), Badhi (1.49) and Koli (1.42) as shown in Table 4.25. The gross reproductive fertility rate of total (pooled) is 1.33. The reason for high gross reproductive fertility rate is due to high infant and childhood deaths.

Table 4.25: Gross Reproductive Fertility Rate

	Kanet	*Koli*	*Badhi*	*Lohar*	*Total (Pooled)*
Gross Reproductive Rate Per Women	2.15	1.42	1.49	1.50	1.33

The gross reproductive fertility rate of total (pooled) is lower than Buddhist (1.50), Hindus (1.60), Bhutias (1.5), Lepchas(1.5), Sherpas (1.5), Tamangs (1.5) of Sikkim; Mao (2.4) of Manipur; Bodhs (1.7), Baltis(1.5), and Brokpas (1.5) of Jammu and Kashmir; Gaddi (1.7), Brahmin (1.6), Rajput (1.6), Scheduled Caste (2.9), Bods (1.8) of Himachal Pradesh; Johari Bhotia (3.0), Rang Bhotia (2.8), Marchha Bhotia (3.1), Rajis (3.6) of Uttaranchal and higher than Arghuns (0.3) of Jammu and Kashmir; Himachal Pradesh (1.05) India (1.70) as given in Table 4.30.

Mean Age of Childbearing

The mean age of childbearing indicates the average age of childbearing. A population having a high mean age of childbearing has a large fraction of total fertility relating to the latter years of childbearing whereas a population with a low mean age at childbearing has a small fraction of total fertility in the latter ages of childbearing.

It has been observed that mean age at child bearing has been found to be maximum in Kanet (27.9 years) followed by Koli (27.8 years), Badhi (27.5 years) and Lohar (26.7 years) as shown in Table 4.26. The reason for high mean age of child bearing is high infant and child mortality and early marriage.

Table 4.26: Mean Age of Child Bearing

	Kanet	*Koli*	*Badhi*	*Lohar*	*Total (Pooled)*
Mean Age of Child Bearing	27.9	27.8	27.5	26.7	27.7

The mean age of child bearing of total (pooled) is 27.7 years which is lower than scheduled caste (32.92 years) of Himachal Pradesh and higher than India (27.40 years).

Marital Fertility Rate

In most of the societies, marital fertility is socially and legally recognised. It is preferred to consider to only those female who are actually exposed to the risk of child bearing i.e the married female only.

General Marital Fertility Rate

The General Marital Fertility rate is expressed as the number of live births per thousand married women in the reproductive age group 15-49 years. However, as the probability of childbearing is not uniform over all the ages, it is also affected by the distribution of married females in various age groups within 15-49 years. It has been observed that General Marital Fertility Rate has been found to be maximum in Lohar (148.76) followed by Badhi (153.85), Kanet (118.14) and Koli (100.00) as shown in Table 4.27. The General Marital Fertility Rate of total (pooled) is 116.35 which is lower than India (173.6), scheduled caste of Himachal Pradesh (236.8) and Gaddis (145.1).

Table 4.27: General Marital Fertility Rate

	Kanet	*Koli*	*Badhi*	*Lohar*	*Total (Pooled)*
General Marital Fertility Rate	118.14	100.00	153.85	148.76	116.35

Age Specific Marital Fertility Rate

The Age Specific Marital Fertility Rate (ASMFR) describes the fertility experience of married females by age at a specified time. The Age Specific Marital Fertility Rate like Age Specific Fertility Rate also showed varied patterns and levels in and across all study ethnic groups. It has been observed that highest age specific marital fertility rate has been found in the age groups of 25-29 years (272.73) followed 20-24 years(153.85), 30-34 years(133.89), 35-39 years (71.43), 15- 19 years (52.02) as shown in Table 4.28. It has been observed that age specific fertility rate is very high in women in age group of 15-29 years. It can be attributed to marriage able of women, lesser use of family planning methods etc.

Table 4.28: Age Specific Marital Fertility Rate

Age Group	*Kanet*	*Koli*	*Badhi*	*Lohar*	*Total (Pooled)*
15-19	47.62	48.78	58.82	90.91	52.02
20-24	142.86	125.00	238.10	250.00	153.85
25-29	271.60	240.00	368.42	350.00	272.73
30-34	144.58	102.80	208.33	160.00	133.89
35-39	70.42	72.29	90.91	50.00	71.43
40-44	36.36	28.17	0.00	0.00	25.64
45-49	0.00	0.00	0.00	0.00	0.00
Total	**713.45**	**617.04**	**964.58**	**900.91**	**709.56**

Total Marital Fertility Rate

Total Marital Fertility Rate summaries the pattern of marital fertility exhibited by the ASMFRs and presents a single index of total marital fertility. In other words, total marital fertility rate is cumulative value of age specific marital fertilities at the end of the reproductive period. It indicates the average number of children expected to be born per married women during the entire span of reproductive period, if the age specific marital fertility rates (on which the rate is based) continue to be same and there is no mortality.

It has been observed that total marital fertility rate has been found to be maximum in Badhi (4.82) followed by Lohar (4.50), Kanet (3.57) and Koli (3.09) as shown in Table 4.29. The reason for high total marital fertility rate in Badhi and Lohar can be attributed to illiteracy, early age of marriage, male sex preference etc. The total marital fertility rate of total (pooled) is 3.55 which is lower than India (4.90).

Table 4.29: Total Marital Fertility Rate

	Kanet	*Koli*	*Badhi*	*Lohar*	*Total (Pooled)*
Total Marital Fertility Rate	3.57	3.09	4.82	4.50	3.55

The Crude Birth Rate, General Fertility Rate, Total Fertility Rate, Gross Reproduction Rate, Child Women Ratio of Tribal and Non-Tribal Population Groups of India and Himalayan Region is given in Table 4.30.

Table 4.30: Crude Birth Rate, General Fertility Rate, Total Fertility Rate, Gross Reproduction Rate, Child Women Ratio of Tribal and Non-Tribal Population Group of India and Himalayan Region

Regions/ State	*Population Groups*	*CBR*	*GFR*	*TFR*	*GRR*	*Child Women ratio*	*Source*
Eastern Himalayan							
Sikkim							
	Buddhist	21.80	92.60	3.10	1.50	553.4	Bhasin & Bhasin (2000)
	Hindus	28.30	108.60	3.20	1.60	516.1	Bhasin & Bhasin (2000)
	Bhutias	22.2	93.2	3.1	1.5		Bhasin & Bhasin (1995)
	Lepchas	20.8	92.1	3	1.5		Bhasin & Bhasin (1995)
	Sherpas	19.6	90.9	3	1.5		Bhasin & Bhasin (1995)
	Tamangs	24.3	92.5	3.1	1.5		Bhasin & Bhasin (1995)

(Table 4.30 Contd...)

Regions/ State	Population Groups	CBR	GFR	TFR	GRR	Child Women ratio	Source
Manipur							
	Mao	12	65.3	4.97	2.4		Maheo 1999
West Bengal							
Kalimpong	Sherpa			6.6	3.2	50.4	Gupta et al (1989)
Kalimpong	Lepcha			5.4	2.6	32.4	Gupta et al (1989)
Rango	Sherpa			6.1	3	85.3	Gupta et al (1989)
Echhay	Sherpa			5.0	2.5	46.3	Gupta et al (1989)
Lava	Sherpa			6.5	3.2	51.2	Gupta et al (1989)
Munsong	Sherpa			7.1	3.5	51.6	Gupta et al (1989)
Labbah (Darjeeling)	Sherpa			5.9	2.9	35.3	Gupta et al (1989)
Mungpoo (Darjeeling)	Sherpa			4.4	2.2	70.5	Gupta et al (1989
	Total	20.7					SRS (2002)

(Table 4.30 Contd...)

Regions/ State	*Population Groups*	*CBR*	*GFR*	*TFR*	*GRR*	*Child Women ratio*	*Source*
Western Himalayan							
Jammu and Kashmir	**Ladakhs**	22.4	88.9	2.69	1.3	563.9	Bhasin & Nag (2002)
	Bodhs	24.4	95.7	2.7	1.7	510.1	Bhasin & Nag (2002)
	Baltis	23.5	98.5	3.1	1.5	689.3	Bhasin & Nag (2002)
	Brokpas	27.1	119	3	1.5	296.2	Bhasin & Nag (2002)
	Arghuns	14.2	49.8	1.6	0.3	360.1	Bhasin & Nag (2002)
	Total	19.7					SRS (2002)
Himachal Pradesh							
	Gaddi	28.4	110.5	3.5	1.7		Bhasin & Bhasin (1993)
	Brahman	25.8	93.3	3.2	1.6		Bhasin & Bhasin (1993)
	Rajput	21.7	83.9	3.2	1.6		Bhasin & Bhasin (1993)
	Scheduled Caste	48.8	201.5	5.9	2.9		Bhasin & Bhasin (1993)
	Bods			3.6	1.8		Bhasin & Bhasin (1993)
	Bods (High Altitude)			4.1	2		Prakesh and Malik (1990)

(Table 4.30 Contd...)

Regions/ State	Population Groups	CBR	GFR	TFR	GRR	Child Women ratio	Source
	Total	30.8	126.3	3.9	1.9		Bhasin & Bhasin (2000)
	Total	15.7		2	0.9		1998-99 NFHS-2 (2000)
	Hindu			2.11	1.03		1998-99 NFHS-2 (2000)
	Buddhist			1.95	0.96		1998-99 NFHS-2 (2000)
	Total	28.2		2.14	1.05		1998-99 NFHS-2 (2000)
	Total	22.1					SRS (2002)
Kinnaur							
	Kanet	25.97	101.5	3.15	2.15	480.47	Present Study
	Koli	23.56	83.81	2.59	1.42	397.85	Present Study
	Badhi	28.53	117	3.67	1.49	407.64	Present Study
	Lohar	29.56	116.1	3.4	1.5	408.45	Present Study
	Total	25.63	96.71	2.98	1.33	428.86	Present Study

(Table 4.30 Contd...)

Regions/ State	*Population Groups*	*CBR*	*GFR*	*TFR*	*GRR*	*Child Women ratio*	*Source*
Central Himalayans							
Uttaranchal		20.2					SRS (2002)
	Johari Bhotia	48.21	175.2	6.2	3		Kapoor (1996)
	Rang Bhotia	53.6	186.4	5.8	2.8		Kapoor (1996)
	Marchha Bhotia	35.98	162.4	6.3	3.1		Kapoor (1996)
	Marchha Bhotia	24.1	128.4	4.2	2.1	481.4	Chachra & Bhasin(1998)
	Dharchula Bhotia	15.6	95.2	2.6	1.3	662.6	Chachra & Bhasin(1998)
	Johar Bhotia	29.6	56.4	1.9	0.9	325.1	Chachra & Bhasin(1998)
	Raji	49.4	198	6.5	3.2		Samal et al (2000)
	Raji	46	213.2	7.4	3.6		Patra (2001)
	Jaunsaris	32	125	3.8	1.9		Kshatriya et al (1997)
India				2.85			1998-99 NFHS-2 (2000)

MEASURES OF MORTALITY

Mortality levels not only affects the demographic profile of a population/region, but are also considered as excellent indicators of the state of development and human well being of a region. Unlike fertility, man's attempts to regulate mortality has been relatively successful and as a consequence, population pressure is building up everywhere, creating a state of disequilibrium. Mortality is an event that occurs over time. Several current measures of mortality provided such understanding and in the present thesis, some of the measures have been studied. All types of mortality causing persistent concern, infant and early childhood mortality (all under 5 mortalities) are considered very significant, since these form a major chunk of all deaths and also influence the fertility and fertility control behaviour of parents. The under five mortalities, like fertility, are affected by a host of demographic, economic, socio-cultural, physical environmental determinants as well as attitude, behaviour of parents related to sex composition of children and family size.

Total Deaths in Different Age Groups

The total deaths in different age groups are given in Table 4.31. It has been observed that total 60 males and 42 females deaths have been recorded and in almost every age groups, the deaths have been taken place except 15-19 years age group in last one year. The maximum percentage of deaths have been found in age group 0-4 years followed by 5-9 age groups, 60-64 and 65 + years. It has been noticed that in Kinnauris, maximum number of death occurred in very early age and old age. During the early stage of life, child is very vulnerable and requires intensive care. It has been reported that due to illiteracy, harsh environmental conditions, far away distance of health institutions and unhygienic living conditions led to high child death.

Table 4.31: Total Death in Different Age Groups

Age	*Sex*	*Kanet*	*Koli*	*Badhi*	*Lohar*	*Total (Pooled)*
0-4	Male	4	5	2	2	13
	Female	6	4	2	3	15
	Total	10	9	4	5	28
5-9	Male	3	3	1	1	8
	Female	3	2	1	1	7
	Total	6	5	2	2	15
10-14	Male		2			2
	Female	1	1			2
	Total	1	3	0	0	4
15-19	Male					0
	Female					0
	Total	0	0	0	0	0
20-24	Male	1	2			3
	Female	2				2
	Total	3	2	0	0	5
25-29	Male		1	1	1	3
	Female	1	2	2	1	6
	Total	1	3	3	2	9
30-34	Male	3				3
	Female					0
	Total	3	0	0	0	3
35-39	Male			1		1
	Female		2	1		3
	Total	0	2	2	0	4
40-44	Male	2			1	3
	Female					0
	Total	2	0	0	1	3
45-49	Male	3	3	1		7
	Female					0
	Total	3	3	1	0	7

(Contd...)

Age	*Sex*	*Kanet*	*Koli*	*Badhi*	*Lohar*	*Total (Pooled)*
50-54	Male		2			2
	Female		1			1
	Total	0	3	0	0	3
55-59	Male	1				1
	Female					0
	Total	1	0	0	0	1
60-64	Male	2	3	2	1	8
	Female	1	1			2
	Total	3	4	2	1	10
65 +	Male	2	2	1	1	6
	Female	1	1	1	1	4
	Total	3	3	2	2	10
Total	**Male**	**21**	**23**	**9**	**7**	**60**
	Female	**15**	**14**	**7**	**6**	**42**
	Total	**36**	**37**	**16**	**13**	**102**

Infant (upto 28 days) Deaths in Different Age Groups

It has been observed that the highest infant deaths are in below age group of 7 days followed by age group of 15-21 days, 8-14 days and 22-28 days. The infant is a period of considerable delicacy and care. After birth, few days are very critical for their survival. The environment conditions of Kinnaur make it more difficult for infant to adopt themselves to new physical environment. In such case, the chances of infant survival become difficult. The infant death can be attributed to illiteracy, harsh environmental conditions, non institutional deliveries, far away distance of health institutions and unhygienic living conditions. The death of male infant is higher than female infants as shown in Table 4.32.

Table 4.32: Infant (upto 28 days) Death in Different Age Groups

Age (in days)	*Sex*	*Kanet*		*Koli*		*Badhi*		*Lohar*		*Total (Pooled)*	
		No.	*%age*	*No.*	*%age*	*No.*	*%age*	*No.*	*%age*	*No.*	*%age*
< 7	Male	1	50.0	1	100.0	1	100.0		0.0	3	60.0
	Female	1	50.0		0.0		0.0	1	100.0	2	40.0
	Total	2	100.0	1	100.0	1	100.0	1	100.0	5	100.0
8-14	Male	0	0.0			1	100.0			1	100.0
	Female	0	0.0				0.0			0	0.0
	Total	0	0.0	0		1	100.0	0		1	100.0
15-21	Male	0		1	50.0				0.0	1	33.3
	Female			1	50.0			1	100.0	2	66.7
	Total	0		2	100.0	0		1	100.0	3	100.0
22-28	Male	1	100.0							1	100.0
	Female		0.0							0	0.0
	Total	1	100.0	0		0		0		1	100.0
Total	Male	2	66.7	2	66.7	2	100.0	0	0.0	6	60.0
	Female	1	33.3	1	33.3	0	0.0	2	100.0	4	40.0
	Total	3	100.0	3	100.0	2	100.0	2	100.0	10	100.0

Infant Deaths in Different Age Groups of Last One Year

Infant death indicates the death which has been occurred below the age of one year of child. It has been observed that the highest infant deaths are in below age group of less than 2 months followed by age group of 2-4 months, 4-6 months and 6-8 months. The death of male and female infant show the equal percentage of death as shown in Table 4.33.

Table 4.33: Infant Deaths in Different Age Groups of Last One Year

Age (in months)	*Sex*	*Kanet*		*Koli*		*Badhi*		*Lohar*		*Total (Pooled)*	
		No.	*%age*	*No.*	*%age*	*No.*	*%age*	*No.*	*%age*	*No.*	*%age*
< 2	Male	2	66.7	2	66.7	2	100.0	1	33.3	7	63.6
	Female	1	33.3	1	33.3		0.0	2	66.7	4	36.4
	Total	3	100.0	3	100.0	2	100.0	3	100.0	11	100.0
2-4	Male		0.0	1	100.0					1	50.0
	Female	1	100.0		0.0					1	50.0
	Total	1	100.0	1	100.0	0		0		2	100.0
4-6	Male						0.0		0.0	0	0.0
	Female					1	100.0	1	100.0	2	100.0
	Total	0		0		1	100.0	1	100.0	2	100.0
6- 8	Male	1	100.0							1	100.0
	Female		0.0							0	0.0
	Total	1	100.0	0		0		0		1	100.0

(Table 4.33 Contd...)

Age (in months)	Sex	Kanet		Koli		Badhi		Lohar		Total (Pooled)	
		No.	%age	No.	%age	No.	%age	No.	%age	No.	%age
8-10	Male				0.0					0	0.0
	Female			1	100.0					1	100.0
	Total	0		1	100.0	0		0		1	100.0
10-12	Male		0.0							0	0.0
	Female	1	100.0							1	100.0
	Total	1	100.0	0		0		0		1	100.0
Total	**Male**	**3**	**50.0**	**3**	**60.0**	**2**	**66.7**	**1**	**25.0**	**9**	**50.0**
	Female	**3**	**50.0**	**2**	**40.0**	**1**	**33.3**	**3**	**75.0**	**9**	**50.0**
	Total	**6**	**100.0**	**5**	**100.0**	**3**	**100.0**	**4**	**100.0**	**18**	**100.0**

Crude Death Rate

The crude death rate is widely and commonly used measure for mortality, describing the frequency with which deaths occur in a population at specified time/period. The crude death rate for Badhi (22.82) found to be maximum followed by Lohar (21.35), Kanet (16.70) and Koli (14.78). The crude death rate of total (pooled) is 17.09. The high crude death rate can be attributed to harsh environmental condition, isolation of villages from the main urban centres, far away distance from health institutions, non availability of health personal, unhygienic living conditions and poor food habits etc. The male, female and total crude death rate is given in Table 4.34.

Table 4.34: Crude Death Rate

Ethnic Group	*Crude Death Rate*		
	Male	*Female*	*Total*
Kanet	18.70	14.52	16.70
Koli	17.72	11.61	14.78
Badhi	23.87	21.60	22.82
Lohar	21.54	21.13	21.35
Total (Pooled)	**19.21**	**14.75**	**17.09**

The crude death rate of total (pooled) is higher than crude death rate of Mao (5.6) of Manipur; Bodhs (14.3), Baltis (16.8), Arghuns (13.2) of Jammu and Kashmir; Himachal Pradesh (8.3); India (8.5); Brahmins (12.9), Rajputs of HP (11.8) and lesser than Brokpas (21.7) of Jammu and Kashmir; Hindus (19.8) of Sikkim; scheduled caste of HP (26.0) as given in Table 4.42.

Age Specific Mortality Rate

To understand the overall death rates, age specific mortality rate is studied in detail. The mortality risks faced by human beings vary from one age group to other. The age specific mortality rate (ASMR) is calculated separately for males and females, as levels of age pattern of the risk of dying vary substantially by sex.

The age specific mortality rate is the broad U pattern with the mortality high among the young and old age group. The mortality is high for age group of 0-4 years which reduce subsequently and again increases at the age group of 60 years and above. The death rate for age group 65 + and children (0-4 years) among each of the study ethnic group appeared high. The death rates in this age groups for the Kinnauris (pool) are 104.17 and 44.66 respectively. The male mortality is found to be higher than female mortality as shown in Table 4.35.

Table 4.35: Age Specific Mortality Rate

Age		*Kanet*	*Koli*	*Badhi*	*Lohar*	*Total (Pooled)*
0-4	M	29.85	36.76	57.14	64.52	38.69
	F	53.57	32.52	68.97	111.11	51.55
	Total	40.65	34.75	62.50	86.21	44.66
5-9	M	23.26	21.28	17.86	21.28.	21.45
	F	24.79	15.50	22.22	24.39	20.83
	Total	24.00	18.52	19.80	22.73	21.16
10-14	M	0.00	16.00	0.00	0.00	5.87
	F	8.70	8.00	0.00	0.00	6.37
	Total	4.24	11.54	0.00	0.00	6.11
15-19	M	0.00	0.00	0.00	0.00	0.00
	F	·0.00	0.00	0.00	0.00	0.00
	Total	0.00	0.00	0.00	0.00	0.00
20-24	M	9.09	15.50	0.00	0.00	9.71
	F	19.42	0.00	0.00	0.00	6.87
	Total	14.08	7.91	0.00	0.00	8.33
25-29	M	0.00	7.87	34.48	35.71	10.87
	F	11.24	16.81	76.92	38.46	23.08
	Total	5.52	12.20	54.55	37.04	16.79
30-34	M	33.33	0.00	0.00	0.00	10.87
	F	0.00	0.00	0.00	0.00	0.00
	Total	16.76	0.00	0.00	0.00	5.64

(Contd...)

Age		*Kanet*	*Koli*	*Badhi*	*Lohar*	*Total (Pooled)*
35-39	M	0.00	0.00	40.00	0.00	4.46
	F	0.00	22.22	45.45	0.00	14.71
	Total	0.00	10.70	42.55	0.00	9.35
40-44	M	28.57	0.00	0.00	58.82	15.87
	F	0.00	0.00	0.00	0.00	0.00
	Total	14.81	0.00	0.00	31.25	8.31
45-49	M	68.18	50.00	62.50	0.00	51.85
	F	0.00	0.00	0.00	0.00	0.00
	Total	35.71	26.55	33.33	0.00	27.45
50-54	M	0.00	33.90	0.00	0.00	13.99
	F	0.00	18.18	0.00	0.00	7.63
	Total	0.00	26.32	0.00	0.00	10.95
55-59	M	29.41	0.00	0.00	0.00	11.49
	F	0.00	0.00	0.00	0.00	0.00
	Total	15.38	0.00	0.00	0.00	6.17
60-64	M	57.14	88.24	181.82	100.00	88.89
	F	32.26	33.33	0.00	0.00	25.97
	Total	45.45	62.50	105.26	55.56	59.88
65 +	M	68.97	133.33	142.86	250.00	109.09
	F	47.62	83.33	200.00	333.33	97.56
	Total	60.00	111.11	166.67	285.71	104.17
Total	**M**	**18.70**	**17.72**	**23.87**	**21.54**	**19.21**
	F	**14.52**	**11.61**	**21.60**	**21.13**	**14.75**
	Total	**16.70**	**14.78**	**22.82**	**21.35**	**17.09**

M= Male, F= Female

Still Birth Rate

Still birth is a term which is accounted in reproductive wastage. Still birth is sometimes referred as foetal death. It represents the death which is occurred before the complete expulsion or extraction from the womb of mother. From 28 weeks of conception to the exact child bearing period, the embryonic loss is accounted as still birth or foetal death. Due to separation from gestation, the foetal is unable to practice breathing or shows the existence of life which indicates heart beat, pulsation, movement of voluntary muscles etc. The Still Birth Rate for Lohar (52.63) found to be maximum followed by Badhi (47.62), Kanet (34.48) and Koli (32.79) as shown in Table 4.36. The still birth rate of total (pooled) is 37.74. It can be attributed to malnutrition, unhygienic delivery practices, negligence on maternal care etc. The still birth rate of total (pooled) is higher than of Himachal Pradesh (14.00); Brahmin (10.8), Rajput (11.6) schedule castes (17.8) of HP and India (9.0).

Table 4.36: Still Birth Rate

	Kanet	*Koli*	*Badhi*	*Lohar*	*Total (Pooled)*
Still Birth Rate	34.48	32.79	47.62	52.63	37.74

Perinatal Mortality Rate

Early neonatal deaths (deaths of infants with first week of life), late fetal deaths (after 28 weeks of gestation) and still births together constitute perinatal mortality. It has been reported that Perinatal Mortality rate is sensitive index reflecting standards of mother and child health care prior to and during pregnancy and child birth and during the first week after birth, as well as the availability and effectiveness of other social development measures. The perinatal mortality rates are found to be highest in Lohar (100.00) followed by Badhi (90.91), Kanet (66.67), Koli (48.39) as shown in Table 4.37. The perinatal mortality rate of total (pooled) is 67.07. The high perinatal mortality rate can be attributed to unsafe deliveries, malnutrition, ignorance in maternal and child health care, lack of knowledge about RCH in mothers, frequent child bearing etc. The perinatal mortality rate of total (pooled) is higher than Himachal Pradesh (52), India (43), Brahmins of HP (63.9), Rajputs of HP (56.2) and lower than scheduled caste of HP (92.3).

Table 4.37: Perinatal Mortality Rate

	Kanet	*Koli*	*Badhi*	*Lohar*	*Total (Pooled)*
Perinatal Mortality Rate	66.67	48.39	90.91	100.00	67.07

Neo Natal Mortality Rate and Post Neonatal Mortality Rate

Infant mortality consists of two main components (a) Neonatal mortality—infant deaths upto 28 days of life and (b) Post Neonatal Mortality—infant deaths from 1 month to 1 year of life. It has been reported that deaths occurring during neonatal period are primarily caused by biological factors which indicate neo natal mortality can be attributed to endogenous or ante-natal causes. The neonatal mortality rate is found to be highest in Lohar (111.11) followed by Badhi (100.00), Kanet (53.57), Koli (50.85) as shown in Table 4.38. The neonatal mortality rate of total (pooled) group is 65.36. The high neonatal mortality rate can be attributed to ignorance of mother during antenatal period, hygienic weaning practices, poor living conditions, unhygienic delivery practices etc. The neonatal mortality rate of total (pooled) is higher than neonatal mortality rate of Khasi (17.2), Garo (33.9) and Mizo (3.5) of Meghalaya; Bodhas (24.4) of Jammu and Kashmir; Himachal Pradesh (22.1); India (46.00) and lower than Hindus (66.7) of Sikkim; Baltis (81.6) and Brokpas (100) of Jammu and Kashmir; Brahmins (75.8), Rajputs (69.0) scheduled caste of HP (94.4) as given in Table 4.42.

Table 4.38: Neo Natal Mortality Rate

	Kanet	*Koli*	*Badhi*	*Lohar*	*Total (Pooled)*
Neo Natal Mortality Rate	53.57	50.85	100.00	111.11	65.36

The post neonatal deaths are most sensitive to socio-economic factors as compared with biological factors. Post neonatal mortality is also attributed to exogenous or environmental conditions. The post neonatal mortality rate is found to be highest in Lohar (166.67) followed by Badhi (100.00), Kanet (53.57), Koli (50.85) as shown in Table 4.39. The post neonatal mortality rate of total (pooled) group is 71.90. The high post neo natal mortality rate can be attributed to paucity of health facilities, frequent child bearing, non institutional deliveries, malnutrition, illiteracy etc.

Table 4.39: Post Neo Natal Mortality Rate

	Kanet	*Koli*	*Badhi*	*Lohar*	*Total (Pooled)*
Post Neo Natal Mortality Rate	53.57	50.85	100.00	166.67	**71.90**

The post neonatal mortality rate of total (pooled) is lower than Hindu (111.1) of Sikkim; Bodhs (73.1), Baltis (81.6)and Brokpas (100.0), Arghuns (76.9) of Jammu and Kashmir; scheduled caste of Himachal Pradesh (112.2) and higher than Khasi (24.8), Garo (18.1), Mizo (19.5) of Meghalaya; Himachal Pradesh (4.0); Brahmin (40.2), Rajput (36.6) of Himachal Pradesh and India (25.0) as given in Table 4.42.

Infant Mortality Rate

The infant mortality rate is an important measure of mortality. It is one of most sensitive indicator of the medical and public health facilities and together with literacy rate and life expectancy at birth estimates, it indicates the state of development and human well being in a region. It also accounts for a large chunk of all death and affect the attitude, behaviour of parents towards family size and structure. The infant mortality rate is found to be highest among Lohar (111.11) followed by Badhi (100.00), Koli (67.80) and Kanet (53.57) as shown in Table 4.40. This difference may partly be attributed to the overall disadvantage with respect to the availability and effectives of medical and other basic facilities as well as several other factors. The infant mortality rate of total (pooled) is 71.90. The high infant mortality rate can be attributed to ignorance toward to reproductive and child health care, poor ante and post natal health care, lack/ignorance of immunization, malnutrition, harsh environmental conditions etc.

Table 4.40: Infant Mortality Rate

Ethnic Group	*Infant Mortality Rate*
Kanet	53.57
Koli	67.80
Badhi	100.00
Lohar	111.11
Total (Pooled)	**71.90**

The high infant mortality rate can be attributed to ignorance toward to reproductive and child health care, poor ante and post natal health care, lack/ignorance of immunization, malnutrition, harsh environmental conditions etc. The infant mortality rate of total (pooled) is higher than Himachal Pradesh (34.4), Hindus (177.8) of Sikkim; Bodhs (97.6), Baltis (163.3), Brokpas (200.0) of Jammu and Kashmir; Johari Bhotia (152.8) Rang Bhotia (212), Marchha Bhotia (158.7), Rajis (344.8) of Uttaranchal; India (67.6) and lower than Mao (23.0) of Manipur; Arghuns (76.9) of Jammu and Kashmir; Brahmins (125.4), Scheduled Caste (263.2) of H P as given in Table 4.42.

Under Five Mortality Rate

The under 5 mortality rate is the number of deaths of children under five years of age per thousand live births. It is another important measure of mortality and principal indicator of human as well as economic progress and the state of nation's children. In low mortality populations, mortality risks between age 1 year and 5 years are also low and the infant mortality rate can be used as a reasonable single index of child mortality but in populations with higher mortality, relatively high proportions of deaths occur between age 1 year and 5 years as well as and so the under five mortality rate is a more important indeed of overall child mortality, although the infant mortality rate remains a crucial measure. The Lohar (277.78) shows the highest under five mortality rate followed by Badhi (200.00), Kanet (178.57) and Koli (152.54) as shown in Table 4.41. The total (pooled) under five mortality rate is 183.01. The higher under five mortality rate can be attributed to malnutrition, harsh environmental conditions, negligence towards child care, non availability of doctors etc.

Table 4.41: Under Five Mortality Rate

Ethnic Group	*Under Five Mortality Rate*
Kanet	178.57
Koli	152.54
Badhi	200.00
Lohar	277.78
Total (Pooled)	**183.01**

It is found to be higher than under 5 mortality rate of population groups Mizo (31.91), Khasi (57.96), Garo (79.01) of Meghalaya; Arghans (153.9) of Jammu and Kashmir; Himachal Pradesh (42.2); India (94.9) and lower than Bodhs (195.1), Baltis (244.9), Brokas (300.0) of Jammu and Kashmir; Rajis (724.1), Johar Bhotia (350) of Uttaranchal as given in below mentioned table. The Crude Death Rate, Infant Mortality Rate, Under 5 Mortality Rate, Neonatal Mortality Rate and Post Neonatal Mortality Rate of Tribal and Non Tribal Population Groups of India and Himalayan Region is given in Table 4.42.

Table 4.42: Crude Death Rate (CDR), Infant Mortality Rate (IMR), Under 5 Mortality Rate (Under 5 MR), Neonatal Mortality Rate and Post Neonatal Mortality Rate of Tribal and Non Tribal Population Group of India and Himalayan Region

Region/ State	*Population Groups*	*CDR*	*IMR*	*Under 5-MR*	*Neonatal Mortality*	*Post Neonatal Mortality*	*Source*
Eastern Himalayan							
Sikkim							
	Buddhist						Bhasin & Bhasin (2000)
	Hindus	19.8	177.8		66.7	111.1	Bhasin & Bhasin (2000)
	Total	17.8	70.2		35.1	35.1	Bhasin & Bhasin (2000)
Manipur							
	Mao	5.6	23				Maheo (1999)

(Table 4.42 Contd...)

Region/ State	Population Groups	CDR	IMR	Under 5-MR	Neonatal Mortality	Post Neonatal Mortality	Source
West Bengal							
Kalimpong	Sherpa						Gupta et al (1989)
Kalimpong	Lepcha		6.58				Gupta et al (1989)
Rango	Sherpa		6.62				Gupta et al (1989)
Echhay	Sherpa		13.83				Gupta et al (1989)
Lava	Sherpa		6.08				Gupta et al (1989)
Munsong	Sherpa		6.81				Gupta et al (1989)
Labbah (Darjeeling)	Sherpa		6.61				Gupta et al (1989)
Mungpoo (Darjeeling)	Sherpa		4.51				Gupta et al (1989)
	Total		6.14				SRS (2002)
Meghalaya							
	Khasi			57.96	17.2	24.8	Adak (2001)
	Garo			79.01	33.9	18.1	Adak (2001)
	Mizo			31.91	3.5	19.5	Adak (2001)

(Table 4.42 Contd...)

Region/ State	Population Groups	CDR	IMR	Under 5-MR	Neonatal Mortality	Post Neonatal Mortality	Source
Western Himalayan							
Jammu and Kashmir	Ladakhs	15.7	221.2	132.8			Bhasin & Nag (2002)
	Bodhs	14.3	97.6	195.1	24.4	73.2	Bhasin & Nag (2002)
	Baltis	16.8	163.3	244.9	81.6	81.6	Bhasin & Nag (2002)
	Brokpas	21.7	200	300	100	100	Bhasin & Nag (2002)
	Arghuns	13.2	76.9	153.9		76.9	Bhasin & Nag (2002)
	Total	6.2	50				SRS (2002)
Himachal Pradesh							
	Brahman	12.9	125.4		75.8	40.2	Bhasin & Bhasin (1993)
	Rajput	11.8	113.3		69.0	36.6	Bhasin & Bhasin (1993)
	S. Caste	26	263.2		94.4	112.2	Bhasin & Bhasin (1993)
	Total	8.3	34.4	42.4	22.1	12.3	1998-99 NFHS-2 (2000)
	Total	7.2		60			SRS (2002)

(Table 4.42 Contd...)

Region/ State	Population Groups	CDR	IMR	Under 5-MR	Neonatal Mortality	Post Neonatal Mortality	Source
Kinnaur							
	Kanet	16.7	53.57	178.57	53.57	53.57	Present Study
	Koli	14.78	67.8	152.54	50.85	50.85	Present Study
	Badhi	22.82	100	200	100	100	Present Study
	Lohar	21.35	111.11	277.78	111.1	166.67	Present Study
	Total	17.09	71.9	183.01	65.36	71.9	Present Study

(Table 4.42 Contd...)

Region/ State	Population Groups	CDR	IMR	Under 5-MR	Neonatal Mortality	Post Neonatal Mortality	Source
Central Himalayan							
Uttaranchal		6.9	50				SRS (2002)
	Johari Bhotia	12.79	152.8				Kapoor (1996)
	Rang Bhotia	16.11	212				Kapoor (1996)
	Marchha Bhotia	13.12	158.7				Kapoor (1996)
	Marchha Bhotia	8.22	142.8				Chachra & Bhasin (1998)
	Dharchula Bhotia	18.12	120	240			Chachra & Bhasin (1998)
	Johar Bhotia	10.01	150	350			Chachra & Bhasin (1998)
	Raji	29	192				Samal et al (2000)
	Raji	47.6	344.8	724.1			Patra (2001)
	Jaunsaris	11.85	81				Kshatriya et al (1997)
India			67.6	94.9			1998-99 NFHS-2 (2000)

REPRODUCTIVE AND CHILD HEALTH CARE

Attention to Diet

It has been observed that during pregnancy, the pregnant women of all the four ethnic groups are provided with nutritious diet including green vegetables, fruits, milk etc. It has been reported that pregnant women are provided with IFA Tablets by the ANM/ Doctors.

Consultation When Suspect Pregnancy

It has been observed that in total (pooled), 87.2 per cent of women consult somebody when they suspect pregnancy. Further, majority of women (38.7 per cent) consulted their mother-in law in the family followed by consultation with any other (sister, sister in law, neighbour etc) (23.0 per cent), LHV/ANM (21.5 per cent) as shown in Table 4.43.

Table 4.43: Consultation When Suspect Pregnancy

	Kanet		*Koli*		*Badhi*		*Lohar*		*Total (Pooled)*	
	No.	*%age*	*No.*	*%age*	*No.*	*%age*	*No.*	*%age*	*No.*	*%age*
Yes	245	89.1	264	88.0	73	85.9	66	79.5	648	87.2
No	30	10.9	36	12.0	12	14.1	17	20.5	95	12.8
Total	**275**	**100.0**	**300**	**100.0**	**85**	**100.0**	**83**	**100.0**	**743**	**100.0**

If yes, whom did you consult?

	Kanet		*Koli*		*Badhi*		*Lohar*		*Total (Pooled)*	
	No.	*%age*	*No.*	*%age*	*No.*	*%age*	*No.*	*%age*	*No.*	*%age*
Mother in law	87	35.5	114	43.2	27	37.0	23	34.8	251	38.7
Husband	7	2.9	8	3.0	6	8.2	3	4.5	24	3.7
LHV/ ANM	49	20.0	63	23.9	15	20.5	12	18.2	139	21.5
TBA	34	13.9	27	10.2	9	12.3	15	22.7	85	13.1
Any other	68	27.8	52	19.7	16	21.9	13	19.7	149	23.0
Total	**245**	**100.0**	**264**	**100.0**	**73**	**100.0**	**66**	**100.0**	**648**	**100.0**

Status of ANC and Place for ANC during Last Pregnancy

Ante Natal Care (ANC) refers to pregnancy related health care provided by doctors or a health worker in medical facility or at home. The safe motherhood initiative proclaims that all pregnant women must receive basic but professional antenatal care. The highest number of women had gone for ANC are from Kanet (96.4 per cent) followed by Koli (96.3 per cent), Badhi (89.4 per cent) and Lohar (85.5 per cent) as shown in Table 4.44.

Table 4.44: Status of ANC and Place for ANC during Last Pregnancy

Status of ANC	*Kanet*		*Koli*		*Badhi*		*Lohar*		*Total (Pooled)*	
	No.	*%age*	*No.*	*%age*	*No.*	*%age*	*No.*	*%age*	*No.*	*%age*
Yes	265	96.4	289	96.3	76	89.4	71	85.5	701	94.3
No	10	3.6	11	3.7	9	10.6	12	14.5	42	5.7
Total	**275**	**100.0**	**300**	**100.0**	**85**	**100.0**	**83**	**100.0**	**743**	**100.0**

It has been reported that maximum number of women had visited local dai (43.7 per cent) for ANC followed by visit to PHC/ SC (38.0 per cent). The lack of awareness and poor transportation facilities are main reason for irregular ANC at Sub Centre/Primary Health Centre. Few of women think that pregnancy is natural phenomenon and did not require any special care. Still due to high education level and awareness among the people, they prefer to visit nearby health centre for ANC. It has been observed that maximum number of women from Kanet (78.9 per cent) followed by Koli (65.7 per cent), Badhi (46.1 per cent) and Lohar (38.0 per cent) visited nearby SC/PHC as shown in Table 4.45 *(See on next page)*.

Visited and Advised for ANC Check up, Immunization, IFA Tablets

In the present study, majority of women of all ethnic groups are visited by ANM/ VHW and advised for ANC Check up, immunization and IFA Tablets during the pregnancy. The highest number of women visited by ANM are from Koli (74.0 per cent) followed by Badhi (72.9 per cent), Kanet (67.6 per cent) and Lohar (66.3 per cent) as shown in Table 4.46.

Table 4.45: Place for ANC during Last Pregnancy

Place for ANC	*Kanet*		*Koli*		*Badhi*		*Lohar*		*Total (Pooled)*	
	No.	*%age*	*No.*	*%age*	*No.*	*%age*	*No.*	*%age*	*No.*	*%age*
PHC/SC	209	78.9	190	65.7	35	46.1	27	38.0	461	65.8
Pvt. doctors	39	14.7	29	10.0	15	19.7	13	18.3	96	13.7
Local Dai	17	6.4	70	24.2	26	34.2	31	43.7	144	20.5
Total	**265**	**100.0**	**289**	**100.0**	**76**	**100.0**	**71**	**100.0**	**701**	**100.0**

Table 4.46: Visited and Advised for ANC Check up, Immunization, IFA tablets

Visited and Advised for ANC Check up, Immunization, IFA tab	*Kanet*		*Koli*		*Badhi*		*Lohar*		*Total (Pooled)*	
	No.	*%age*	*No.*	*%age*	*No.*	*%age*	*No.*	*%age*	*No.*	*%age*
No Body	56	20.4	41	13.7	12	14.1	9	10.8	118	15.9
ANM/VHW	186	67.6	222	74.0	62	72.9	55	66.3	525	70.7
TBA	12	4.4	19	6.3	10	11.8	18	21.7	59	7.9
Any Other	21	7.6	18	6.0	1	1.2	1	1.2	41	5.5
Total	**275**	**100.0**	**300**	**100.0**	**85**	**100.0**	**83**	**100.0**	**743**	**100.0**

Immunization with Two Doses of TT during the Pregnancy

In the present study, majority (more than 97 per cent) of women of all ethnic groups immunised for two doses of TT during the pregnancy. This high frequency of immunization for two doses of TT during the pregnancy could be due to high awareness level among the community people and better out reach of PHC staff as shown in Table 4.47.

Table 4.47: Immunization with Two Doses of TT during the Pregnancy

Immunization with 2 Doses of TT during Pregnancy	*Kanet*		*Koli*		*Badhi*		*Lohar*		*Total (Pooled)*	
	No.	*%age*	*No.*	*%age*	*No.*	*%age*	*No.*	*%age*	*No.*	*%age*
Yes	272	98.9	292	97.3	80	94.1	78	94.0	722	97.2
No	3	1.1	8	2.7	5	5.9	5	6.0	21	2.8
Total	**275**	**100.0**	**300**	**100.0**	**85**	**100.0**	**83**	**100.0**	**743**	**100.0**

In the present study, majority of people (48.0 per cent) of all ethnic groups consult PHC staff (i.e ANM, Doctor, Nurses) for problem during the pregnancy. This high frequency of consulting PHC staff could be due to high awareness level among the community people and better out reach of PHC staff as shown in Table 4.48.

Table 4.48: Person Consulted for Any Problem during Pregnancy

Person Consulted for Problem during Pregnancy	*Kanet*		*Koli*		*Badhi*		*Lohar*		*Total (Pooled)*	
	No.	*%age*	*No.*	*%age*	*No.*	*%age*	*No.*	*%age*	*No.*	*%age*
Nobody	45	16.4	34	11.3	7	8.2	6	7.2	92	12.4
PHC Staff	130	47.3	146	48.7	42	49.4	39	47.0	357	48.0
TBA	69	25.1	87	29.0	23	27.1	28	33.7	207	27.9
Tribal practitioner	12	4.4	9	3.0	8	9.4	6	7.2	35	4.7
Any other	19	6.9	24	8.0	5	5.9	4	4.8	52	7.0
Total	**275**	**100.0**	**300**	**100.0**	**85**	**100.0**	**83**	**100.0**	**743**	**100.0**

Place of Last Delivery

In present study majority of last deliveries (80.5 per cent) are conducted at their respective houses only. The highest number of deliveries at home are conducted in Lohar (95.2 per cent) followed by Badhi (91.8 per cent), Koli (77.7 per cent) and Kanet (75.6 per cent) as shown in Table 4.49. The reason for maximum number of deliveries at home in Lohar and Badhi ethnic group is due to lack of awareness about the importance institutional deliveries, non availability of health staff, far away distance of health institutions, lack of transportation facilities etc.

Table 4.49: Place of Last Delivery

Place for Delivery	*Kanet*		*Koli*		*Badhi*		*Lohar*		*Total (Pooled)*	
	No.	*%age*	*No.*	*%age*	*No.*	*%age*	*No.*	*%age*	*No.*	*%age*
At home	208	75.6	233	77.7	78	91.8	79	95.2	598	80.5
At PHC	45	16.4	33	11.0	5	5.9	3	3.6	86	11.6
Dist Hospital	18	6.5	29	9.7	1	1.2	1	1.2	49	6.6
Pvt Hospital	4	1.5	5	1.7	1	1.2		0.0	10	1.3
Total	**275**	**100.0**	**300**	**100.0**	**85**	**100.0**	**83**	**100.0**	**743**	**100.0**

Person who Conducted the Delivery

It has been observed that majority of deliveries are conducted by TBA (68.9 per cent) followed by ANM(18.4 per cent), doctors (9.3 per cent), Elderly women in the village(4.4 per cent), untrained TBA (2.3 per cent) and Family member (1.6 per cent) as shown in Table 4.50. Though ANM's services are easily available most of the deliveries take place at home and that too under the guidance of TBA, elderly women and Dai (traditional) who have acquired their professional knowledge through family tradition. There is a provision of arranging training of such traditional Dai in safe delivery and provision of giving them a kit for the same. However, in most cases, the traditional dais are not trained and seldom do they have proper equipments for safe delivery. The round the clock availability of TBA in far away villages is one of main reason for maximum deliveries conducted by TBA.

Table 4.50: Person who Conducted the Delivery

Person who conducted Delivery	*Kanet*		*Koli*		*Badhi*		*Lohar*		*Total (Pooled)*	
	No.	*%age*	*No.*	*%age*	*No.*	*%age*	*No.*	*%age*	*No.*	*%age*
TBA	189	68.7	175	58.3	59	69.4	52	62.7	475	63.9
Untrained TBA	4	1.5	6	2.0	2	2.4	5	6.0	17	2.3
Elderly lady in the village	12	4.4	16	5.3	3	3.5	2	2.4	33	4.4
Family Member	3	1.1	2	0.7	3	3.5	4	4.8	12	1.6
ANM	42	15.3	62	20.7	15	17.6	18	21.7	137	18.4
Doctors	25	9.1	39	13.0	3	3.5	2	2.4	69	9.3
Total	**275**	**100.0**	**300**	**100.0**	**85**	**100.0**	**83**	**100.0**	**743**	**100.0**

Place for Child Immunization

As part of National Health Programme, immunization is given due emphasise. As the mandate of the programme, all children under the age of six years are to be immunizsed against diphtheria, whooping cough, tetanus, poliomyelitis, measles and tuberculosis. This programme is carried out all government hospitals, CHC and PHC. In the present study, health worker visited their home for immunisation of children. 73.5 per cent of Lohar had immunised their children at their home by health worker followed by 65.9 per cent of Badhi, 58.7 per cent of Koli and 56.7 per cent of Kanet as shown in Table 4.51. It has been observed that due to lack of convergence between ICDS and Health functionaries, lack of awareness about the immunization and far away distance of health institutions, the immunization is generally carried out at home.

Table 4.51: Place for Child Immunization

Place for Child Immunization	*Kanet*		*Koli*		*Badhi*		*Lohar*		*Total (Pooled)*	
	No.	*%age*	*No.*	*%age*	*No.*	*%age*	*No.*	*%age*	*No.*	*%age*
PHC	39	14.2	32	10.7	3	3.5	2	2.4	76	10.2
CHC	12	4.4	9	3.0	4	4.7	3	3.6	28	3.8
Distt. Hospital	56	20.4	63	21.0	4	4.7	2	2.4	125	16.8
Health Worker visited home	156	56.7	176	58.7	56	65.9	61	73.5	449	60.4
Anagan-wadi	12	4.4	20	6.7	18	21.2	15	18.1	65	8.7
Total	**275**	**100.0**	**300**	**100.0**	**85**	**100.0**	**83**	**100.0**	**743**	**100.0**

FAMILY PLANNING

Family planning can favourably influence the health development and well being of the family and has striking impact on the health of mother and children. The health benefits of family planning results from avoidance of unwanted pregnancy, a change in the total number of children born to mother, achievement of an optimum interval between pregnancies and changes in the time at which birth occur particular the first and the least in relation to the age of the parents and especially that of the mother. The successful limitation of family size by the couples depends not only on their small family ideals but also on their psychological acceptance of family limitation, knowledge of birth control methods, availability of contraceptives, psychological and economic costs and environemental favourable to the practice of birth control. Certain motivational aspects moderated through social and cultural consideration are presumed to curtail fertility.

Knowledge about Family Planning Methods

The knowledge about family planning methods is found to be universal among all ethnic groups. The Knowledge of family planning methods has attained the threshold level of social acceptability and has been disseminated successfully by the government to meet the needs of eligible women and men.

Knowledge about Family Planning Methods

In the present study, the most of people had replied with multiple choices about knowledge of family planning method i.e Natural–coitus interrupts/abstinence, IUD (intra uterine device), Oral pills, Sterilization TT/VT, Nirodh. In majority, Lohar (22.9 per cent) followed by Badhi (21.2 per cent), Kanet (20.4 per cent), and Koli (20.3 per cent) had knowledge about Natural-coitus interrupts/abstinence as shown in Table 4.52. It has been observed that 37.3 per cent of people have knowledge about more than 2 family planning methods. The anganwadi workers and ANM play an important role in imparting knowledge about family planning methods.

Table 4.52: Knowledge About Family Planning Methods

Knowledge about Family Planning Method	*Kanet*		*Koli*		*Badhi*		*Lohar*		*Total (Pooled)*	
	No.	*%age*	*No.*	*%age*	*No.*	*%age*	*No.*	*%age*	*No.*	*%age*
Natural-coitus interrupts/ abstinence	56	20.4	61	20.3	18	21.2	19	22.9	154	20.7
IUD (intra uterine device)	24	8.7	31	10.3	7	8.2	4	4.8	66	8.9
Oral pills	12	4.4	17	5.7	9	10.6	5	6.0	43	5.8
Sterilization TT/VT	21	7.6	23	7.7	10	11.8	8	9.6	62	8.3
Nirodh	34	12.4	37	12.3	12	14.1	16	19.3	99	13.3
More than 2	128	46.5	131	43.7	29	34.1	31	37.3	319	42.9
Total	**275**	**100.0**	**300**	**100.0**	**85**	**100.0**	**83**	**100.0**	**743**	**100.0**

Ever Seen any Material on Family Planning

To assess the influence of mass media propagation on family planning, the most of people had replied with multiple choices about seeing any material on family planning method i.e posters, radio, TV etc. Still Badhi (21.2 per cent) followed by Lohar (19.3 per cent), Koli (17.3 per cent), and Kanet (17.1 per cent) had seen family planning method from Television. Only 5.9 per cent of Lohar, 3.0 per cent of Badhi, 4.3 per cent of Kanet and 4.0 per cent of Koli had never heard about any material related to family planning as shown in Table 4.53. In total, 30.1 per cent of people have seen 2 or more than 2 material related to family planning methods. The Information Education and Communication activities carried out various government and non government agencies is main source of family planning related material in the villages. At the same time, Anganwadi workers and ANM plays an important role in displaying the family planning material at their respective centres.

Table 4.53: Ever Seen Any Material on Family Planning

Ever Seen any Material on Family Planning	*Kanet*		*Koli*		*Badhi*		*Lohar*		*Total (Pooled)*	
	No.	*%age*	*No.*	*%age*	*No.*	*%age*	*No.*	*%age*	*No.*	*%age*
No	11	4.0	9	3.0	5	5.9	7	8.4	32	4.3
Posters	28	10.2	30	10.0	17	20.0	24	28.9	99	13.3
Radio	19	6.9	23	7.7	13	15.3	11	13.3	66	8.9
TV	47	17.1	52	17.3	18	21.2	16	19.3	133	17.9
2 or more than 2	170	61.8	186	62.0	32	37.6	25	30.1	413	55.6
Total	**275**	**100.0**	**300**	**100.0**	**85**	**100.0**	**83**	**100.0**	**743**	**100.0**

Discussion about the Family Planning Methods

In the present study, the most of people had replied with multiple choice about discussion about family planning method with spouse, village dai, PHC staff, neighbour, relative and friends. Still Badhi (28.2 per cent) followed by Lohar (25.3 per cent), Koli (17.3 per cent), and Kanet (16.4 per cent) discussed with their spouse about the family planning methods as shown in Table 4.54. In total, 36.1 per cent of people discuses about family planning methods with 2 more than 2 individuals.

Table 4.54: Discussion about the Family Planning Methods

Discussion about Family Planning Method	*Kanet*		*Koli*		*Badhi*		*Lohar*		*Total (Pooled)*	
	No.	*%age*	*No.*	*%age*	*No.*	*%age*	*No.*	*%age*	*No.*	*%age*
Spouse	45	16.4	52	17.3	24	28.2	21	25.3	142	19.1
Village Dai	2	0.7	6	2.0	3	3.5	3	3.6	14	1.9
PHC Staff	36	13.1	31	10.3	13	15.3	16	19.3	96	12.9
Neighbor/ friends/ relatives	3	1.1	20	6.7	11	12.9	13	15.7	47	6.3
2 or more than 2	189	68.7	191	63.7	34	40.0	30	36.1	444	59.8
Total	**275**	**100.0**	**300**	**100.0**	**85**	**100.0**	**83**	**100.0**	**743**	**100.0**

Attitude Towards Family Planning

Attitude Towards Restricting the Number of Children and Adopting Small Family Norm

In the present study, in all ethnic groups had positive attitude towards restricting the number of children and adopting small family norms. In Koli (89.0 per cent) followed by Kanet (88.7 per cent), Badhi (69.4 per cent) and Lohar (66.3 per cent) have positive attitude towards restricting the number of children and adopting small family norms as shown in Table 4.55. In total, 84.1 per cent have positive attitude towards restricting the number of children and adopting small family norm. The reasons for positive attitude towards restricting the number of children and adopting small family norm are high education level, knowledge about importance of small family size, low socio-economic conditions etc.

Table 4.55: Attitude Towards Restricting the Number of Children and Adopting Small Family Norm

Attitude Towards Restricting the Number of Children and Adopting Small Family Norm	*Kanet*		*Koli*		*Badhi*		*Lohar*		*Total (Pooled)*	
	No.	*%age*	*No.*	*%age*	*No.*	*%age*	*No.*	*%age*	*No.*	*%age*
Yes	244	88.7	267	89.0	59	69.4	55	66.3	625	84.1
No	31	11.3	33	11.0	26	30.6	28	33.7	118	15.9
Total	**275**	**100.0**	**300**	**100.0**	**85**	**100.0**	**83**	**100.0**	**743**	**100.0**

Believe that Children Are?

The ethnic groups consider sterility as a misfortune and disfavour the ideas of limiting the size of the family. However, among the young generation, some favour this idea. They regard children as gift of God. Fear of high mortality may be the other reason of their willingness to restrict the number of children. The ideas behind reproduction are:

(a) For the perpetuity of the family as well as the continuity of race;

(b) For the continuity of parents name after death;

(c) To help and make take care of parents in old age.

In the present study, the most of people had replied with multiple choice for their believe in children as good's gift or they are helping hand in the family or necessary for looking after in old age. Still think that Kanet (20.4 per cent) followed by Koli (15.0 per cent), Badhi (12.9 per cent) and Lohar (9.6 per cent) thinks that children are necessary for looking after in old age as shown in Table 4.56.

Table 4.56: Believe that Children Are

Believe that Children Are	*Kanet*		*Koli*		*Badhi*		*Lohar*		*Total (Pooled)*	
	No.	*%age*	*No.*	*%age*	*No.*	*%age*	*No.*	*%age*	*No.*	*%age*
God's Gift	26	9.5	21	7.0	9	10.6	12	14.5	68	9.2
They are helping hand in the family	35	12.7	57	19.0	18	21.2	16	19.3	126	17.0
Necessary for looking after in old age	56	20.4	45	15.0	11	12.9	8	9.6	120	16.2
Multiple choice more than one	147	53.5	171	57.0	47	55.3	45	54.2	410	55.2
Other	11	4.0	6	2.0		0.0	2	2.4	19	2.6
Total	**275**	**100.0**	**300**	**100.0**	**85**	**100.0**	**83**	**100.0**	**743**	**100.0**

Person Who Takes Decision for Contraceptive Practices

In the present study, in all ethnic groups generally husband and women (both) (53.3 per cent) take decision for adopting any contraceptive. In Kanet (57.8 per cent) followed by Koli (57.3 per cent), Lohar (43.4 per cent) and Badhi (34.1 per cent) husband and wife both participate in decision making process for contraceptive usage as shown in Table 4.57 *(See on next page)*. The joint decision of husband and wife for adopting the contraceptive practices can be attributed to the equal importance to women in family planning.

Place for Availing Family Planning Services

In the present study, most of women in all ethnic groups are availing family planning services from PHC staff (85.3 per cent). The highest number of women availing family planning services from PHC staff are from Koli (87.0 per cent) followed by Kanet (86.9 per cent), Badhi (80 per cent) and Lohar (79.5 per cent) as shown in Table 4.58. The reasons for PHC staff as source of family planning services are extensive outreach of PHC in the district, absence of private health institutions, far away distance of district hospital etc.

Table 4.57: Person who Takes Decision for Contraceptive Practices

Person who Takes Decisions for Contraceptive	*Kanet*		*Koli*		*Badhi*		*Lohar*		*Total (Pooled)*	
	No.	*%age*	*No.*	*%age*	*No.*	*%age*	*No.*	*%age*	*No.*	*%age*
Husband	69	25.1	78	26.0	27	31.8	23	27.7	197	26.5
Wife	26	9.5	21	7.0	21	24.7	21	25.3	89	12.0
Both	159	57.8	172	57.3	29	34.1	36	43.4	396	53.3
Elderly female	21	7.6	29	9.7	8	9.4	3	3.6	61	8.2
Total	**275**	**100.0**	**300**	**100.0**	**85**	**100.0**	**83**	**100.0**	**743**	**100.0**

Table 4.58: Place For Availing Family Planning Services

Place for Availing Family Planning Services	*Kanet*		*Koli*		*Badhi*		*Lohar*		*Total (Pooled)*	
	No.	*%age*	*No.*	*%age*	*No.*	*%age*	*No.*	*%age*	*No.*	*%age*
Traditional Practitioner	4	1.5	10	3.3	4	4.7	6	7.2	24	3.2
PHC staff	239	86.9	261	87.0	68	80.0	66	79.5	634	85.3
District Hospital	22	8.0	29	9.7	11	12.9	9	10.8	71	9.6
Pvt. Practitioner	10	3.6		0.0	2	2.4	2	2.4	14	1.9
Total	**275**	**100.0**	**300**	**100.0**	**85**	**100.0**	**83**	**100.0**	**743**	**100.0**

Family Planning Practices

The practice of family planning method depends on the knowledge of contraceptives techniques and the attitude in their social environment towards birth control.

Women Using any Family Planning Method Currently

In the present study, most of women in all ethnic groups are using any family planning method currently. The highest number of women using any family planning method currently are from Kanet (64.7 per cent) followed by Koli (62.3 per cent), Lohar (55.4 per cent) and Badhi (51.8 per cent) as shown in Table 4.59. Male female differences imply that the burdens of fertility control are born by women but this is a function of contemporary contraceptive technology where the most of the efficient method are geared to women. Generally the husband s try to avoid to use contraceptive devices whereas women are enforced to adopt it because the low status in the family.

Table 4.59: Women Using any Family Planning Method Currently

Women Using any Family Planning Method Currently	*Kanet*		*Koli*		*Badhi*		*Lohar*		*Total (Pooled)*	
	No.	*%age*	*No.*	*%age*	*No.*	*%age*	*No.*	*%age*	*No.*	*%age*
Yes	178	64.7	187	62.3	44	51.8	46	55.4	455	61.2
No	97	35.3	113	37.7	41	48.2	37	44.6	288	38.8
Total	**275**	**100.0**	**300**	**100.0**	**85**	**100.0**	**83**	**100.0**	**743**	**100.0**

Motivation for Family Planning Method

In the present study, majority of women of all ethnic groups are motivated by medical person (ANM/ Doctors) (36.6 per cent) for using family planning method due to extensive out reach of PHC/Sub Centre in the district, availability of technical knowledge about various family planning method and main source of family planning methods. The highest number of women motivated by medical person are from Kanet (39.3 per cent) followed by Badhi (41.2 per cent), Koli (34.0 per cent) and Lohar (32.5 per cent) as shown in Table 4.60. It has been reported that husband and other family members also play a crucial role in motivating for family planning method. After acquiring the knowledge about method, the eligible women discuse it with her husband and other member of the family to learn about their experience and to seek their advice on suitability of family planning method.

Table 4.60: Motivation for Family Planning Method

Motivation for Family Planning Method	*Kanet*		*Koli*		*Badhi*		*Lohar*		*Total (Pooled)*	
	No.	*%age*	*No.*	*%age*	*No.*	*%age*	*No.*	*%age*	*No.*	*%age*
Medical Person (ANM/ Doctors etc.)	108	39.3	102	34.0	35	41.2	27	32.5	272	36.6
Husband	85	30.9	95	31.7	21	24.7	19	22.9	220	29.6
Family Member	82	29.8	103	34.3	29	34.1	37	44.6	251	33.8
Total	**275**	**100.0**	**300**	**100.0**	**85**	**100.0**	**83**	**100.0**	**743**	**100.0**

Satisfaction Level with Family Planning Method

In the present study, majority of women (49.7 per cent) of all ethnic groups are somewhat satisfied with family planning method (which they are using or used in the past). The highest number of women satisfied with family planning method are from Koli (40.3 per cent) followed by Kanet (39.6 per cent), Badhi (37.6 per cent) and Lohar (33.7 per cent) as shown in Table 4.61. The reasons for not completely satisfaction and dissatisfaction with family planning method are non availability/irregular supply of contraceptives, side affects associated with contraceptive, myths associated with usage of contraceptives etc.

Table 4.61: Satisfaction Level with Family Planning Method

Satisfaction Level with Family Planning Method	*Kanet*		*Koli*		*Badhi*		*Lohar*		*Total (Pooled)*	
	No.	*%age*	*No.*	*%age*	*No.*	*%age*	*No.*	*%age*	*No.*	*%age*
Satisfied	109	39.6	121	40.3	32	37.6	28	33.7	290	39.0
Somewhat satisfied	135	49.1	147	49.0	44	51.8	43	51.8	369	49.7
Dissatisfied	31	11.3	32	10.7	9	10.6	12	14.5	84	11.3
Total	**275**	**100.0**	**300**	**100.0**	**85**	**100.0**	**83**	**100.0**	**743**	**100.0**

HEALTH SEEKING BEHAVIOUR

It has been reported that illed Kinnauris with better socio-economic condition and education, consult doctors for their ailment as they perceive that modern clinical services are much effective and cure them in shorter time period. In case of un satisfaction with modern health services, people shift to other magico-religious practices in consultation with village temple devta or herbal medicines in consultation with local herbalist.

Medical Attention Given in Case of Illness

The maximum medical attention given by government doctors as reported by Kanet (80.4 per cent) followed by Koli (79.7 per cent), Lohar (74.7 per cent) and Badhi (68.2 per cent) as shown in Table 4.62. The minor illness among the people is ignored initially and not given proper attention. But if the suffering due to illness increases, the people contact nearby health institutions depending on the condition of patients.

Table 4.62: Medical Attention Given in Case of Illness

Medical Attention given in case of Illness	*Kanet*		*Koli*		*Badhi*		*Lohar*		*Total (Pooled)*	
	No.	*%age*	*No.*	*%age*	*No.*	*%age*	*No.*	*%age*	*No.*	*%age*
Tribal Healer	26	9.5	32	10.7	9	10.6	8	9.6	75	10.1
Pvt. Doctors	28	10.2	29	9.7	18	21.2	13	15.7	88	11.8
Govt.	221	80.4	239	79.7	58	68.2	62	74.7	580	78.1
Total	**275**	**100.0**	**300**	**100.0**	**85**	**100.0**	**83**	**100.0**	**743**	**100.0**

Medical Facilities Available in the Village

In the present study, the Kanet (85.1 per cent) followed by Koli (82.0 per cent), Lohar (80.7 per cent) and Badhi (74.1 per cent) responded positive about the medical facilities available in the village as shown in Table 4.63. It has been reported that due to far away village, unapproachability of few of the villages, the medical facilities in terms of presence of PHC/SC in their respectively village is not feasible due to high infrastructural cost.

Table 4.63: Medical Facilities Available in the Village

Medical Facilities Available in the Village	*Kanet*		*Koli*		*Badhi*		*Lohar*		*Total (Pooled)*	
	No.	*%age*	*No.*	*%age*	*No.*	*%age*	*No.*	*%age*	*No.*	*%age*
Yes	234	85.1	246	82.0	63	74.1	67	80.7	610	82.1
No	41	14.9	54	18.0	22	25.9	16	19.3	133	17.9
Total	**275**	**100.0**	**300**	**100.0**	**85**	**100.0**	**83**	**100.0**	**743**	**100.0**

Table 4.64: Frequency of Immunization Programme Carried Out in the Village

Frequency of Immunization Programme Carried Out in the Village	*Kanet*		*Koli*		*Badhi*		*Lohar*		*Total (Pooled)*	
	No.	*%age*	*No.*	*%age*	*No.*	*%age*	*No.*	*%age*	*No.*	*%age*
Once in month	163	59.3	151	50.3	13	15.3	15	18.1	342	46.0
Once in 2 month	101	36.7	89	29.7	30	35.3	26	31.3	246	33.1
Once in 3 month		0.0	43	14.3	26	30.6	29	34.9	98	13.2
Thrice in year	10	3.6	15	5.0	14	16.5	12	14.5	51	6.9
Once in year	1	0.4	2	0.7	2	2.4	1	1.2	6	0.8
Total	**275**	**100.0**	**300**	**100.0**	**85**	**100.0**	**83**	**100.0**	**743**	**100.0**

Frequency of Immunization Programme Carried Out in the Village

In the present study, the Kanet (59.3 per cent) followed by Koli (50.3 per cent), Lohar (18.1 per cent) and Badhi (15.3 per cent) responded that the immunization of programme is carried out once in month in the village as shown in Table 4.64. Being a part of national health programme, immunization is essentially to be carried out regular. It has been reported that due to non availability of ANMs, far away places and wider geographical coverage, the frequency of immunization carried out in the village is less. In such a condition, parents have to take their children to nearby Hospital/ CHC/PHC for immunisation of their children.

Adequacy of Medical Facilities

In the present study, most of the respondent replied complained about inadequacy of medical facilities in nearby health institutions. In Koli (55.7 per cent) followed Badhi (54.1 per cent), Kanet (48.0 per cent) and Lohar (42.2 per cent) respondent with inadequacy of medical facilities as shown in Table 4.65. The adequacy of medical facilities has been reported in the form of non availability of medical officer, ANM, non availability of medicines and other essential health services, non friendly attitude of the doctors etc.

Table 4.65: Adequacy of Medical Facilities

Adequacy of Medical Facilities	*Kanet*		*Koli*		*Badhi*		*Lohar*		*Total (Pooled)*	
	No.	*%age*	*No.*	*%age*	*No.*	*%age*	*No.*	*%age*	*No.*	*%age*
Yes	143	52.0	133	44.3	39	45.9	48	57.8	363	48.9
No	132	48.0	167	55.7	46	54.1	35	42.2	380	51.1
Total	**275**	**100.0**	**300**	**100.0**	**85**	**100.0**	**83**	**100.0**	**743**	**100.0**

Regular Visit of Dai/ANM/Doctors

In the present study, the regular visit to Dai/ANM/Doctors are made by Kanet (59.3 per cent) followed by Koli (51.3 per cent), Badhi (47.1 per cent) and Lohar (41.0 per cent) as shown in Table 4.66 *(See on next page)*. It has been reported that patients/ pregnant women does not make regular visit to the Dai/ANM/ Doctors. Only in case of extreme emergency or as per requirement, Dai/ANM/Doctors are visited frequently.

Person who Take Decisions for Consulting Doctor

In the present study, the decision for consulting doctor is taken collectively at household level in Koli (55.7 per cent) followed by Kanet (53.5 per cent), Lohar (42.2 per cent) and Badhi (40.0 per cent) as shown in Table 4.67. It has been reported that consulting a doctor is mostly a collective decision in the family. It can be attributed to high family values, equal participation of each member in the family, unity in the family etc.

Table 4.66: Regular Visit of Dai/ANM/Doctors

Regular Visit to Dai/ ANM/ Doctors	*Kanet*		*Koli*		*Badhi*		*Lohar*		*Total (Pooled)*	
	No.	*%age*	*No.*	*%age*	*No.*	*%age*	*No.*	*%age*	*No.*	*%age*
Yes	163	59.3	154	51.3	40	47.1	34	41.0	391	52.6
No	112	40.7	146	48.7	45	52.9	49	59.0	352	47.4
Total	**275**	**100.0**	**300**	**100.0**	**85**	**100.0**	**83**	**100.0**	**743**	**100.0**

Table 4.67: Person who Take Decisions for Consulting Doctor

Person who Takes Decisions for Consulting Doctors	*Kanet*		*Koli*		*Badhi*		*Lohar*		*Total (Pooled)*	
	No.	*%age*	*No.*	*%age*	*No.*	*%age*	*No.*	*%age*	*No.*	*%age*
Head of the Family	32	11.6	48	16.0	24	28.2	20	24.1	124	16.7
Husband	49	17.8	34	11.3	9	10.6	13	15.7	105	14.1
Elderly male	47	17.1	51	17.0	18	21.2	15	18.1	131	17.6
Collective Decision	147	53.5	167	55.7	34	40.0	35	42.2	383	51.5
Total	**275**	**100.0**	**300**	**100.0**	**85**	**100.0**	**83**	**100.0**	**743**	**100.0**

Ever Visited/Availed Health (Apart from Family Planning) Facilities at PHC/CHC

In the present study, in all ethnic groups, majority of people had visited/availed health facilities at PHC/ CHC. In Badhi (78.8 per cent) followed by Koli (72.7 per cent), Lohar (68.7 per cent) and Kanet (67.3 per cent) had ever visited/availed health facilities at PHC/CHC as shown in Table 4.68. It has been observed that 31.3 per cent of total (pooled) have not ever visited/availed facilities at PHC/CHC. It can be attributed to the far away distance of the PHC/

CHC, any other member of family might have collected the medicines on the behalf of patient, frequent visit of ANM, free medical services during health camp organised in the village etc.

Table 4.68: Ever Visited/Availed Health (apart from Family Planning) Facilities at PHC/CHC

Ever Visited/ Availed Facilities at PHC/CHC	*Kanet*		*Koli*		*Badhi*		*Lohar*		*Total (Pooled)*	
	No.	*%age*	*No.*	*%age*	*No.*	*%age*	*No.*	*%age*	*No.*	*%age*
Yes	185	67.3	218	72.7	67	78.8	57	68.7	527	70.9
No	90	32.7	82	27.3	18	21.2	26	31.3	216	29.1
Total	**275**	**100.0**	**300**	**100.0**	**85**	**100.0**	**83**	**100.0**	**743**	**100.0**

HEALTH DEVELOPMENT AND HEALTH CARE SYSTEM

The different health system are organised under two broad sectors namely, government and private. It is important to note that the two major medical systems, viz modern medicine and Ayurveda are recognised and instituted under the government sector in Kinnaur District. The organisation of the health care services under the government sector is one of two parallel organisational structures: Parallel because there is hardly any meeting point among them in terms of training, research or organisation. The private sector has two parallel systems and number of small traditional systems.

Role of Health Department in Modern Medicine

Centre Level

Ministry of Health and Family Welfare is instrumental and responsible for implementation of various programs on national scale in the areas of Family Welfare and Prevention and Control of major communicable diseases. Apart from these, the ministry also assists the state in preventing and controlling the spread of seasonal disease outbreaks and epidemic through technical assistance.

To tackle the menace of communicable and non communicable diseases, the Department of Health is implementing National Health Programmes, through out the country for Malaria, Tuberculosis, Leprosy, Blindness, AIDS, and Cancer etc.

State Level

The offices of Health and Family Welfare functioning under Ministry of Health and Family Welfare located in the state capitals. These offices have been established to liaison and coordinate various national health and family welfare programmes with the states, monitor the centrally sponsored health and family welfare schemes, provide on the spot technical guidance, test checking of records in the state, maintain the malaria clinic, review and analysis of monthly technical reports, supervise health information and field units and associate in various implementation and grant sanction. Director, Shimla is on the strength of Department of Health and Family Welfare.

At the State Level, the Department of Health and Family Welfare is the apex body to cater to the health services in the whole state. The main function of the Department of Health and Family Welfare is to manage the functioning of the government health institutions/hospitals in the state. Department of Health and Family Welfare is primarily responsible for providing the health services in the state. The Director of Health and Family Welfare is the Chief Executive of the department and is responsible for providing health services. The Director is assisted by the Deputy Directors who are given the responsibility of the various national health programmes.

The Department of Health and Family Welfare, Himachal Pradesh has taken an initiative of developing the Himachal Health Vision 2020. The concept of Vision 2020 for the State of Himachal Pradesh is the brain wave of Prof. Prem Kumar Dhumal, Hon'ble Chief Minister of Himachal Pradesh when he delivered an address at " The Partnership Summit-1999" at Confederation of Indian Industries (CII) Conference at Jaipur and his speech in Himachal Pradesh Vidhan Sabha in the budget session of 1999-2000. He reiterated his resolve to adopt Vision 2020 for the State while presenting the Budget Estimates for 1999-2000 and declared that Himachal Pradesh should be one of the top five States in the country in the next 20 years. According to Vision 2020, the state should attain the status equivalent to that of the developed countries.

The Government of HP consequently issued directions to all the departments to fulfill the budget assurances and to constitute task forces for drafting Vision 2020. The Health and Family Welfare

Department constituted a Task Force comprising eight members for developing Himachal Health Vision 2020. The main objective of preparing the document is to spell out the long term goals with clearly defined strategies and propose concrete actions to achieve Health for All within 20 years through situation analysis of current and future health scenario.

District Level

At district level, the Chief Medical Officer (CMO) is the administrative head to take care of the health needs. Deputy/ Additional CMO is the second person in-charge. Each Deputy CMO is given the responsibility of particular national programmes/zones. District Malaria Officer, District Leprosy Officer and District Immunization Officers look after the concerned programmes in the district. Kinnaur has district hospital where the curative services are provided. In district hospital, there are specialist cares available like paediatric, ENT, Ophthalmology, Obstetrics and Gynaecology, General Surgery and General Medicines.

Rural Health Infrastructure

The Community Health Centers (CHC) are established and maintained by the state government under MNP/ BMS Programme. It is manned by four medical specialists supported by 21 paramedical and other staff. It has 30 indoor beds with X-ray, labour room, operation theatre and laboratory facilities. It serves as a referral center for four PHCs and also provides facilities for obstetric care specialized consultation.

The Primary Health Center (PHC) is the first contact point between the village community and the medical officer. These are established and maintained by the state government under the Minimum Needs/Basic Minimum Services Programme. A PHC is manned by a Medical Officer and is supported by 14 other paramedical and other staff. It acts as a referral unit for six sub centers and has 4-6 beds. The activities of PHCs involve curative, preventive, promotive and family welfare services.

The Sub Centers are the most peripheral contact point between the primary health care system and the community and have mainly promotional and educational function relating to Maternal and Child Health, Family Welfare, Nutrition, Immunization, Diarrhoeal Control and Control of Communicable Diseases Programmes. The

Sub-Centers are also provided with basic drugs for minor ailments needed to take care of essential health needs of women and children. It is manned by one Multi-purpose Worker (Male) and One Multi-purpose worker (Female)/ANM. The ICDS Anganwadi Worker assists the ANM in health services delivery by growth monitoring, maintaining the health records, organizing immunization and health day, collecting and motivating women for immunization etc. She also distributes Iron Folic Acid tablets and Vitamin A solution to the pregnant women and children. Beside she also gives medicines for minor ailments such as fever, ear and eye infection.

To provide health services in the villages, ANM fixes the date of her visit to the village and inform the date of visit to the AWW. Further, the AWW informs the villagers about the visit of ANM. On the appointed day, ANM carries out immunization of pregnant women and children, ANC/PNC, motivates the women for using the family planning methods, etc. ANM provides some medicines like ORS, Tab. IFA, Bleaching powder, Mala-D, Condoms, Tab. Paracetamol to the AWW for distribution to the villagers. (i.e. AWW acts as Depot Holder).

Medicines Supply Logistics in Department of Health and Family Welfare

Budgeting

While estimating the budget for the Health and Family Welfare Department, certain points are considered, which are number of Blocks in the district, SC/ST Population, number of Health Institutions etc. At the district level, the Chief Medical Officer (CMO) prepares the Annual Budget for the District and submits it to the State-Department of Health and Family Welfare in the month of December. Normally the budget for the last year is considered as the basis for estimation of current year's budget. The last year's budget is augmented by approximately 10 per cent to estimate current's budget. While preparing a budget, the number of the patients treated is also considered. At the Block level, there is no provision for preparing the budget. Similarly at sector and Sub Center level, no budget is prepared. There is a fixed budgetary allocation for CHC/PHC. While preparing the budget, least consideration is given to the Annual Action plan prepared at each level. A top down approach is followed for estimating the budget in

Health Department. The ANM/Supervisors/Medical Officer- PHC/ BMO do not play any role in the preparation of the budget.

Procurement

Medicines under National Programmes: The drugs and supplies under various National Programmes such as Leprosy, Tuberculosis, Family Planning, Malaria, etc are procured from the Ministry of Health and Family Welfare (GOI). After receiving the requirement from each state, Ministry of Health and Family Welfare (GOI) places the order to the contracted supplier/manufacturer for procurement. The CMO, procures medicines from the State Store at Shimla and distributes to the PHCs/CHCs from where the medicines are further distributed to Sub Centers.

Sub Center's Medicines Kit A and B: For the procurement of Medicines Kits A & B, the CMO sends the requirement based on number of Sub Centers in the district to the Ministry of Health and Family Welfare (GOI) with a copy to Director, Department of Health and Family Welfare, Shimla. After receiving the requirement from each state, the Ministry of Health and Family Welfare places the order to the contracted supplier/manufacturer for the procurement. BMO procures the medicines kits from the District HQ and ANM procures the Medicines Kits A & B from the CHC.

Medicines under State Supply: For the purchase of other essential medicines, the State- Department of Health and Family Welfare, Shimla allocates 80 per cent of the total budget to the District CMO for making the local purchase. The 20 per cent of the budget is kept at the State for procurement of emergency medicines (epidemic, natural calamities etc). These emergency medicines for the District are procured on the basis of demand forwarded by the CMO. For procurement of medicines at District level, CMO calls for the tender from various Manufactures/Distributors on the basis of rate contract. At CHC level, there is no provision for local level procurement. BMOs do not have any role in procurement of CHC's medicines. The BMOs have no purchasing power at CHC level and have to depend entirely on District supply. It ultimately leads to shortage and irregular medicines supply at the CHC/PHC/SC.

The annual action plan of various levels flow upward from Sub Center level to Director of Health and Family Welfare. Although these action plans are supposed to be the basis for budgeting and material projection, but they are not properly considered for its purpose. The main reason for this is push down system of supply of medicine kits from GOI to Sub Centers. The flow of action plan is shown in following diagram:

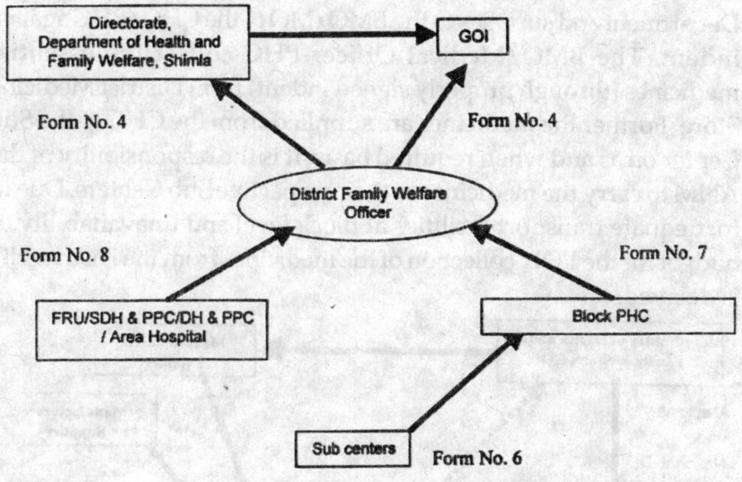

Fig. 4.1: Flow Chart of Action Plans

Distribution System of Medicines

Distribution Channel for Medicines under National Programme In RCH programme, the Ministry of Health and Family Welfare places the order for Medicine Kits A & B to the manufacturer. The manufacturer supplies the Kits directly to the District HQ as per the order. The District HQ then distributes the Kits, normally once a year to the CHC. BMO collects the Medicine kit A and B from the district HQ and subsequently distributes it to the Sub Centers. The multi layer system of distribution of Medicine Kits (A&B) creates irregularities and delay in supply to the Sub Center.

The medicines under various other national programmes, such as TB, Malaria, Leprosy, Blindness are supplied to the state store at Shimla by the manufacturer/supplier contracted by the Ministry of Health and Family Welfare. The State Medicine Store supplies the medicines to the District HQ at Recong Peo. The BMO collects the medicines by issuing an Indent. The ANM collects supplies from the CHC. The ANM does not make any indent for getting the medicines.

Distribution Channel for Medicines under State Supply The CMO purchases medicines with the allocated fund from Department and supplies to the BMO/MO of the CHC/PHC against indent. The BMO/Medical Officer-PHC collect the essential medicines (through properly signed indent) from District Medicine Store. Further the medicines are supplied from the CHC to the Sub Center on as and when required basis. It is the responsibility of the ANM to carry the medicines to their respective Sub Centers. Due to inadequate transport facilities at Block level and unavailability of budget for the POL, collection of the medicines from the District HQ is delayed.

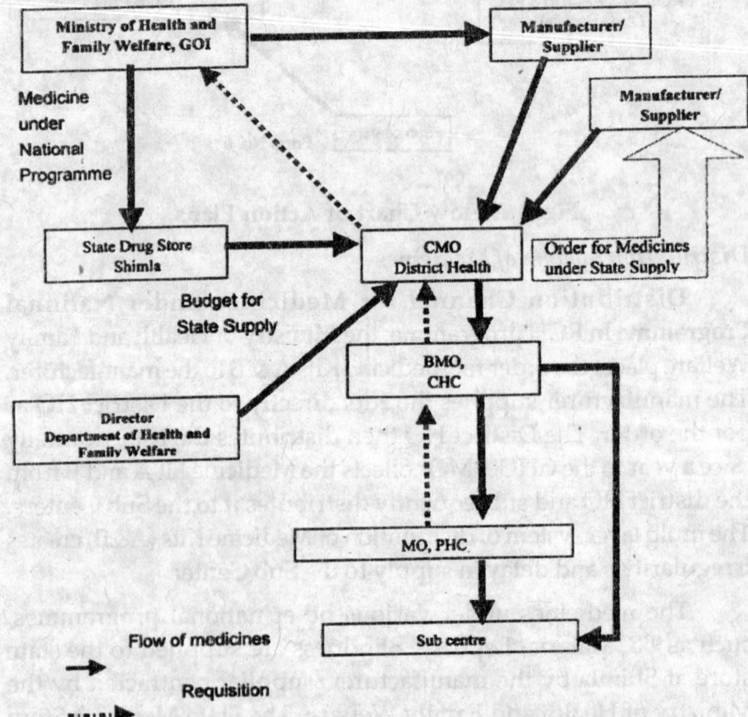

Fig. 4.2: Supply System (For Health and Family Welfare Department) Order for Medicine Kit A & B

Transportation of Medicines and Equipment

The medicines under various National programmes are transported to the State Store by truck hired by the Manufacturer. From State Store, the Medicines/items are transported to District HQ either by department vehicle or by hired truck arranged by the Department of Health and Family Welfare. It is the responsibility of the Director, Department of Health to arrange for the transportation of the medicines to the District HQ. Only the Medicine Kits A & B are directly transported to CMO by Supplier or Manufacturer, (contracted by Ministry of Health and Family Welfare, Govt. of India) through hired private transport.

The medicines under State Supply are transported to the District stores through hired truck arranged by the Manufacturer/ supplier. Transportation of medicines from District HQ to CHC/ PHC is made either by departmental vehicle or by hired vehicle. It is the responsibility of the BMO/MO to arrange vehicle for collection of medicines. Because of very limited availability of vehicle with BMO/MO, jeeps or medium sized trucks are often hired for collection of medicines. Transportation of medicines is most difficult between BPHC and Sub Center. Normally the ANM has to carry the medicines and kits from the BPHC to sub center on her own, she carries the medicines by public transport, if her sub center is connected by a proper road. Otherwise she has to walk a long way with the kits. The distance of sub center and the terrain in many parts of district are always the demotivating factors for an ANM to bring medicines to her Sub Center.

It has been observed that in spite of availability of the vehicle at the CHC, the BMOs are unable to improve the status of supply of the medicines due to lack of budget for the vehicle maintenance and Petrol/diesel. Even for minor repair of the vehicle, it takes months to receive the approval/budget from the District HQ. Vehicle is a major constraint in medicine distribution system. The number of vehicles available with the District, Health Department is not at all sufficient to make any arrangement so that the medicines can be supplied from District to Block and from Block to Sub Center.

Storage

Due to less availability of the almirahs and racks, the medicine boxes are kept on the floor. It has been observed that in most of

CHC, the medicines are stored at any available room. Due to non-availability of any separate store room at Sub Center level, the medicines and other items are stored at the corner of a room. In fact, the volume of supplies available at a time at Sub Center level is so low that it does not require a separate storeroom. At CHC level, due to lack of proper storage room and utilizing any "available space" in the building, the so-called "Store Room" lacks the storage infrastructure like racks, almirahs, ventilation etc. It has been also observed that the rooms, which are either not fit for general use (like OPD/Indoor Patients Department etc) or have some faulty structure (like leaking roof, small and dinghy etc), are used for storing the medicines and other equipments. At sub center level, lack of building structure or non availability of racks/almirah or bad condition of storing structure (like locking systems, broken windows), pose a threat for storage for Medicines/equipment.

Store Keeping Practices

At the district HQ, the compounder/Pharmacist look after the storeroom under overall supervision of Deputy CMO/Medical Officer. Similarly at the Block level, the compounder/Pharmacist looks after the storeroom under the supervision of BMO. It has been observed that proper record about the expiry date of the medicines is maintained at district and block level. Although Pharmacist follows the practice of issuing the 'first expiry-first out' system, but due to lack of proper infrastructure and storage arrangement, the medicines can not be arranged according to expiry date leading to incongruity in the issuing pattern. The Pharmacist because of their involvement in other affairs of Hospital finds it difficult to keep the check on the expiry dates and other records on regular basis.

At the district, as per the norms, there should be a buffer stock for three months in the stock and also sufficient quantity of medicines to tackle the epidemic in the district. Sometimes due to delay and shortfall in the supply, it becomes difficult to keep the buffer stock at optimum level. Formal training on drug logistics and store management has not been arranged for any level of staff. The ANMs/Supervisors are not at all trained in stores handling or logistic management. Lack of training leads to poor record maintenance and inefficient handling of stores.

Vaccine Supply Logistics

Procurement of Vaccines: The Department of Health and Family Welfare sends the requisition to Ministry of Health and Family Welfare for procurement of vaccines. The Ministry of Health and Family Welfare, GOI place the order to selected manufacturer. The contracted manufacturer sends the vaccines as per the order to the regional vaccines warehouse through flights equipped with cold chain equipments. From the regional stores the vaccines are distributed to the District. At District level, the Vaccines are procured from the Regional Vaccine HQ through proper Indent.

Distribution of Vaccines On receiving the order from Ministry of Health and Family Welfare (Govt. of India), the Manufacturer supplies the Vaccines directly to the Regional Vaccines warehouse in the State. At the District level, District Immunization Officer/ Deputy CMO is responsible for collecting the vaccines from the Regional Vaccine HQ to the District level Cold Chain Storage. District Immunization Officer/DMO places formal indent to the Regional Vaccines warehouse and collect the vaccines through the District Vaccine Van or Cold Boxes.

District Vaccine Van or Cold Boxes supplies the vaccines from District Cold Storage to CHC. From CHC, the vaccines are transported to sector PHCs in cold boxes/vaccine carriers. The ANM collect the vaccines from CHC either a day before or at early hours on fixed appointed immunization day in the village. Due to distance of Sub Center from PHC/CHC and lack of transportation facilities, some of the villages in the sub center area are often neglected.

The PHCs and Sub Centres, however, provide the infrastructure for universal health services. They have made almost the whole (including the illiterate ones) the client population aware of the modern medical facilities (howsoever meagre may be the quantum of services readily available). Some effective measures, though no doubt inadequate, are being taken against debilitating and decapacitating diseases.

Health Institutions in Kinnaur District

Initially, Medical allopathic system was introduced by at Pooh and Chini Village. Later on Forest department maintained sub assistant surgeon in the dispensary at Kilba. The period of merger and formation of district in 1961 has witnessed a significant development of Medical facilities by the State Social Welfare board. At present total medical facilities available is as shown Table 4.69 and Table 4.70.

Table 4.69: Health Institutions in Kinnaur District

1. District Hospital	1 (100 bedded)
2. Civil Hospital	1 (10 bedded)
3. Community Health Centers	3 (60 bedded) but in actual 11 only
4. Primary Health Center	17 (36 bedded)
5. State Special Dispanceries	2 (20 bedded)
6. Dental Clinics	3 (Peo, Sangla, Pooh)
7. Health Sub centers	32
8. Ayurvedic Hospital	1
9. Ayurvedic Dispensers	39
10. Total Number of Doctors at Hospital	13 out of 19
11. Doctors in PHC	15 out of 22
12. Doctors in CHC	12 out of 12
13. S/Nurses in Hosp	12 out of 15
in CHC	4 out of 12
in PHC	2 out of 17
14. W/Nurses in Hosp	3 out of 4
in CHC	0 out of 2
15. FHW in Hosp.	4 out of 6
in CHC	2 out of 3
in PHC	10 out of 17
16. Total no. of Dai in Hospital	2 out of 2
in CHC	3 out of 3
in PHC	5 out of 5

Table 4.70: Block-wise Health Institutions in Kinnaur

Name of Block	*Hospital/ Health Institution*	*Name of the Primary Health Centre*	*Name of the Sub-Centre*	
POOH BLOCK	CHC Pooh Civil Hospital, Chango	PHC, Spillow PHC, Giabong PHC, Moorang PHC. Ribha PHC, Kanam PHC, Lippa PHC, Hango	Chuling Malling Asrang Shalkher Dubling Tashingang	Nesang Ropa Labrang Karla Rispa Akpa Charang
KALPA BLOCK	Distt. Hospital, Reckong-Peo CHC, Sangla I.T.B.P. Hospital, Reckong-Peo	PHC, Kalpa PHC, Kilba PHC, Sapni PHC, Tangling	Ralli Chansu Rakcham Baturi Kanai	Baturi Shong Brun
NICHAR BLOCK	CHC, Nichar Project Hospital, Bhawa-Nagar	PHC, Tapri PHC, Rupi PHC, Katgoan PHC. Urni PHC, Meeru PHC, B.Kamba	Yulla Karba Ponda Sungra Tranda Chaurra	Runang Yangpa Panvi Barri Kachrang

Ayurvedic System

As many other systems, the indigenous health care system also declined in importance during the long period of western colonial rule in India. However, the system survived this hostile atmosphere and in this the princely rulers played some role. However, it was the demand from the people which nourished it even on a small scale. Though the system of Ayurveda has a sub continental tradition, in its early stages it seemed to have developed with certain regional characteristics.

Indian System of Medicine and Homoeopathy play a vital role in the Health Care system of the state. Because of the variety of climatic conditions it has traditionally been rich repository of medicinal plants use in various Indian System of Medicine namely Ayurveda, Tibetan, Bhot System, Unani and Yoga flourished in its cradle since time immemorial. In the tribal areas, Tibetan System of Medicine under the name of 'Bhot Chikitsa Padhati' continues to the popular. There are several areas which remain cut off from other parts of the state and country on account of heavy snow fall and rain fall and the people have to depend upon the traditional local system of medicine for their health requirement. Incidentally these very areas are a natural habit of several rare and very useful medicinal plants. The people of Himachal Pradesh traditionally have immense faith in Ayurvedic and other Systems of Medicine and in this context the Government of Himachal Pradesh embarked upon a programme to serve the people living in the remotest corners through these systems. Accordingly, a large number of dispensaries and hospitals were opened and at present 1/3 of the total patients (3 lakhs) in the State are meeting their health requirements through the network of these hospitals and dispensaries. These systems are popular in rural and remote areas where the practitioners of modern medicine are not really available and are often the only available alternative in the remote tribal areas.

The Department of Ayurvedic (Indigenous medicine) functions under the control of a Director for overall administration and programmes implementation. Kinnaur district has one Ayurvedic hospital and 39 Ayurvedic dispensaries.

Traditional Health System

Belief in the supernatural being is one of the aspects of all religions and there are systems and rituals in the tribal religion, which establish direct relationship between the human needs and the supernatural forces, both benevolent and malevolent, to obtain their grace or to mitigate their wrath. It acts as a binding force amongst the individuals too. Religion has in its central core, a set of systematic belief- patterns and practices, which uphold such beliefs. These religious practices, which uphold such beliefs. These religious practices or beliefs are variable in nature from group to group. Some consider these practices as techniques for achieving success or a means to make adjustment of human beings with nature and other situations beyond their calculation and control.

Other aspects of man's life, illness, accident, epidemic, ill fortune etc. have their appropriate rites, performed only by those who need or want them and directed to the spirits evil or angered by omissions as diagnosed by the priest by divination.

Buddhist Health System

The contribution of Buddhist monasteries in inculcating humanistic values in the social system of the Lamaistic western Himalaya, comprising Ladakh, Lahul, Spiti and Kinnaur, has been overwhelming. These institutions ushered in an era of socio-cultural resurgence in a region which had long remained infested with innumerable primitive beliefs and superstitions under the generic name of *lung-pa'l chos*.

Extraneous influence notwithstanding, Lamaism still holds ground in the upper part of Kinnaur where all the social, cultural and religious activities are conducted according to the monastic canons. The Lamas there perform all the religious rites for their lay patrons. The lama is the astrologers (*rtis-pa*) for them and performs various religious ceremonies. Interestingly, the lama has no place in marriage ceremonies. The reason for this may be that the lama would not be a party to the establishment of a household as a result of the canonic restriction imposed on them. In lower Kinnaur, where Lamaism is losing ground, the lama invariably solemnizes the marriage and blesses the newly weds.

The socio-religious impact of the monasteries has been relative to the religious following which these institutions command. This is at its greatest in upper Kinnaur, sparse in the Lower part of Kinnaur. The Buddhist monasteries have helped dispel many primitive beliefs, superstitions and dogmas by popularising the written language and literacy. In the remote Lamaist society of the Kinnaur, modern medical facilities were unheard of till a few decades ago and the people depended entirely upon the lama doctors in the monasteries for their medicines. Those are based on the age-old Ayurvedic system of Indian and local herbs therapy which have already been incorporated into *Tan- gyur*, providing a sound base for the Tibetan system of medicine. The monasteries thereby not only assumed the role of 'spiritual healer' for the laity but also provided better medical facilities for them. In the more orthodox Lamaist pockets, the pre-Lamaist totemic methods of treatment also continued. In fact in the tantric Buddhism with which Lamaism identifies itself, tantric methods of healing play a significant role and even advanced Lamaist societies depend upon tantric treatment for ailments.

In earlier time, the Tibet Medicines system was greatly followed in the Kinnaur. According to one of the responded Lama in Kinnaur, there are about forty books or works written in Tibet, on medicine, besides the five volumes in the Stan-gyur collection, and the scattered occasional instructions on medicament in the Kah-gyur.

According to this system, the theory of human constitution is illustrated by a similitude taken from the Indian fig-tree. Thus, there are three roots or trunks; thence arise nine stems; thence spread 47 boughs or branches: two blossoms, and three fruits. The explication of the simile as applied to the states of the body. The single root or basis of diseases: the stems, branches, and leaves arising thence, taken or considered in a healthy and in a diseased state. Distinction; with respect to wind; ditto, with respect to bile: as also to phlegms their respective offices, operations or influences.

There are seven supports of the body on which life depends; the chyle, blood, flesh, fat, bone, marrow, and semen. Description of the three sorts of excretions or sordes of body, ordure, urine and sweat.

The three generative causes of disease are: lust or areent desire; passion or anger; dullness or ignorance. By the first is caused wind; by the 2nd. Bile; by the last, phlegm. The accessory causes of disease are four: 1, season with respect to cold and heat; 2, any evil spirit; 3, wrong use of food; and 4, ill conduct of life.

The parts of the body, commonly subject to diseases, are six: the skin, the flesh, the veins, the bones, the viscera, and the bowels.

The proper places of the three humours are: that of the phlegm in the upper parts of the body, as the proper place of dullness, in the brain or skull ; that of the bile, in the middle part of the body, which is appropriate to anger; and the wind resides in the lower part of the trunk, in the waist and loins, as in its proper place.

There are 15 ways or channels through which disease spreads itself. The channels of the motion of wind are, the bones, the ear, skin, heart, artery, and the guts. The blood, sweat, the eye, the liver, the bowels, are the ways or vehicles of bile. The chyle, flesh and fat, marrow and semen, ordure and urine, the nose and the tongue, the lungs, the spleen, and the kidneys, the stomach, and the bladder, are the vehicles for the conveyance of the phlegmatic humour.

With respect to the three humours, this further distinction is made: wind is predominant in the diseases of old people ; bile, in those of adolescents or youths; and phlegm, in children.

With respect to place (or part of the body); wind occurs in the cold parts of the body; bile in the dry and hot parts; phlegm abides in the moist and unctuous parts.

The several seasons, in which the diseases caused by any of these three humours prevail, are thus stated: diseases, caused by wind, arise commonly during the summer season, before the dawn, and about mid-day. Those caused by bile, in autumn, about midday and midnight. Phlegm prevails during the spring season, and in the morning and evening.

There are specified nine sorts of diseases, in which there is no hope of recovery. On the twelve causes by which any of the diseases caused by any of the three humours, is changed into another, as wind into bile and phlegm etc.

All diseases are classed under two heads: heat and cold. Those, in which wind and phlegm prevail; being of natural water, belong to cold. Blood and bile, being of natural fire, belong to heat. The diseases caused by the worms and the serum, belong both to cold and heat.

According to one of the respondent, on finding the certain symptoms of diseases, he approached traditional healer in other village for remedy. The traditional healer examined the tongue and urine and felt the pulse. The traditional healer took the oral history of disease from respondent. He asked questions like how the disease first arose, and its progress, what pain is felt, what sort of food has been useful or noxious?

Further the traditional healer explained the following condition of tongue, urine and pulse help him in diagnosing the diseases

- **With respect to the tongue:** If the tongue is red, dry and rough, it is the sign of prevailing wind: if covered with a yellowish white thick substance, it is the sign of bile ; if covered with a dim, white, soft, and moist substance, it is the sign of phlegm.

- **With respect to the urine:** If the urine of the patient is blue, clear like spring-water, and has much spume or froth, it is the symptom of wind ; if yellowish red and thick, steaming or vapouring greatly, and diffusing a smell, it is the sign of bile ; if white, with little smell and steam or vapour, it is the sign of phlegm.
- **With respect to the pulse:** When the traditional healer feels the pulse, if beating greatly upwards it somewhat stops (if irregular), it is the sign of wind ; a quick full beating is the sign of bile: a sunk, low, and soft beating is the sign of phlegm.

The physician's asked 29 questions to the patient about his food, exercise, and the pains or relief felt after having taken such and such food, made such and such an exertion. Traditional healer suggested the patient the hot bath at regular interval with completely avoiding the wind exposure.

Village Gods and Goddess Health System

Village are generally self- sustained units. Every village has a temple where people congregate for common worship. The village gods are carried in palanquins, on number of occasions, to places of religious interest or village fairs. When in trouble, the people go to the deities to seek their guidance and help. The village god is suspected to watch over the destiny of the village. He protects, rewards, threats, threatens and punishes the people, while they in turn, worship him/her by singing and dancing.

It is believed that the village gods are benevolent spirits who never fed displeased with their devotees. However, these spirits are accompanied by attendants known as ganas, who inflict curse on deviants. These curses are called dosh (displeasure) and the evildoer has to correct himself in line with the wishes of the god.

PRIVATE HEALTH CARE SYSTEM

The private health care system in Kinnaur district is limited to presence of RMPs and other local non registered medical practitioners. Most the people prefer to visit government health institutions in the district. The following reason has been reported for visit to private health practitioners:

- Absence of doctors s in government health institutions
- non availability of medicines
- far away distance of government health institution.

The contribution of private health care system is very negligible in Kinnaur district. It has been reported that most of the people prefer to visit government health institution or consult village devta or monasteries (in upper Kinnaur area). In case of extreme emergency, the people take the patent to Rampur or Shimla for further treatment.

HEALTH IN GOVERNMENT RECORDS

After a study of data pertaining to Ante-natal cases, one can clearly observe an increasing trend in the number of institutional deliveries in 1996-97 was 66, which increased to 202 in 1997-98 and showed a further increases to reach 234 in the year 98-99.

This trend clearly shows the increased acceptance level of various institutions for the delivery of a child. This trend is further increasing year after year and the people of the district are slowly realizing the importance of institution for childbirth. They have recognized slowly that institutional deliveries tend to be safer. Here, I would like to mention the case of one of my informant Prakash. His wife gave birth to a still child after her first pregnancy in 1996. The local women conducted the delivery at home. In 1998 when his wife was pregnant for a second time, he took her to the PHC for delivery. This has happened because increase in level of awareness about the health care system.

As far the success of various family planning methods such as vasectomy, tubectomy, IUD etc. mixed results. In 1996-97 there were a total number of 186 operations of vasectomy and tubectomy put together. This number swelled to 406 during 1997-98 but further decreased to just 162 in 1998-99. The main reason for this is that during the year 1997-98 there was a huge campaign for family planning by the Government. A large number of people were convinced to undergo operation for family planning. The reason behind this was that once they have realized the importance of this, they will readily accept it. Sizeable amount of resources was spent on this effort. But the result has been quite different from what was desired.

Due to this drive a large number of operations were done sometimes in unhygienic conditions. This led to large number of people developing infection or some other problems. Here again a case study is quite interesting Hari of Chini village underwent the operation during this period but after this he development on infection. He had to suffer for 3 months because of this. Also there was not much postoperative care provided. Today Hari severely criticize this method to everyone who comes for advice. This sort of case causes a bad publicity for otherwise safe system. The lacks of post-operative care and follow up only shown the inefficiency of the system. This work together to produce a counter effect and all the positive aspects are thus overlooked.

It has also been noticed that in record years i.e. 1998-99 the number of people going in for family planning measure such as Vasectomy and tubectomy is more among those who are literate. This number is maximum at 77 among those who have done primary schooling. This trend clearly indicates that education is in reality spreading awareness among the people and more and more educated people are realizing the importance of family planning.

This is also a general trend to use the family planning methods more when one has 2 or more children. The people having 2 or more children are one's who use the various family planning methods in larger numbers.

About the morbidity pattern of the diseases, the disease of tuberculosis is quite prevalent in this area. But over the year this has been showing a declining trend. Malignant diseases are really found in this region. A very few cancer patients are reported and almost 40 per cent out of total number of cancer patients suffer from the cancer of stomach.

After studing the hospital records, it was found out that earlier in 1996, there used to be a large number of patients suffering from diseases due to nutritional deficiencies such as Kwashiorkor (392). This has been controlled over the year though the no. of patients suffering from Masasmus has gone up to 122 in 1998.

Disorders of the eye are also quite widely found and especially the infectious disease like conjunctivitis. There have been efforts to control it and this has been slowing a declining trend.

Respiratory diseases like bronchitis have been growing at an alarming rate. People suffering from bronchitis reached 22,453 during the year 1998, which has been the highest number reached so far. The most prevalent disease in the district is the acute respiratory disease including influenza. They have shown an increasing trend and the number of patients suffering from them has increased to 51,005 in 1998 showing an increasing trend. Gastric diseases are the second most prevalent group and also show an increasing trend. The number of patients suffering from them was 22,438 in 1998.

In the last, each ethnic group is a distinctive unit with its own socio-cultural background and values. As a matter of fact the ethnic group differ from each other ethnically and socio- culturally. Among the members of the same ethnic group the local problems have different dimensions. Similarly the pace of development, not to speak of different ethnic groups, is not at all uniform among the members of the same ethnic group inhabiting different regions what, therefore, hold good in one case may not necessarily apply to other cases.

Administrative procedural problems, lack of knowledge about the selection of proper area and time of implementation of the schemes, lack of coordination among implementing agencies, difficult and inaccessible tribal areas, lack of missionary zeal and helping attitudes in many cases among the implementing agencies, peculiar psychology of the tribal people, their reaction etc. are the main reasons for the slow growth and failure.

The problem of development in the Himalayan region is not only mutli-directional but also of such a high magnitude which may be compared with the height of the Himalaya itself. In spite of it, a number of programmes and projects have been formulated and launched for conservation and development of hill resources. Maintenance of ecological balance, improved water utilisation, power development, development of livestock, forestry, sericulture, land use, marketing, processing and transportation are some of the worth mentioning programmes which has been initiated with an ultimate objective to raised the standard of living of the local population.

One would try to visualize what could be the level of development—individual, familial, village and societal? Because of the diversities, the idea of individual approach will be hard to materialize. It could be either familial of societal approach which can be used. The plan of development on caste groups needs a familial approach, while the village level plans refer to societal development. In other words, the concept, planning and approach have been adopted to plan programmes of development of Kinnauries.

When we see the relationship between man and environment we find that whenever man inhabited a new area he chose for his settlement a site that supplied him with fuel, fodder, food and timber and protected him from enemies. Water remained essential for his very survival. Yet, in more insecure times the environment of the Himalayas with its ridges, forests, stream, height and its climate sometimes gave protection to refugees from the subcontinent and beyond and taught them self-reliance. The environment both protected and governed these communities.

As the Himalayas are relatively less accessible, this break has come later and the rupture is still in the process of taking place. Nature is still very dominant in the Himalayas and its resistance to this break is still very strong. Technology and concepts introduced in the Himalayas from the plains are facing resistance from the natural environment. All the technology and infrastructure introduced in the hills has failed substantially to improve the quality of people's lives and has resulted in the devastation of natural resources, the disintegration of society and an increase in out migration.

The obstacles to development are clearly located in ecology, ideology, lethargy, lack of capital, expertise, traditional practices and ideas of values.

5

Summary, Conclusion and Strategies

SUMMARY

The relationship between health and social conditions is a constantly changing one. It has even been said that social conditions primarily determine the health status of populations. The present study focuses on health profile and health development of four ethnic groups of district Kinnaur, Himachal Pradesh.

For the present study, the Kinnaur district of Himachal Pradesh has been selected. The Kinnaur valley is stretched over about 80 kilometers in length and 64 kilometers in breadth. The area lying within the district territory is nearly 6679 sq. kilometers. The Kinnaur district is divided into three blocks, Kalpa, Nicher and Puh. The tehsil in Kalpa, Nichar and Puh blocks are Kalpa and Sangla, Nichar and Morang and Puh respectively. (Hangrang is sub teshil in Puh Block).

The inhabitants of Kinnaur district; i.e. Kinnauri has been spelt out with many variants such as Konawri, Koonauri, Kanauri, Kannauri, Kinnaura and Kinnauri (presently in use). The ethnic population of Kinnaur district, Kinnauri are divided into five groups i.e. Kanet, Koli, Badhi, Lohar and Nagloo. The entire population is classified as Scheduled Caste or Scheduled Tribe. The population of the district comprises of five distinct classes of Kanets (Rajputs), Chamangs {(Kolis, preparing shoes, weaving, tailoring and music drum beating)} and Domangs { Lohar (Blacksmith) and Badhi (carpentry), Nagloo (Basket makers)}. The Kanet constitute the majority of Kinnaur population. They are declared as Scheduled tribe people. The other four ethnic groups i.e Koli, Badhi, Lohar and Nagloo constitute the scheduled caste population of district. The Koli, Badhi, Lohar and Nagloo mainly comprises of the artisan class and are considered untouchables whether they be Koli having the occupation of preparing shoes, weaving, tailoring and music drum beating, the Badhi having the carpentry and mason, Lohar—Blacksmith and Nagaloo–Basket making.

In the context of present study, four ethnic groups were selected i.e Kanet, Koli, Badhi and Lohar. The rationale for selection of these four ethnic groups is uniform and wider distribution in the district.

In first chapter, the various aspects of health definition, role of health care system in India, health policy, ecological and health dimension, role of non government agencies, Ethnic groups: definition and perspectives, tribes and caste system in Himachal Pradesh and brief review of literature is discussed in details. The study area, Kinnaur have been declared as Scheduled Areas under the Fifth Scheduled of the Constitution by the President of India as per the Scheduled Areas (Himachal Pradesh) Order , 1975 (CO 102) dated 31st November 1975.

Defining Health

Development of the notion of social responsibility for health and the duty of individuals for the care of their health was espoused by Dr Andrija Stampar, who was to become President of the First World Health Assembly of WHO. It was he who played a crucial role in drafting the definition of health that was to be incorporated into the first paragraph of the preamble to the WHO Constitution and subsequently into the International Covenant on Economic, Social and Cultural Rights. Thus it was that over half a century ago, the founders of the World Health Organization defined health as "a state of complete physical, mental and social well-being and not merely the absence of disease or infirmity". The Constitution further recognized "the enjoyment of the highest attainable standard of health ... as one of the fundamental rights of every human being" (WHO, 1948) This "right to health", as it became expressed in an abbreviated version in many subsequent documents, includes the right to adequate food, water, clothing, housing, health care, education, security in the event of unemployment, sickness, disability, old age or lack of livelihood in circumstances beyond an individual's control.

Optimal health is defined as a balance of physical, emotional, social, spiritual, and intellectual health. Lifestyle change can be facilitated through a combination of efforts to enhance awareness, change behaviour and create environments that support good health practices. Of the three, supportive environments will probably have the greatest impact in producing lasting change". Anthropology of health is of recent origin. As subject for specialized anthropological study, health and ill health is as yet only in its adolescence. Anthropology of Health is application of anthropology to the health field. The anthropology has its genesis in certain fundamental aspects of human life whereas health and health care services possess certain distinctive characteristics.

Various health policies not only provide guidance with regard to the meaning and importance of the definitions of health, but also gives explicit emphasis to the need to move from policy to action. In doing so, we recognize how important it will be to harness the will and action of diverse sectors and partners for health at all levels.

Heath Care System

Health care system is intended to delivery the health care services to the different segment of the population. It constitutes the management sector and involves organizational matters. In India, five major sectors or agencies represent health care system, which differs from each other by the health technology applied and by sources of fund for operation. These five major sectors are public sector, private sector, indigenous sector of medicine, voluntary health agencies, and National Health programme. (Park, 1991). The major objective of any health care system is to provide health care services at the maximum to the people in order to prevent and cure diseases by proper utilization of Knowledge, skill and technology. Although there is no specifically isolated designed Health care system but other social institution, cultural norms, value system in the society greatly influenced the utilization pattern of available knowledge, technology, man-resources involved, it is essential to understand the existence of various health care system existing/

present in any community, people involved in it, accessibility of the system, cultural practices of the people, infrastructure available etc.. After independence, India dedicated herself to the creation of the new social order based on social equity, freedom to justice, economic disparity, illiteracy and dismal health condition prevailing in the country were the main focus in the initial stage and subsequent 5 year plan.

Much before the Alma Ata Conference the Constitution of India provided (Under Directive Principle of State Policy: Article 36-51 Part IV) as follows: " The State shall in particular direct the policy towards securing that the health and strength of workers, men and women and the tender age of children are not abused and that citizens are not forced by economic necessity to enter avocations unsuited to their age or strength, that childhood and youth are protected against exploitation and against moral and material abandonment. The State within the limits of its economic capacity and development make effective provision for securing the right to work, to education and to public assistance in case of unemployment, old age, sickness and disablement and in other cases of underserved want. The State shall make provision for securing just and humane conditions of work and maternity relief. The State shall regard the raising of standard of living of its people and the improvement of public health as among the primary duties."

In the 30th World Health Assembly in 1977 refereed to Health for all slogan as " a level of health that will permit people to live a socially and economically productive life". India as a signatory to the Alma Ata Declaration is committed to the achievement of Health for all by 2000 AD. India is also signatory to the South-East Asian Charter on Health. Against this background, the current objective of state and national health plans in the health sectors is to organize and provide universal primary health care to all sections of the society with special attention to the needs of those living in the tribal, hilly and remote area. Greater equity in the access to health care, better utilization of limited resources, appropriate policy development, active participating of the community in achieving self reliance in the health and sustaining the health services and inter- sectoral action are the key components of the Global Strategy of health for all.

The Indian Health Care System operates at the national level by Ministry of Health and Family Welfare, under which all the states and union territories comes. States control the district health institutions. Chief Medical Officer/Chief Health and Medical Officer is the head of the district health institution, sub-district hospitals comes which controls to the community health centres which is present in 1/1,00,000 population. Under community health centres, primary health centres comes which present 1/30,000 population for plains and 1/20,000 for tribal and hilly region. Primary health centres (PHC) are present in block. Under PHC, sub centres comes which are present 1/5000 population plain and 1/ 3000 population for tribal and hill region.

Population policy in India has had a long and somewhat chequered history. Starting in the 1950s as a largely urban clinic-based programme, the family planning programme increasingly became, in the 1960s, rural in its focus and more community oriented in its approach. By the late 1960s, however, contraceptive method-specific targets had become a key instrument in the programme, and in the 1970s, the programme and the policy within which it was embedded became increasingly target-driven, centralized and top-down. It operated largely in isolation from other sectors, such as education, which might plausibly have been argued to have some impact on demographic behaviour and patterns. Indeed, population and health policies were usually only nominally linked to each other even though both had a home in the Ministry of Health and Family Welfare.

Altitude being one of the significant determining factors the life and living of the people of these areas distinctly differ from the rest of Indian living in the plane. In almost all respects including roads communications and other infrastructural facilities these areas are deficient and are consequently lagging behind the average level of development of the rest of the country. Health Ecology is a broad, interdisciplinary approach to health and wellness. It is based on the dynamic interplay among diverse environmental, political, spiritual, philosophical, psychological, and personal factors and it incorporates an ecological perspective.

Role of NGOs in Improving the Health Status

The past five decades have witnessed the difficult problems encountered in providing health care services to poor people, the majority of whom live in more than half-a-million villages and in the proliferating slums of cities. Charitable and voluntary organizations since time immemorial have been contributing significantly towards the health care of the community. With the passage of time, Non-Governmental Organizations (NGOs) have equipped themselves adequately and come up enthusiastically in providing services like relief to the blind, the disabled and disadvantaged and helping the government in mother and child health care including family planning programmes. As a result, now all concerned have realized the potentiality of NGOs and their considerable merit compared to the public/private health sectors because of their staffs motivation, dedication and sympathy for the deprived sections of society and their personalized approach towards the solution of the problems.

As per the National Population Policy (NPP) 2000 and National Health Policy (NHP) 2002, there should be greater involvement' of NGOs in the implementation of different health and family welfare programmes in the country. In order to utilize the high motivational skills of NGOs on an increasing scale, it has even directed that the national health programmes should earmark a definite portion of the budget in respect of identified programme components, to be exclusively implemented through the NGOs. Consequently, the perception of strengthening linkages between the Government and NGOs has become very crucial for India's health sector reforms, especially for the decision-making, planning and management procedures.

It is very interesting and encouraging to note the evolution of NGOs' activities in the health sector in India. In recognition of the crucial role played by them, Government of India started granting financial aids to NGOs for various schemes. The laudable role played by the various national and regional level NGOs is briefly documented in the various documents, where special mention has been made of such organizations like All India Blind Relief Society, Family Planning Association of India (FPAI), Indian Medical Association, Indian Red Cross Society, National Society for the

Prevention of Blindness, Sant Parmanand Blind Relief Mission, T.B. Association of India, Bombay Mothers and Children Welfare Society; to name a few. After the Alma Ata Declaration in 1978, greater role for the NGOs was seen to ensure Health for All through the primary health care approach. Their role was also considered as most crucial to translate the concept of 'People's Health in People's Hands' into action.

Planning Commission and the Central Council of Health and Family Welfare emphasized the role of NGOs in health care delivery system. Subsequently a very large number of NGOs took the health and family welfare field and made notable contributions all over the country. There are a large number of voluntary organizations, like Voluntary Health Association of India, Hind Kusht Nivaran Sangh, Christian Medical Association of India, Paediatrics Association of India, Federation of Obstetrician and Gynaecologist of India, Lions Club, Rotary Club, Family Planning Foundation of India (which are mainly funding research and training), that have shown their sustainable contribution to the community. Some of these organizations are working for control of specific diseases such as Tuberculosis, Leprosy, Blindness, Family Planning, while many others are providing Maternal and Child Health and General Health Care services. Of late, the Government of India is encouraging NGOs to take up contraceptive and immunization services in a big way including training of nurses, ANMs, AWWs, Dais etc. The government has tried to streamline the system by identifying National NGOs, Mother NGOs and Field NGOs with a view to facilitate flow of funds and supervision of work output.

Defining Ethnic Groups

The word ethnic comes from the Greek ethnos, originally meaning nation. In its earliest English usage, about 1470 B.C., the word referred to culturally different heathen countries or nations (those not Christian or Jewish). Apparently, the first usage of ethnic group to denote national origin developed in the period of heavy immigration from southern and eastern European nations to the United States in the early twentieth century. Since the 1930s and 1940s, a number of prominent social scientists have suggested that the narrower definition of ethnic group, more in line with the original Greek meaning of nationality, makes the term more useful.

The term ethnic means of or pertaining to a group of people recognized as a class on the basis of certain distinctive characteristics such as religion, language, ancestry, culture or national origin. The ethnic group is a set of individuals whose identity as such is distinctive in terms of common cultural traditions or heritage.

In this thesis, the broader term ethnic group is extensively used. An ethnic group is a group of people who identify with one another, or are so identified by others, on the basis of either presumed cultural or biological similarities, or both.

The term 'Scheduled Tribes' first appeared in the Constitution of India. Article 366 (25) defined scheduled tribes as "such tribes or tribal communities or parts of or groups within such tribes or tribal communities as are deemed under Article 342 to be Scheduled Tribes for the purposes of this constitution". According to Article 342 of the Constitution, the Scheduled Tribes are the tribes or tribal communities or part of or groups within these tribes and tribal communities which have been declared as such by the President through a public notification. The essential characteristics of these communities are Primitive Traits, Geographical isolation, Distinct culture, shy of contact with community at large and economically backward.

There is no agreed definition of a tribe. The word, as such, has dictionary meaning of "a race or family descended from the same ancestor: an aggregate of families forming a community usually under the government of a chief." The Hindi equivalent 'Adivasi' has a clearer connotation. It means those who are the earliest inhabitants of the country.

The Constitution of India incorporates several special provisions for the promotion of educational and economic interest of Scheduled Tribes and their protection from social injustice and all forms of exploitation. These objectives are sought to be achieved through a strategy known as the Tribal Sub-Plan (TSP) strategy, which was adopted at the beginning of the Fifth Five Year Plan. The strategy seeks to ensure adequate flow of funds for tribal development form the State Plan allocations, schemes/programmes of Central Ministries/Departments, financial and Developmental Institutions. The cornerstone of this strategy has been to ensure earmarking of funds for TSP by States/UTs in proportion to the ST population in those State/UTs. Besides the efforts of the States/UTs and the Central Ministries/Departments to formulate and implement Tribal Sub-Plan for achieving socio-economic development of STs, the Ministry of Tribal Affairs is implementing several schemes and programmes for the benefits of STs.

In order to give more focused attention to the development of Scheduled Tribes, a separate Ministry, known as the Ministry of Tribal Affairs was constituted in October 1999. The new Ministry carved out of the Ministry of Social Justice and Empowerment, is the nodal Ministry for overall policy, planning and coordination of programmes and schemes for the development of Scheduled Tribes.

The Indian society is highly stratified and is divided into schedule castes, scheduled tribes etc. Hindu caste system is highly complex institutions, though social institutions resembling caste in one respect or another are not difficult to find elsewhere, but caste as we know it in India, is an exclusively Indian phenomenon. The word caste comes from the Portuguese word 'casta' signifying breed, race or kind. Risley (1915) defines it as ' a collection of families or groups of families bearing a common name; claiming a common descent from a mythical ancestor, human divine; professing to follow the same hereditary calling; and regarded by those who are competent to give an opinion as forming a single homogenous community' is generally associated with a specific occupation and that a caste is invariably endogamous, but is further divided as a rule, into a smaller of smaller circles each of which is endogamous. (called Jati). The internal exogamous division of the endogamous caste is Gotra. The main features of caste are hierarchy, endogamy and hyergamy (male of higher caste marrying a.female of lower caste) occupational association; consciousness of caste membership and restriction on food, drink and smoking; distinction in dress and speech and confirmation to peculiar customs of particular caste ritual and other privileges and disability; caste organisation and caste mobility.

Demographic Analysis

In anthropology, due emphasis is given to demographic analysis in health studies of any particular society. Demography is considered as the interdisciplinary science and policy oriented concerned with understanding and measuring population change in any society. Demography is a professional and global interdisciplinary science during 1930s. The practitioners of demography are originated from diverse field such as sociology, economic and statistic. Anthropology is considered as the holistic study of man irrespective of time, space and culture and in demographic research, fertility, mortality and migration of human beings are studied in detail. Therefore, both demography and anthropology are concerned with the study of human beings in various dimensions.

In anthropological sense, one type of qualitative data embody the own ideas of the local respondents and it is referred to as 'emic' data and another type of qualitative data is based on own constructs and categories of researchers and it is also referred to as 'etic' data (Pike, 1954). Demographers generally follow 'etic' method for data collection. The formal surveys generally collect the as usual data where qualitative methods specially collect the better quality data (Bleek, 1987). The demographic and anthropology is an important and substantial for understanding the theoretical and methodological orientation. The researchers of anthropology and demography emerge the new consideration of the related literature, apply the suitable method for data collection and then prepare the survey report.

Benjamin defined (1965), 'Fertility measures the rate of which a population adds to itself by births and is normally assessed by relating the number of births to the size of some section of population, such as the number of married couples of the numbers of women of child bearing age, i.e. an appropriate yard stick of potential fertility'. Thompson and Lewis (1965),'Fertility is generally used to indicate the actual reproductive performance of a woman or groups of women. The crude birth rate (number of births per 1000 population per year) is only one measure of fertility'. Fertility indicates the number of children which were produced by the women. Various attempts have been made to study the association of culture with fertility behaviour among certain tribes through social and demographic variables. It has been observed that among tribals, every aspect of life from birth to death is being influenced by the prevalence of customs, beliefs and notions. Though fertility is a biological phenomenon there are a number of other factors influencing the levels and differentials of fertility among tribals. Demographers usually measure the fertility differentials by taking into account women's income, occupation, education, family type, age at menarche, age at marriage, etc. (Dandekar & Dandekar, 1953; Dandekar, 1959; Roy Burman, 1961; Nag, 1962; Das, 1973; Vidyarthi & Rai, 1977; Sahu, 1983; Basu & Kshatriya, 1989).

The mortality is another important demographic indicator. The United Nations (UN) and World Health Organization (WHO) (1970) define death as 'the permanent disappearance of all evidence of life at any time after birth has taken place (post natal) cessation of vital functions without capacity of resuscitation'. Mortality indicates permanent extinction of all signs of life from living body and it is occurred at any time after birth. The poor food habits contribute the malnutrition specially among children and pregnant mothers which leads to increased susceptibility to morbid conditions. In India, maternal care is largely neglected and the expectant mother to a great extent is not inoculated against tetanus. The consumption of iron, calcium and vitamins during pregnancy is poor. From the inception of pregnancy to its termination, no specific nutritious diet is consumed by a woman. Vaccination and immunization of infant and children are inadequate. In addition, extremes of magico- religious beliefs and taboos tend to aggravate the problems which influence high mortality (Dandekar & Dandekar, 1953; Hauser & Duncan,1959; Dandekar, 1959; Thomas & Lewis, 1965; Ghosh, 1970; Sharma, 1978; Sirajuddin & Basu, 1984; Murthy, 1987; Bhasin & Bhasin, 1990; Kshatriya, 1993; Pandey & Goel, 1999).

Household income indicates the total income of the earning member of a particular family. The house hold income determines the socio-economic status of the household. It has been reported that fertility is affected by socio-economic factors whereas socio-economic factors are generally depended on quantity of income and sources of income. (Bhasin & Bhasin, 1990; Basu, 1999).

Education plays an important role for the development society. Education is considered as an development indicator which also effects the demographic behaviours of any particular population group. In earlier studies, it was reported that non availability of schools and lack of basic infrastructure is common phenomenon in tribal areas. (Tiwari, 1984; Patnaik, 1989; Singh, 1995).

In various anthropological studies and writing, early age at marriage is reported among the tribal communities. Further the age at marriage is linked with the demography of any population group. The lower fertility is presented by those women who married late (Majumdar, 1962; Agarwala, 1966; Nag, 1974; Khan, 1976; Bhasin & Bhasin, 1990; Roy, 1995).

A few comparative studies indicate that with the progress of civilization the trend of menarcheal age is becoming lower. It has been reported that characteristics like caste; sub-caste; nature of the area studies, i.e. rural, semi-rural and urban; socio-economic status (like family income, occupation of the head of the family and education family size), birth rank and climatic conditions of the geographical area also influence the menarcheal age. (Shah, 1958; Banerjee & Mukharjee, 1961; Biswas, 1967; Chattopahdyay & Khullar, 1969; Mukherjee, 1972; Ghosh & Kumari, 1973; Mukherjee, 1974; Beall, 1983; Malik & Hauspie, 1986; Bhasin & Bhasin, 1990). Menopause is regarded as the symbolic end of womanhood. The menstruation and Ovulation stopped and ceased the reproductive function of women. Age at menopause is an important biological phenomenon plays a great role in reproductive performance of women. Menopause marks the end point of a transition of up to several years during which female fecundity gradually approaches zero. According to various researchers mean age at menopause of the Indian population is between 40 and 50 years age. However, the mean age at menopause ranges from 44 to 50 years. The age at menopause is influenced by nutrition, physical environmental factors, etc. (Agarwala, 1966; Gogoi, 1972; Gunasundaramma, 1980; Beall, 1983; Sidhu & Sindhu, 1985; Mastana, 1987; Bhasin & Bhasin, 1990).

The mothers and children's health have been considered as most important factors for the improvement of national health status. For this cause, maternal and child health aspects have been included in family welfare programme of India. Nowadays, these health service activities have been continued as child survival programme and safe motherhood programme. (Kantikar, 1979; Bhasin & Basin, 1990; Basu, 1994, Prasad & Nagaraj 2001).

In a particular environment, health and disease are depended on the resources of biological and cultural factors. It has been reported that in tribal societies, health and disease is totally related with socio-cultural factors. (Pandey & Saxena 1988; Bhatia, 1990; Rajyalakshmi & Greevani, 1992; Kar & Gogoi, 1993; Dutta, 1999).

Aim and Objectives

Till date various social scientists and quite a few anthropologists have studied the social cultural dimension of Kinnauris. But no attempt has been made to get the insight of demographic dimensions and health system of the ethnic groups. The anthropological interpretation of health system and comparative analysis of health situation of various ethnic groups reflects knowledge, belief and practices related to health among the ethnic populations of Kinnaur district. The aims and objectives of the present study are as follows:

1. To study the socio-cultural dimensions of Kinnauris
2. To study the health profile and health care system of four ethnic groups of district Kinnaur, Himachal Pradesh
3. To study the drug logistic system and health development in district Kinnaur and its impact
4. To suggest the appropriate strategy for improvement of health status in district Kinnaur

Scope of The Study

There is an urgent need to investigate into the health issues in order to help the planners and administrators in planning and administering health related programmes for the community. Moreover health planner and administrators need feedback of operational information for decision making. The present study explore into health profile of four ethnic groups, drug logistic system and health development in the district and to suggest appropriate strategy to be followed for improvement of health status of Kinnauris.

Limitations of the Study

- Non uniformity of sample size among ethnic groups. Due to lesser population of three ethnic groups i.e Koli,

Lohar and Badhai in the village, the sample size of these groups is lesser than the sample size of Kanet ethnic group.

- Due to flood situation in August 2000, the field work was stropped due to lack of proper transportation services to the district. The field work was reinstated after few months. This natural calamities had unnecessary delayed the field work in the district.

Area and People

The Chapter 2 discus about the area and people, Geographical Location, Climate, Forest and Flora/Fauna, Lakes and Springs, Rivers and Water Resources, Housing Structure, Population, Language, Food, Dress etc. It also discus in details about the People, historical perspective, Ethnic Composition of Kinnaur District, Tribe-Caste Classification in Himachal Pradesh etc. The ethnographic profile of the four ethnic groups in Kinnaur district is described in the chapter.

The inhabitant of the Kinnaur, i.e. Kinnauri has been referred at several places in Sanskrit literature. There are many variants of Kinnauri such as Koonauri, Kannaura, Koonawri, Kinnaura, Kinnaur, Kinnauries, Kinauri, and Kinnauri.

The entire population is classified as Scheduled Caste or Scheduled Tribe. The population of the district comprises of four distinct classes of Kanets (Rajputs), Chamangs {(Kolis, preparing shoes, weaving, tailoring and music drum beating)} and Domangs { Lohar (Blacksmith) and Badhi (carpentry), Nagloo (Basket makers)}. The Kanet constitute the majority of Kinnaur population. They are declared as Scheduled tribe people. The other four ethnic groups i.e Koli, Badhi, Lohar and Nagloo constitute the scheduled caste population of district.

Kinnauri are polyandrous people. All brothers usually share one common wife that makes the polyandry of fraternal type. In the polyandrous family the relationship between son and fathers is not difficult to establish. Normally the practice is that the all the husbands are recognized as the fathers of each child. The eldest among the husbands is called Teg Boda (elder father) and others as Jigich Boba or Goto (Young Fathers). For day to day use the eldest among the living brothers is spoken of as the father of all the children born to a polyandrous wife. These days the institution of polyandry is gradually on its way out.The joint family system prevails in Kinnaur. Under one roof not only father, sons, brothers but also uncles, nephews of the same descent live and sometimes even own property in common. The system is under serious strain due to the advent of modern civilization in the area.

The economic structure of people is primarily agrarian and essentially rural in character with 68.7 per cent of the total working population engaged in agriculture. In the earliest time, the Kinnauri have primarily been an agro-pastoral tribe. This occupation had been followed by the people since time immemorial. Forest is very important for the Kinnauris. They are dependent on the forest for their day to day need of timber and firewood. The livestock depend on it for fodder and grazing. Besides producing rain (ecological balance), the forest produce varieties of medicinal plants which also serves as income for the people. Sizeable parts of wooded area are covered by Chilgoza trees which bears edible nuts and fetch handsome price.

Village are generally self- sustained units. Every village has a temple where people congregate for common worship. The village gods are carried in palanquins, on number of occasions, to places of religious interest or village fairs. When in trouble, the people go to the deities to seek their guidance and help. The village god is suspected to watch over the destiny of the village. He protects, rewards, threats, threatens and punishes the people, while they in turn, worship him/her by singing and dancing. Now a day, the village gods, lamaistic Buddhism and a local version of Hinduism exist side by side. Most of the ceremonies are performed by lamas both in Hindu and Buddhist families. Despite early Christian mission at Puh and other parts of the region, Christianity has not been able to spread very much in the district. The missionary influence can be seen in the habits of cleanliness and the use of refined form of knitting was acquired by the people from the missionaries.

Methodology

The Chapter 3 focus on main focuses of the chapter is on the various methods that were chosen for collecting data, during the progress of this research. It deals with the different dilemmas that were faced by the researchers in the initial stages of the study. This chapter further delineates the reasons for choosing Kinnaur as the locale for study. It talks about the selection of sample, how and where it was selected? The chapter describes the various methods that were chosen and the reasons why they were opted for? It also makes an attempt to portray the difficulties faced in interviewing the respondents, because of restricted time they had. Being totally engrossed in their family and work, they sparingly gave time to the researcher.

Village profile schedule was developed to collect the detailed information about the village. The household scheduled was prepared to elucidate the information from the each household. The information from the Households was obtained regarding Demography, Morbidity, Maternal and Child Health, Family Planning and Health Utilization and Awareness. The elaborate interview schedule designed to elicit the information from the respondents.

To ensure the authenticity of the interview schedule it was essential to put it to a pretest, which helped in developing the continuity during conduction of the interview. It was advantageous in improving the efficiency of smooth flow from one topic to another, without any interruptions. A sample of 20 household was selected during the pilot study (from June 1999 to August 1999). The main reason being the researcher had to get acquainted with a tribal area, the people and their language. Long periods, were spent with the respondents, as they were enthusiastic to talk to the researcher. Many a times they had to be interrupted gently as they used to get carried away with their talks, case studies required longer duration, as in-depth information had to be collected and along with them home visits had to be made. The intensive field work was conducted from July 2000 to August 2000, January 2001 to June 2001, December 2001-March 2002 and final validation of data and information from March 2003-July 2003.

Health Profile and Health Development of Four Ethnic Groups

The chapter 4 discuses the various demographic indicators like fertility, mortality, maternal and child health care, health seeking behaviour and family planning system of four ethnic groups and pooled (Total) population. Further, an attempt has been made to understand the Health Care System in the Kinnaur district of Himachal Pradesh in order to get insight view of the existence of traditional and other health system in Kinnaur, health institutions, infrastructure availability, supply logistic system etc. The present study has been undertaken among the four population groups namely Kanet (Schedule Tribe), Koli (Schedule Caste), Badhi (Schedule Caste) and Lohar (Schedule Caste) constitute the major share of total population. In total 743 households were surveyed from 15 villages of Kinnaur district. The total population of the sampled villages is 5970 (3123 males and 2847 female). The maximum population covered in Koli group (41.9 per cent) followed by Kanet (36.1 per cent), Koli (11.7 per cent) and Lohar (10.3 per cent). Under the study due to numerical dominance, the Koli and Kanet population have been recorded maximum in the sample. The number of Badhi and Lohar is less in total population studies due their low population in the village as compare to other population.

The majority of Kanet (40.0 per cent) had land holding between 10-30 hectares where Koli (42.0 per cent), Badhi (40.0 per cent) and Lohar (47.0 per cent) had land holding less than 10 hectares. Majority (80.5 per cent) of people in all four population groups reside in their own house. It was observed that 84.3 per cent of Koli owned their house followed by Kanet (82.5 per cent), Lohar (72.3 per cent) and Badhi (68.2 per cent). Majority of Kanet and Koli group stay in Pucca House whereas majority of Badhi and Lohar stay in Semi Pucca and Kachha houses respectively. It has been observed that 90.2 per cent of the Kanet population stay in Pucca house followed by Koli (89.7 per cent), Badhi (27.1 per cent) and Lohar (22.9 per cent). Majority of people (pooled) (39.6 per cent) reside in houses with 3-5 rooms. The maximum number of Kanet (44.7 per cent) and Koli (41.0 per cent) reside in house with 6-8 rooms whereas Badhi (49.4 per cent) and Lohar (48.2 per cent) reside in house with 3-5 rooms and 1-2 rooms respectively. The main source of water is stored water at common place and used later on through small water channels. It was observed that majority of people Kanet (64.0 per cent) followed by Koli (63.3 per cent), Lohar (47.0 per cent) and Badhi (44.7 per cent) used store water as main source of water. The housing of animals had been kept separately / at the corner of house / lower floor of the house as reported by Kanet (88.7 per cent) followed by Koli (92.0 per cent), Badhi (68.2 per cent) and Lohar (61.4 per cent).

The population composition is taken as the internal structure of a population group with regards to one or more demographic traits such as age, sex, marital status and economic activities, education characteristics etc at a given period of time.

The age distributions of complied population (pooled data of four population groups) have shown 33.4 per cent of population under 15 years and only 4.4 per cent in the age group of 60+. It shows that more or less young population composition with moderately high fertility which can be well compared with growing population structure (high proportion of young and childbearing population and less proportion of old people) of any developing region. On comparing the inter population variation, the child population in the age group 0-14 years is highest in Badhi (36.7 per cent) followed by Lohar (35.0 per cent), Kanet (34.0 per cent) and lowest value among Koli (31.5 per cent). On comparing the population (36.1 per cent) of age group 0-14 years of India, it was observed that percentage of population of Badhi between age group 0-14 years is higher than percentage of population between age group 0-14 years in India where percentage of population between age group 0-14 years of Kanet, Badhi, Lohar is lesser than percentage of population between age group 0-14 years in India.

The next broad age group of 15-49 years, comparing of population at fertile and peak productive ages, the highest per cent is found in Koli (57.9 per cent) followed by Lohar (54.0 per cent), Kanet (52.9 per cent) and lowest value among Badhi (51.6 per cent). The Koli (929) showed the highest sex ratio closely followed by Kanet (920), Lohar (874) and Badhi (859).

Further, among pooled population, it is observed that 52.0 per cent population (25.5 per cent of male and 26.5 per cent female), 44.0 per cent population (24.8 per cent of male and 19.2 per cent female), 4.0 per cent population (2.1 per cent of male and 2.0 per cent female) are under married, unmarried and widow / divorced category receptively.

The total dependency ratio was found to be highest among Badhi followed by Kanet, Lohar and Koli. The total dependency ratio for total (pooled) population worked out to be 60.66. the total dependency ratio is found to be lower than Brahmins (77.94), Rajputs (75.44) and Schedule caste (91.16) of HP, India (85.30). In study population groups, it has been observed that the per cent of family comprises of two members among the Kanet, Koli, Badhi and Lohar is 4.4 per cent, 4.7 per cent, 2.4 per cent and 4.8 per cent respectively.

The lowest Household income was observed Lohar (25.3 per cent) followed by Badhi (24.7 per cent), Kanet (18.2 per cent) and Koli (16.3 per cent) whereas the highest income grade was found out in Kanet (33.5 per cent) followed by Koli (28.7 per cent), Badhi (24.7 per cent) and Lohar (14.5 per cent).

The literacy rate (as per cent of particular population group) was observed the highest among Kanet (51.3 per cent) followed by Koli (51.2 per cent), Badhi (50.6 per cent) and Lohar (50.2 per cent). The literacy rate of total (pooled) population is found to be 51.1 per cent.

The menarche is considered as an important physiological episode. It represents the specific stage of first periodical regular flow of blood from womb in all the health normal females. The first menstruation start is taken into account as the age of puberty and maturity and from this stage, females are biologically capable to conceive. It is observed that age at menarche in Total (Pooled) varies from 12 to 15 years. The maximum per cent of women in Total (Pooled) population menstruated at the age of 14 years (43.6 per cent) and 13 years (38.0 per cent) whereas only 11.7 per cent and 6.7 per cent women achieved menstruation cycle at the age of 12 and 15 years respectively.

It is observed that in Total (Pooled), maximum menopausal age varies from the age of 44 years (37.8 per cent) to 45 years (24.6 per cent). The menopausal age of 17.4 per cent women was started at 43 years whereas only 10.2 per cent attained menopause at the age of 47 years. The mean age of menopause is found to similar (44.6 years) in four ethnic groups and Total (Pooled) except 44.5 years in Koli group.

The maximum mean age at marriage of women is found in Kanet (21.3 years) followed by Koli (21.2 years), Badhi (20.9 years) and Lohar (20.1 years). The mean age at marriage of total (pooled) is found to be 21.1 years. The maximum mean age at marriage of men is found in Koli (24.0 years) followed by Kanet (23.8 years), Badhi (22.7 years) and Lohar (22.6 years). The mean age at marriage of total (pooled) is found to be 23.6 years.

In the present study, the crude birth rate highest among Lohar (29.56) followed by Badhi (28.53), Kanet (25.97) and Koli (23.56). The Crude Birth Rate of Koli is lower because more number of Koli were adapting to family planning as compared to the other groups. The general fertility rate was found to be highest among Badhi (116.96) followed by Lohar (116.13), Kanet (101.45) and Koli (83.81).

Total fertility rate has been found to be maximum in Badhi (3.67) followed by Lohar (3.40), Kanet (3.15) and Koli (2.59). Mean age at child bearing has been found to be maximum in Kanet (27.9 years) followed by Koli (27.8 years), Badhi (27.5 years) and Lohar (26.7 years).

The highest infant deaths have been reported below age group of less than 2 months followed by age group of 2-4 months, 4-6 months and 6-8 months. The crude death rate for Badhi (22.82) found to be maximum followed by Lohar (21.35), Kanet (16.70) and Koli (14.78). The crude death rate for total population worked out to be 17.09. The neonatal mortality rate is found to be highest in Lohar (111.11) followed by Badhi (100.00), Kanet (53.57), Koli (50.85) The infant mortality rate was found to be highest among Lohar (166.67) followed by Badhi (150.00), Kanet (107.14) and Koli (101.69).

In pooled population, 87.2 per cent of women consult somebody when they suspect pregnancy. Further, majority of women (38.7 per cent) consulted their mother-in-law in the family followed by consultation with any other (sister, sister-in-law, neighbour etc) (23.0 per cent), LHV/ANM (21.5 per cent) The highest number of women had gone for ANC were from Kanet (96.4 per cent) followed by Koli (96.3 per cent), Badhi (89.4 per cent) and Lohar (85.5 per cent). It is observed that maximum number of women from Kanet (78.9 per cent) followed by Koli (65.7 per cent), Badhi (46.1 per cent) and Lohar (38.0 per cent) visited nearby SC/ PHC. Majority of people

(48.0 per cent) of all population groups consult PHC staff (i.e ANM, Doctor, Nurses) for problem during the pregnancy. Majority of last deliveries (80.5 per cent) were conducted at their respective houses only. The highest number of deliveries at home were conducted in Lohar (95.2 per cent) followed by Badhi (91.8 per cent), Koli (77.7 per cent) and Kanet (75.6 per cent). It has been reported that majority of deliveries were conducted by TBA (68.9 per cent) followed by ANM(18.4 per cent), doctors (9.3 per cent), Elderly women in the village(4.4 per cent), untrained TBA (2.3 per cent) and Family member (1.6 per cent).

In majority, Lohar (22.9 per cent) followed by Badhi (21.2 per cent), Kanet (20.4 per cent), and Koli (20.3 per cent) had knowledge about Natural-coitus interrupts/abstinence. The most of people had replied with multiple choice about discussion about family planning method with spouse, village dai, PHC staff, neighbour, relative and friends. Still Badhi (28.2 per cent) followed by Lohar (25.3 per cent), Koli (17.3 per cent), and Kanet (16.4 per cent) discussed with their spouse about the family planning methods. In all population groups had positive attitude towards restricting the number of children and adopting small family norms. In Koli (89.0 per cent) followed by Kanet (88.7 per cent), Badhi (69.4 per cent) and Lohar (66.3 per cent) have positive attitude towards restricting the number of children and adopting small family norms. The most of people had replied with multiple choice for their believe in children as god's gift or they are helping hand in the family or necessary for looking after in old age. Still think that Kanet (20.4 per cent) followed by Koli (15.0 per cent), Badhi (12.9 per cent) and Lohar (9.6 per cent) thinks that children are necessary for looking after in old age.

In Kanet (57.8 per cent) followed by Koli (57.3 per cent), Lohar (43.4 per cent) and Badhi (34.1 per cent) husband and wife both participate in decision making process for contraceptive usage The highest number of women availing family planning services from PHC staff were from Koli (87.0 per cent) followed by Kanet (86.9 per cent), Badhi (80 per cent) and Lohar (79.5 per cent).

The highest number of women using any family planning method currently were from Kanet (64.7 per cent) followed by Koli (62.3 per cent), Lohar (55.4 per cent) and Badhi (51.8 per cent). The highest number of women motivated by medical person were from

Kanet (39.3 per cent) followed by Badhi (41.2 per cent), Koli (34.0 per cent) and Lohar (32.5 per cent). The highest number of women satisfied with family planning method were from Koli (40.3 per cent) followed by Kanet (39.6 per cent), Badhi (37.6 per cent) and Lohar (33.7 per cent). The maximum medical attention given by government doctors as reported by Kanet (80.4 per cent) followed by Koli (79.7 per cent), Lohar (74.7 per cent) and Badhi (68.2 per cent).

The Kanet (85.1 per cent) followed by Koli (82.0 per cent), Lohar (80.7 per cent) and Badhi (74.1 per cent) responded positive about the medical facilities available in the village. The Kanet (59.3 per cent) followed by Koli (50.3 per cent), Lohar (18.1 per cent) and Badhi (15.3 per cent) responded that the immunization of programme was carried out once in month in the village. Most of the respondent replied complained about inadequacy of medical facilities in nearby health institutions. In Koli (55.7 per cent) followed Badhi (54.1 per cent), Kanet (48.0 per cent) and Lohar (42.2 per cent) respondent with inadequacy of medical facilities.

Role of Health Department

As the role of Health Department in Modern Medicine, Ministry of Health and Family Welfare is instrumental and responsible for implementation of various programmes on national scale in the areas of Family Welfare and Prevention and Control of major communicable diseases. Apart from these, the ministry also assists the state in preventing and controlling the spread of seasonal disease outbreaks and epidemic through technical assistance.

To tackle the menace of communicable and non communicable diseases, the Department of Health is implementing National Health Programmes through out the country for Malaria, Tuberculosis, Leprosy, Blindness, AIDS, and Cancer etc.

The offices of Health and Family Welfare functioning under Ministry of Health and Family Welfare located in the state capitals. These offices have been established to liaison and coordinate various national health and family welfare programs with the states, monitor the centrally sponsored health and family welfare schemes, provide on the spot technical guidance, test checking of records in the state, maintain the malaria clinic, review and analysis of monthly technical reports, supervise health information and field units and associate in various implementation and grant sanction. Director, Shimla is on the strength of Department of Health and Family Welfare.

At the State Level, the Department of Health and Family Welfare is the apex body to cater to the health services in the whole state. The main function of the Department of Health and Family Welfare is to manage the functioning of the government health institutions/hospitals in the state. Department of Health and Family Welfare is primarily responsible for providing the health services in the state. The Director of Health and Family Welfare is the Chief Executive of the department and is responsible for providing health services. The Director is assisted by the Deputy Directors who are given the responsibility of the various national health programmes.

The Department of Health and Family Welfare, Himachal Pradesh has taken an initiative of developing the Himachal Health Vision 2020. The concept of Vision 2020 for the State of Himachal Pradesh was the brain wave of Prof. Prem Kumar Dhumal, Hon'ble Chief Minister of Himachal Pradesh when he delivered an address at " The Partnership Summit-1999" at Confederation of Indian Industries (CII) Conference at Jaipur and his speech in Himachal Pradesh Vidhan Sabha in the budget session of 1999-2000. He reiterated his resolve to adopt Vision 2020 for the State while presenting the Budget Estimates for 1999-2000 and declared that Himachal Pradesh should be one of the top five States in the country in the next 20 years. According to Vision 2020, the state should attain the status equivalent to that of the developed countries.

The Government of HP consequently issued directions to all the departments to fulfill the budget assurances and to constitute task forces for drafting Vision 2020. The Health and Family Welfare Department constituted a Task Force comprising eight members for developing Himachal Health Vision 2020. The main objective of preparing the document was to spell out the long term goals with clearly defined strategies and propose concrete actions to achieve Health for All within 20 years through situation analysis of current and future health scenario.

At district level, the Chief Medical Officer (CMO) is the administrative head to take care of the health needs. Deputy/ Additional CMO is the second person in -charge. Each Deputy CMO is given the responsibility of particular national programmes/zones. District Malaria Officer, District Leprosy Officer and District Immunization Officers look after the concerned programmes in the district. Kinnaur has district hospital where the curative services are provided. In district hospital, there are specialist cares available like paediatric, ENT, Ophthalmology, Obstetrics and Gynaecology, general Surgery and general medicines.

While estimating the budget for the Health and Family Welfare Department, certain points are considered, which are number of Blocks in the district, SC/ST Population, number of Health Institutions etc. At the district level, the Chief Medical Officer (CMO) prepares the Annual Budget for the District and submits it to the State-Department of Health and Family Welfare in the month of December. Normally the budget for the last year is considered as the basis for estimation of current year's budget. The last year's budget is augmented by approximately 10 per cent to estimate current's budget. While preparing a budget, the number of the patients treated is also considered. At the Block level, there is no provision for preparing the budget. Similarly at sector and Sub Center level, no budget is prepared. There is a fixed budgetary allocation for CHC/PHC. While preparing the budget, least consideration is given to the Annual Action plan prepared at each level. A top down approach is followed for estimating the budget in Health Department. The ANM/Supervisors/Medical Officer-PHC/ BMO do not play any role in the preparation of the budget.

Other Health System

Indian System of Medicine and Homoeopathy play a vital role in the Health Care system of the state. Because of the variety of climatic conditions it has traditionally been rich repository of medicinal plants use in various Indian System of Medicine namely Ayurveda, Tibetan, Bhot System, Unani and Yoga flourished in its cradle since time immemorial. In the tribal areas, Tibetan System of Medicine under the name of Bhot Chikitsa Padhati continue to the popular. There are several areas which remain cut off from other parts of the state and country on account of heavy snow fall and rain fall and the people have to depend upon the traditional local system of medicine for their health requirement. Incidentally these very areas are a natural habit of several rare and very useful medicinal plants. The people of Himachal Pradesh traditionally have immense faith in Ayurvedic and other Systems of Medicine and in this context the Government of Himachal Pradesh embarked upon a programme to serve the people living in the remotest corners through these systems. Accordingly, a large number of dispensaries and hospitals were opened and at present 1/3 of the total patients (3 lakhs) in the State are meeting their health requirements through the network of these hospitals and dispensaries. The Department of Ayurvedic (Indigenous medicine) functions under the control of a Director for overall administration and programmes implementation. Kinnaur district has one Ayurvedic hospital and 39 Ayurvedic dispensaries.

Tibetan Medicine System

In earlier time, the Tibet Medicine system was greatly followed in the Kinnaur. According to one of the responded Lama in Kinnaur, there are about forty books or works written in Tibet, on medicine, besides the five volumes in the Stan-gyur collection, and the scattered occasional instructions on medicament in the Kah-gyur. The contribution of Buddhist monasteries in inculcating humanistic values in the social system of the Lamaistic western Himalaya, comprising Ladakh, Lahul, Spiti and Kinnaur, has been overwhelming. These institutions ushered in an era of socio-cultural resurgence in a region which had long remained infested with innumerable primitive beliefs and superstitions under the generic name of lung-pa'I chos. The socio-religious impact of the monasteries has been relative to the religious following which these institutions command. This is at its greatest in upper Kinnaur, sparse in the lower part of Kinnaur.

Other Health Care Systems

Villages are generally self-sustained units. Every village has a temple where people congregate for common worship. The village gods are carried in palanquins, on number of occasions, to places of religious interest or village fairs. When in trouble, the people go to the deities to seek their guidance and help. The village god is suspected to watch over the destiny of the village. He protects, rewards, threats, threatens and punishes the people, while they in turn, worship him/ her by singing and dancing. It is believed that the village gods are benevolent spirits who never fed displeased with their devotees. However, these spirits are accompanied by attendants known as ganas, who inflict curse on deviants. These curses are called dosh (displeasure) and the evildoer has to correct himself in line with the wishes of the god.

The private health care system in Kinnaur district is limited to presence of RMPs and other local non-registered medical practitioners. Most the people prefer to visit government health institutions in the district.

Key Demographic Indicators—At a Glance

Demographic Indicators	*Ethnic Groups*				
	Kanet	*Koli*	*Badhi*	*Lohar*	*Total (Pooled)*
Sex Ratio	920	929	859	874	912
Masculinity Proportion	52.1	51.8	53.8	53.4	52.3
Young Age Dependency Ratio	55.96	48.58	62.23	57.41	53.58
Old Age Dependency Ratio	8.87	5.6	7.51	6.74	7.08
Total Age Dependency Ratio	64.83	54.19	69.73	64.15	60.66
Index of Aging	15.85	11.53	12.06	11.74	13.21
Crude Birth Rate	25.97	23.56	28.53	29.56	25.63
General Fertility Rate	101.5	83.81	117	116.1	96.71
Total Fertility Rate	3.15	2.59	3.67	3.4	2.98
Gross Reproductive Rate	2.15	1.42	1.49	1.5	1.33
Child Women Ratio	480.47	397.85	407.64	408.45	428.86
Crude Death Rate	16.7	14.78	22.82	21.35	17.09
Infant Mortality Rate	53.57	67.8	100	111.11	71.9
Under 5–Mortality Rate	178.57	152.54	200	277.78	183.01
Neonatal Mortality	53.57	50.85	100	111.1	65.36
Post Neonatal Mortality	53.57	50.85	100	166.67	71.9

Condition of Health Service Delivery at Grass Root

(i) Availability of Health Facilities

To cater the health services in 3 blocks of Kinnaur District, there are 1 District Hospital, 1 Civil Hospital, 3 Community Health Centers and 17 Primary Health Centers. 32 Sub Centers provide their services in villages of Kinnaur district. It was observed that most of the sub centers lack the basic amenities like toilets, electricity and water supply. Moreover in few of villages, due to outskirt location of the Sub Center, ANM feels insecure and find difficult to stay in the Sub Center.

(ii) Availability of Staff

It was observed that at the district level, due to vacant posts of Medical office and specialist, the existing staffs shares the workload. The increasing number of patients in the hospital puts a lots of work pressure on the staff. Similarly it was observed that due to remote location of PHC/CHC, the medical officers do not prefer to work there. In Pooh Block, it is observed that in the absence of Medical Officer, the whole affair of PHC is taken care by Pharmacist and ANMs. During focus group discussion with medical staff, they were of opinion that non-availability of staff puts lots of pressure on them. Due to wider and difficult geographical terrain, some time they feel difficulty in providing the service to the people.

Convergence of Services

To care the health status of the community, the normative convergence between ICDS and Health and Family Welfare Department states that AWW will take care of Nutritious Food, Growth Monitoring, Pre School Education, support services to the ANM for organizing the Health/Immunization camp in the village, motivating the women for family planning and immunization. Anganwadi Worker coordinate with ANM for carrying out health related activities like Growth Monitoring, Immunization, Health Check up, treatment of worms in the stomach, Anaemia, Goiter, deficiency of Vit A.

(i) District Level Co-ordination

In the Kinnaur district, there is norm to organize the jointly monthly meeting between CMO and DWCDO. But in actual practice, occasionally the officials of both the departments interact with each other to discus issues related to their functioning. The meeting is participated by Collector, DMO, DWCDO, MOs. Due to their pre-engagement and other reasons, the convergence did not seem to be working out effectively.

(ii) Block Level Co-ordination

In the Kinnaur district, BMO and CDPO meet occasionally to discus the issues related to functioning of ANM and AWW at community level. It was observed that due to shortage of staff in CHC, the BMOs do not get time to attend the meeting timely. Even if the meeting is organized, the agenda and follow-up of the meetings are not very much clear for effective implementation.

(iii) Sector Level Co-ordination

It is observed that the sector supervisors ICDS and Sector Supervisors-Health Department never organize the joint sector meetings to solve the differences between ANM and AWW and formulate the strategies for the improving the health status of the villagers.

(iv) Community Level Co-ordination

Since beginning of the ICDS programme, the ANM and AWW are jointly organizing Immunization camps in the village. In districts, the Nutrition and Health Day is observed on first Tuesday of the each month. During the Nutrition and Health Day, AWW distribute the nutritious food, record growth and assist the ANM in organizing the immunization camp. It was observed that occasionally absence of ANM on Health and Nutrition day without informing the AWW creates problem for Anganwadi Worker and bring a negative reaction among the villagers about health department ANM and AWW jointly maintain the records related to pregnant women, lactating mothers, immunization, birth and death in the village. ANM keeps certain items like ORS, Bleaching powder, family planning items at AWC.

Problems at Each Levels

(i) Problems Faced by Store Keeper

During the course of discussion with the Store Keepers Health Department, problems faced by them are reported which are as follows:

- Lack of proper storage place for medicines and other equipment
- There is no provision of regular electricity and proper ventilation in the store
- Lack of storage infrastructure in the store i.e. almirah/ racks tables/chairs etc.
- Due to other engagement of the store keepers in their departmental affairs, so called "Store Keeper" gets very little time in managing the work of the store room properly

(ii) Problems Faced by ANMs

The problems faced by the ANM are as follows:

- Due negative attitude of the villagers towards the Modern health system, ANM faces lots of problem in providing the health services in the villages;
- In certain places it is reported that due to lack of co-operation from the AWW, ANM faces problems in carrying out the immunization programme in the villages;
- The short and irregular supply of medicines is the biggest problem of the ANM in the village. Due to this they find themselves handicap in providing the health services to the villagers;
- Due to far away distances of the villages from the Sub Centers, the ANM faces problem in carrying the medicines/vaccines to all the villages comes under Sub Center. Due to this, ANM face problem in providing the treatment and medicines to the patient during odd timing;
- The increase in the population size of the villages, now the ANM has to provide the health services to the population of more than 3000 population;
- Due to lack of infrastructure like lack of proper building, damaged building structure, less no. of rooms for Sub Center in the village, ANM faces problem in storage of medicines and conducting deliveries at Sub Center;

- Due to lack of equipments like No Delivery table, No stove for sterilization at Sub Center, ANM faces problem in providing the better health services;
- In the absence of proper stationary, ANM faces problem in maintaining the record at the village level;
- It is also observed and reported that due to co-operation from the supervisors, ANM does not get the proper guidance and assistance in providing the health service at the village level;
- In most of the cases, due to outskirts location of the Sub Center Building in the village, ANMs face security problems;
- ANM always have pressure from the higher authorities to co-ordinate with the AWW in providing the health services at the community level. Due to low level of literacy of AWW, ANM faces problem in co-ordination;
- Due to the absence of residential facilities at the Sub Center/ Villages, most of the ANM faces problem. And it is the one of the major reason for non-staying of the ANM at their HQ.

(iii) Problems Faced by BMO

During interaction with doctors and focus group discussion with them, it has been noticed that villagers still prefer to opt for traditional medical system. When they fail to get good result or illness become critical, only than they approach doctors or modern health system. The problems faced by the BMOs are as follows:

- It is being reported that BMOs at the block level are not given the power to spend the amount for purchasing certain medicines at the block level. They find themselves handicap while handling the epidemic incidence in the block;
- No transportation budget is provided to the BMOs. Due this constrain, they face problem in supplying the medicines and other equipment to the PHC/Sub Center;
- Due to appointment of lesser number of doctors at CHC, BMOs are always pre occupied in managing the daily affairs of the CHC. They face problem in monitoring and inspecting the work of PHC/ supervisors/Sub Center.

Health Development Through NGOs

During the course of visit to Kinnaur, various NGOs were contacted to know their perception about health and development issues in district. It has been observed that 83 major NGOs are operating in the state whereas 3 Kinnaur based active NGOs are working in the district. Apart from these 3 NGOs, occasionally NGOs from other districts and states implement time bound projects in the district Kinnaur. The objectives of the most of these organisations are to create awareness about clean drinking water, health and sanitation, to promote community participation, to plan, to create awareness among the youth and adolescent girls about the health, sanitation etc.

During the course of discussion with NGO personal in Kinnaur, it is observed that NGOs have greater outreach in the villages. These NGOs are directly or indirectly implementing health related programmes like IEC activities or Health Camps in the village. The grassroot level approach and dedication are major factors for NGOs success in the community. On being asked about NGOs role in improving the health status of the community, the representatives of NGOs gave following suggestions:

- NGOs should be involved in organizing the Information, Education and Communication activities in the village to bring about the awareness about the Modern health system and role of Anganwadi worker in providing the health services in the village. According to them, till date no advocacy for the role of Anganwadi Worker is done at the community level. Due to this, villagers always misunderstand and limit Anganwadi Worker's role in providing the nutritious food only;
- According to the NGO personals, it is very essential to provide proper counseling and motivation of the Pregnant and Lactating women for immunization and usage of family planning methods in the villages;
- In order to improve the supply of the Medicine and reduce the transportation cost, the NGOs field office/ community worker should be used as the depot for keeping the Medicines in the Villages. It will not only reduce the transportation cost/burden but also enhance

the availability of the medicines in the community. It will be the responsibility of the recognized NGO in the district/block to collect the medicines from the Health Department/ICDS department and store it either at their Field Office/community worker or distribute it to the AWC/SC;

- It is observed that most of the ANM and AWW lack the knowledge about the usage of IEC material in the community. They are still using the traditional methods of IEC which are not effective in the present context. NGOs can play a major role in organizing training programme for AWW and ANM for use of Modern and cost effective IEC material to bring the awareness among the villagers.

The major threats to the NGOs in the Kinnaur are lack of long-term commitment from the funding organization, isolation from main land, lack of technical skills related to project formulation and implementation, availability of trained manpower and non-recognition from the state government officials. Whatever programme they had implemented in the villages is of short duration with a limited impact on the community. (Kapoor and Singh, 1997)

STRATEGIES FOR IMPROVEMENT OF HEALTH SERVICES IN DISTRICTS

Identification of Underserved Areas

The initiatives for development of ethnic population of Kinnaur can be most effective and optimum, if it is implemented in the areas not properly served by the public health facilities. The problem is to define and identify the underserved areas. Underserved area can be defined on the basis of priority of health sector reform. However, some common indicators can be considered for identifying the undeserved areas, which are:

- Infrastructure conditions
- Population Coverage (Ethnic Group-wise)
- Manpower availability
- By Comparison of Demographic indicators like MMR/ IMR/Death rate/Birth Rate (Ethnic Group-wise)
- Programme Output

Developing Directory of Non Government Organisations Himachal Pradesh (Kinnaur)

The NGOs have a crucial role to play in the process of nation-building and planned socio-economic development everywhere, particularly in developing nations like India. With a view to improving the Health service delivery at peripheral level, Department of Health and Family Welfare, Himachal Pradesh should identify the good and dedicated NGOs working in various parts of the district and involve them in healthcare service delivery.

Methodology to be Followed for Preparation of NGOs Directory

Health Department should collected exhaustive information about the NGOs and their activities from various sources, such as: Department of Social Welfare, Mother NGOs in Himachal Pradesh (Ministry of Health & Family Welfare), Himachal Pradesh State AIDS Control Society, CAPART, District Social Welfare Department.

For collection NGO's details working in Himachal Pradesh (especially Kinnaur), a model questionnaire should be prepared. The data should be collected through:

- **Postal Survey:** The questionnaires should be mailed to NGOs, whose names and addresses are collected from various sources. The NGOs should be given a cut off date for submission of the filled-in questionnaires in the self-addressed stamped envelope provided along with the questionnaires.
- **Personal NGOs Visit:** The BMO/MO should interact with the representative of NGOs at their respective jurisdiction.

(i) Methods for Updating NGO Database

The NGOs database should be regularly updated in order to include the new NGOs and to delete the non-functioning and closed organizations from the database. The updating process cannot be a one time activity. It should be so meticulously planned and deftly executed that the information about the existing and new NGOs flow at a regular interval to the CMO's office where the database is maintained and updated. The following steps may be taken to ensure regular update of NGO database:

- **Appointing NGO Coordinator:** CMO may engage one person for coordinating with the NGOs. His/her main responsibility will be to keep track of all the NGOs working in the district and interact with them regularly, disseminate information about various health projects and programmes.
- **Mailing NGO Information Forms:** The NGO coordinator should mail the NGO information form to all the NGOs, whose names are available in its database at a regular interval (preferably annually). The NGOs will be requested to furnish the latest information about their organization along with requisite documents within a given data. They will also be requested to provide the name and address of any other NGOs working in their project area or district.
- **Involving Block Medical Officer and other district officials:** An executive committee will be constituted at District level to facilitate the process of NGO empanelment. The committee will be headed by the Chief Medical Officer and may include the District Field Officer, District Immunization Officer, representatives of voluntary agency and also other identified district level official like District Social Welfare Officer etc. The NGOs working in the field of Health regularly interact with CMO and other district officials like District Immunization officer, District Health Officer, District Training Officer etc. The CMO or District Officials should be entrusted to provide name and address of NGOs. They should also make it mandatory for all the NGOs to register themselves with CMO office. The other district offices like ADC Office, District Social Welfare Office etc can also be involved for providing the name of NGOs supported by their department. NGOs registration can be facilitated by improving inter departmental co-ordination.
- **Receiving the Information Forms and Updating the Database:** The Information Forms received from the NGOs should be entered into the database. For the

existing NGOs, the database should be updated with the changes, and for the new NGOs, new record will be created.

- **System of Empanelment of NGOs:** The Health department may initiate the system of empanelment of Non-Government Organisations with the department for engaging them in Health programmes/projects. The NGOs which submit the filled-in Information Form to the Health Department may be issued a certificate of empanelment with District Hospital and considered for involvement in health related projects subject to their qualification and experience.
- **Annual District-wise NGOs Meeting:** One-day NGOs meeting can be organized annually at the district level under the chairmanship of CMO and district officials. During the meeting, the NGOs should be advised to update the information provided earlier. This will enhance the understanding between public and private sector and it will also help in understanding the various projects implemented by various NGOs in the district.
- **Involving Mother NGO:** GOI has identified mother NGO for Kinnaur district. The Mother NGO has been given the responsibility to identify the NGOs in the district. As the Mother NGO, it is there responsibility to identify the NGOs, pre-appraisal, project proposal acceptance, grant sanctioning, training of NGOs, monitoring and evaluation of projects etc. The NGOs registration can be facilitated by actively involving the Mother NGO in district Kinnaur.

(ii) Capacity Building Need Assessment for NGOs in Project Formulation

It has been observed that NGOs working in the district are good, dedicated and hard working. The services of these NGOs may be utilized very effectively if they are properly guided and nurtured. NGOs can be very useful in executing a project or different components of a project. But the NGOs often lack in the skill of developing a project proposal or formulating a new project with new ideas. Before initiating any activity for involving the NGOs in Health programs, it is essential to develop their skills in new project formulation and preparation of project proposals, so that they can come ahead with innovative projects in health programmes.

Capacity Building Need Assessment for NGOs is the process of identifying the gaps between what NGOs know in project formulation and can do in relation to their work and what they ought to know and be able to do to perform a project more efficiently. Identification of capacity building needs forms the basis for developing a capacity building training. The principal advantage of making needs the starting point for planning is that this strategy consciously focuses attention on the participants-the NGOs. The broad objective of the Capacity Building Need Assessment is to identify the NGOs and their capacity building needs in project formulation with special reference to Health programme.

It is important to design a workshop for capacity building of the NGOs in project formulation and proposal preparation. The workshop will help in developing the skills among the representatives of Non government organization and voluntary organization working in the field of health and family welfare about the project formulation and to acquaint them with its basic concepts, scope and methodologies to help find feasible, cost effective innovative health projects.

Developing Public-Private Partnership in Health Sector

The state government can take initiative in developing a model of public private partnership in Kinnaur district with financial assistance from international funding agencies. For more than a quarter century, NGOs have played a key role in the provision of health services in Himachal Pradesh. The Department of Health and Family Welfare assumed an increasingly important role in the direct provision of services. NGOs, for which national/ international donors had been the largest source of financial support, face increasing hardship, as did the profit-seeking organisations. Notwithstanding the NGOs and profit-making organisations have continued to seek opportunities to rationalise health services delivery, reduce the cost burden on the population and improve the quality of care. (Khanna and Kapoor, 2004).

Whereas the public health institutions lack professional approach and commitment, the profit-seeking private health organisations fail to reach the poorest of the poor because of their high cost of service and unwillingness to work in the remote areas. The non-profit seeking voluntary organisations, commonly known as NGOs, have the dedication and willingness to work for the poor in the remote areas but most of them lack professional training. Hence, there is a need for developing a mechanism of partnership among the public and private sectors (for-profit and not-for-profit) to supplement the strengths of each other.

(i) Public-Private Partnership—What does it mean?

The term "public-private partnership" is used to refer specifically to the collaborative programmes between the public sector and the not-for-profit/for-profit private sector. Public Private Partnership can achieve positive public health results and at the same time meet individual organisational goals of the partners. Such partnerships allow considerable leveraging of each partners resources and unique strengths and results are often attained in less time, at lower cost, and with greater sustainability than efforts by any single partners

The Public Private Partnership begins when funding agencies decides to take on the catalyst role and bring public and private organisations together to tackle a specific health problem. The catalyst investment makes it possible for public and private organisation to join forces and apply their strengths to achieve health goals.

The health sector can be divided into a number of different organisational groups. It is important to recognise that the terms "public" and "private" is used in loose and ambiguous ways. Some times 'private' is used as the opposite of 'public' to refer to all non-state organisations whether profit-seeking or not, and sometimes the term is used to refer to that set of non-state organisations that seek profit. The characteristics of public and private health service providers are shown in the Tables 5.1 and 5.2 below:

Table 5.1: Characteristics of Public and Private Providers

	Public Sector	Private Sector		
		Voluntary	Profit-Motivated	
			Private for Profit	*Company Health Services*
Objective	Social Promotion of welfare	Social Promotion of welfare	Profit Generation	Productivity increase
Target Group	Broad with emphasis on the undeserved	Specific target groups or locations	High income consumers of health care	Worker and their dependent
Structure and Methods	Frequently rigid and homogenous	Potentially varied and flexible	Varied and flexible	Varied and flexible
Regulatory Mechanism	Provider of regulatory mechanism. Has legal powers	Regulated by outside bodies	Regulated by outside bodies	Regulated by outside bodies
Funding	Social Means (eg. Tax) and service sales	Donations, service sales and government funding	Service sales, government funding is form of subsidies	Company budget, government funding in form of subsidies
Accountability	Electorate service user	Donors, service users trustee	Owners of capital, service users	Company

Table 5.2: Partners' Contribution to PPP and Benefits

Partner	*Contributions*	*Benefits*
Private sector	• Assigns personnel for planning and championing the effort. • Implements strategy using own resources. • Continues to use new techniques and approaches	• Leverages resources to achieve project goals. • Creates new alliances with public sector and other organizations. • Learn new methods of health research. • Shares risks of health development.
Public sector	• Assigns personnel for planning and coordination with the private sector. • Provides health expertise to the private sector. • Assists in implementing the strategy. • Motivates involvement at the local level.	• Improves public health in less time and with less investment. • Improves and strengthens its programmes. • Leverages resources to achieve health goals. • Reinforces healthy behaviour at the household and community level. • Enlists new, "non-traditional" partners in public health.
Catalyst	• Facilitates partnership. • Provides technical assistance. • Guides development of the strategy. • Monitors implementation. • Disseminates information about results.	• Leverages resources to achieve project goals. • Brings about sustainable, public-health oriented changes in private sector approaches. • Demonstrates the benefits of PPP. • Provides a proven approach other organizations can use.

(ii) Role of Private Sector in Health Service Delivery

The private sector and relevance of public private partnership policies in the health care services are perhaps the least studied areas in our country. Notwithstanding all statements made in our plan and the policy documents, it is evident that in health care, it is not the public sector but the private sector which occupies a predominant place. This is clear from the fact that over 80 per cent of the health care expenditure in our country is accounted for by the private sector. The private sector is conspicuous by its absence in our plan documents. The public-private partnership is not even tried out at a smaller scale to understand their relevance. Not only that, certain public private partnership methods used in health care, like contracting out in building construction, involvement of private doctors in the disease control programme, occasional philanthropic donations for public health institutions etc have not been studied and their efficiency compared with the public provision of health care.

The voluntary health effort, as it exists today in Himachal Pradesh, can be broadly classified as follows:

- *Specialized Community Health Programme:* Many of these programmes look beyond health and run other schemes such income-generation schemes so that the poor people can meet their basic nutritional needs;
- *Integrated Development Programme:* Health is a part of these programmes. Their approach towards healthcare may not be as systematic or as effective as that of the previous grou;
- *Health Care for Special Groups of People:* This includes education, rehabilitation and care of the handicapped. As very few government infrastructures exist in this sector, the private sector organizations specialised in this field play an important role;
- *Government Voluntary Organisation:* These are voluntary organisations which play the role of implementing government programmes like Family Planning and Integrated Child Development Services;

- *Health Work Sponsored by Rotary Clubs, Lions Clubs and Chambers of Commerce:* They normally concentrate on organising different camps, such as eye camps for conducting cataract operations in the rural areas on a large scale;
- *Health Researchers and Activists:* The efforts of these groups are usually directed towards writing occasional papers, organizing meetings on conceptual aspects of health care and analyzing government policy through their journals;
- *Campaign Groups:* These groups are working on specific health issues, such as a rational drug policy and amniocentesis, among others.

A good number of NGOs are working in the above areas of the State and are trying to develop alternative models for providing quality healthcare to the poor at a low cost. However, none of these models can be considered as perfect and universally replicable. This is truer in case of partnership with profit seeking private health sector. In fact, no model has been implemented in India where profit-seeking private organisations have truly entered into partnership with public sector for providing health services, particularly clinical health services, for the common people.

Privatization of public health facilities is another tricky issue, which needs policy decision at government level. The implementation of Public Private Partnership mechanism depends on government's willingness to accept private sector as partner in health service provision.

The NGOs may be involved in the following areas of activities:

- Campaigns, advocacy and lobbying both at central and state levels evolving new policies aimed at improving the health status of tribal people;
- Organizing formal and non-formal training programmes and doing follow up to strengthen capacity building of voluntary agencies and their members;
- Strengthening grassroots-level health care delivery by equipping village health workers with training and communication materials;

- Implementing effective communication strategies through various audio-visual and folk mediums;
- Dissemination of information on various health issues among the people at various level.

(iii) Partnership Mechanism

There is no time-tested model for public-private partnership in health sector in India. Some efforts have been made sporadically in some places, but they are very specific to certain socio-economic conditions and not replicable to other places. The model of handing over the management of PHC to SEWA-Rural at Jagadhia in Gujarat, or the model of involving NGOs in providing non-clinical services in one or two hospitals in Himachal Pradesh, or the model of partnership under Andhra Pradesh Urban Slum Healthcare Project are successful to some extent in their respective area and social conditions. None of them is proven as scalable on larger area. Therefore, for Himachal Pradesh, one should be very cautious in selecting any particular model and implementing it at a large scale. It is always recommended to implement one model as a pilot case on one PHC/CHC area, rectify the flaws of the model and then replicate it on a larger area. The following steps may be taken for involving the private/NGO sector in Health programme.

(a) Need Assessment

Before taking up any project, the need for the particular project should be assessed in the target area and among target population.

(b) Project Identification

Partnership with NGOs can be developed for implementing each identified project in the target area. The partnership strategy may vary depending on the need and outcome of the project.

(c) Deciding on Partnership Mechanism

Based on the project that is to be implemented in a particular area, the partnership mechanism has to be selected from a number of options and fine tuned as per the condition of the local area. The most model which seems most suitable for the local condition and the project may be adopted and adapted as per the needs.

(d) Developing Terms of Reference (ToR) for NGOs/ Private Parties

The ToRs outlining the partnership arrangements should be worked out before hand. These may include (a) objectives of the partnership arrangements, (b) tasks required to be carried out by the NGO/private parties, (c) time schedule for completion of preparatory tasks such as recruitment of staff and household survey for identification of eligible couples, (d) facilities that would be provided by the Government, (e) reports required to be submitted by the NGO/private parties and (f) review mechanism that would govern funds flow.

(e) Inviting NGOs/Private Parties to Participate

Based on the partnership arrangements and the ToR, the NGOs/private parties should be invited to participate in the project. The invitation to the private parties may be open or limited as suggested in the partnership model.

(f) Implementation Through Private Participation

The NGO/private parties for a particular project will be selected based on certain selection criteria developed by Health Department, and the project will be implemented as per the partnership mechanism.

(iv) Alternative Models of Partnership

Three models/mechanisms are suggested for involvement of private/NGOs:

(a) Nodal Agency Model

The Department of Health and Family Welfare may engage Mother NGOs* in Himachal Pradesh state as the Nodal Agency for involving private and voluntary sector in Kinnaur District. The Mother NGOs are already selected by the Ministry of Health and FW (GOI) under RCH scheme , and have requisite experience in selection of small NGOs, their capacity building, monitoring and

---

* Mother NGOs are the organisations, selected by Ministry of Health and Family Welfare for identification, pre-appraisal, funding support and monitoring and evaluation of small NGOs and their activities on behalf of the Ministry.

evaluation, fund releasing and submission of progress report to Ministry. They work as extended arms of Ministry of Health and Family welfare for selection of NGOs and routing the central government funds for improving the health status in the state. The Mother NGOs are allotted certain number of districts in the state depending upon their experience in health sector and outreach in the state. These Mother NGOs are selected after thorough screening of their track records, infrastructure availability, and manpower capability, number of projects handled in health sector etc.

Responsibilities of Nodal Agency

The responsibilities of Nodal Agency will be:

- To identify the local grass root level NGOs in Kinnaur district or some other districts;
- Pre-appraisal of NGOs;
- Collection and screening of project proposals on behalf of state government;
- Short listing of NGOs;
- Approval from Health Department for fund sanctioning;
- Releasing the fund to NGOs;
- Capacity building of NGOs for project implementation;
- Developing reporting format and collecting the quarterly reports from NGOs;
- Monitoring and evaluation of project;
- Finally sending the report to state government/ funding agency.

Mechanism for State Government and Nodal Agency Partnership

- Preliminary Interaction with the Mother NGO.
- Briefing them about type of projects to be funded.
- Allocating fund to Nodal agency for realising to the NGOs.
- Providing 15 per cent of the total fund as the administration cost to the Nodal Agency for carrying out their work effectively.

General Principles

As it is primarily the interest and need of the Health Department to enter into such a partnership, it is desirable that the Department identifies appropriate Nodal Agency and invites them to participate in its programmes. It cannot be expected that the NGOs will always approach the Government for financial and technical support.

The relationship should be based on mutual trust and complement, where both state government and Nodal Agency admit, understand and appreciate each other's strength and limitations. Since it would be unrealistic to expect the NGOs to master these intricacies in a short period of time, therefore considerable flexibility in these respects on part of the Nodal Agency / State Health Department is called for.

Functional and financial accountability on part of the Nodal Agency / NGOs will need to be defined differently for partnership project from government department, but will be an essential part of the partnership.

General Issues

The identification of the Nodal Agency (Mother NGO) to be funded should be transparently impartial and liberal and the choice be left to Selection Committee / independent forum of experts of wide representation, including from NGOs. The criteria for identification will be based initially on simple assessment of interest, experience of field work and minimum capability, before selecting.

The minimum period of partnership should be for two years, which is a reasonable time for tangible results to be gained. An additional period of one year or so may be provided initially for the Nodal Agency to make preparation. A period of preparation can also help concerned to judge realistically whether the experiment is likely to succeed, before the actual work begins. A final selection process should take into account the integrity, capability and commitment of the Nodal Agency (Mother NGO).

All levels of personnel in the government hierarchy who are likely to come in contact with the work of the Nodal Agency from time to time be specifically informed about the special importance of the partnership venture and strongly encouraged to minimize impediments and maximize positive response in dealing with the NGO. This may mean inter- sectoral meetings at high levels that can then percolate such positive attitudes in the hierarchy down the grassroots levels.

Programmes

While it may be necessary to hand over "total health care responsibility" of the area to the NGO (selected by Nodal Agency) in keeping with the general profile of a PHC, it would be useful to have the NGO to work with greater emphasis on a given work area of its interest and of national and local importance, such as women's health or child health, with the understanding that such focused work is likely to produce solutions.

Since most community health programmes need a backup curative care facility, it would be desirable to encourage government hospital near NGO projects to form tie-ups with the NGO, which may not be able to put up a hospital on its own.

Manpower

As mentioned above, it would be in the interest of success of partnership venture to provide for the guidance of necessary technical staff to the Nodal Agency for the period of the project.

The staff employed would be totally answerable and responsible to Nodal Agency alone for the period of the project.

Reporting and Monitoring

The purpose of monitoring is to seek and resolve obstacles and problems. A State-level monitoring committee and district level monitoring committee can be formed to deal with all aspects of partnership, beginning from identification of the NGOs, through finalizing the terms of the project, to ongoing monitoring of all aspects of the progress. The committee may be chaired by the DGHS and consist of nominated experts, NGO representatives, and government officers from State and District levels. In general, reporting should be simplified as far as possible. The state government should help the Nodal Agency with reporting procedures where required, by giving timely inputs and anticipating problems, if necessary by deputing appropriate accounts officers for the purpose.

Grants and Finances

The release of the grant should be from state to avoid predictable impediments from the bureaucracy at lower levels, which can be very upsetting to the work of Nodal Agency unwilling to compromise on basic values.

The grants to be given by the nodal organisation to any NGO under this mechanism should not exceed a certain maximum limit, as decided by the Health Department and the apex committee.

Recognizing the limited resources of Nodal agency and NGOs selected by Nodal agencies and possible lack of buffering funds, the initially release of the grant should be for 6 months in advance. Later releases can be quarterly following submission of simple financial statement.

A quarterly statement of expenditure incurred under the particular project will be submitted to the Department by the 30th of the month following each quarter for requirement of funds to the rolling fund, so that it does not fall short of a certain pre-decided level at any stage.

The nodal organisation will be responsible for the proper utilization of the amount advanced to it for disbursement to NGOs and due safeguards will be observed while releasing the amount to NGOs.

Financial aspects of the nodal agency will be monitored by the State level apex committee, which can be accordingly constituted.

Monitoring of financial aspects should be on the spirit and broad objective of the project and not mechanically on the basis of expenditure under standard, narrow "heads". Re-appropriation should be permitted accordingly, the extent of flexibility to be determined by the cell in an on-going manner, thus providing for mid-course changes.

The accounting system being followed by the NGO should be allowed to be retained, and should be based on norms generally acceptable to the auditing system.

Thus the principle to be followed in fiscal matters is to give considerable leverage as long as:

- Audited statement of funds are submitted in time
- There is no evidence to doubt the integrity of the nodal agencies and
- The funds are utilized for the broad purpose for which they were allocated which can be determined by the committee.

(b) Networking/Collaboration Through NGO Forum

NGO Forum for Health will be a collaboration aimed at improving family planning/reproductive health in Himachal Pradesh especially in Kinnaur District by advocacy and resources sharing. The specific goal of the forum is to enhance the capacity of various private voluntary and non-governmental organizations through networking and collaboration and involve them in Health services. The track records of the forum members in communities and the intersectoral nature of their activities give them the scope to respond with innovative, creative solutions to local needs. The Partners' experience in empowering and building capacity in the communities they serve, in mentoring local organizations, and in collaborating with the for-profit, not-for-profit, and public sectors, are additional strengths Forum can draw on to bring about sustainable improvements in Health.

The NGO Forum for Health would have a core staff of its own to oversee overall collaborative efforts. It will be supported by some elected organisations, which would conduct NGO Forum for Health functions at local levels. The core discipline represented will be anthropologist, health management, training, communication, social science, statistics, demography and financial management etc.

The NGO Forum for Health will work with some regional organisations or affiliates and individuals to promote and bring effective collaboration of NGOs at state and district levels. Such regional organisations will be identified and appropriately strengthened. They would help the NGOs to train their staff, to plan, develop, train and advise on the communication activities and to develop management information system for their schemes, help them to evolve supervisory chain and continuously monitor the work of NGOs.

State Government Role in NGOs Forum for Health

State Government role in the NGO Forum will be as a facilitator in the formation of the Forum. State Government will also facilitate the Forum in its regular activities but not implement them directly.

Responsibilities/Activities of NGO Forum

- To function as an interest group for health, for initiating action in this area for sharing experience and providing mutual guidance and support to each other.
- To promote co-operation and co-ordination among NGOs, individuals, Government Departments and Funding Agencies having programmes for Health related issues and to pursue framing of laws for improvement of health status in Himachal Pradesh with special reference to Kinnaur and other tribal areas in state.
- To initiate a joint and co-operative action plan for health programme with NGOs actively involved in health related activities.
- To create awareness on health related issues which needed to be addressed and lobby with all political parties, and to have these issues highlighted in party manifestos, address the problem through questions raised in both Houses of Parliament; Government authorities; Statutory bodies like the Health Department; Pressure Groups.
- To have documentation and also conduct research in the relevant areas to broaden out understanding on health related issues and also to publish Newsletter/ magazines.
- To organise workshops, seminars, conferences on related issues and host or organise orientation programme for different stakeholders like the doctors, Health department functionaries etc., highlighting the health related issues and the need to deal with it.
- To organise training programmes for its members (and other) to build the capacity of the member NGO's.

- Work towards initiating and developing a movement for health sector reform.
- To acquire, receive and held properly both moveable and immovable including securities and negotiable instruments.
- To raise moneys and funds by borrowing or otherwise as may be deemed fit for and on behalf of the forum.
- To purchase, construct, manage and alter building and to equip them suitably for the purpose of the running of the Forum.
- To manage, sell, transfer, ledge or otherwise dispose of or deal with property of any kind which may be at the disposal of the Forum.
- To enter into contracts for and in connection with any of the purpose of the Forum, mentioned herein.
- To perform all such acts as may be necessary or proper for the achievement of any or all the objectives of the Forum.
- To facilitate the flow of information to all member organizations in this regard.

Institutional Arrangement

The functioning of NGO Forum for Health can be overseen by the network committee constituted by State Government consisting of:

- Two representatives of PGI-Shimla
- CMO of Kinnaur district
- One representative (Convenor/Deputy Convenor) of NGOs Forum
- One representative of intermediary organisation
- One representative of IMA-Himachal Pradesh
- One independent expert–anthropologist

The Director General, Department of Health and Family Welfare will be its Chairman.

(c) Direct Selection Model

A large number of NGOs are institution-based, running dispensaries/hospitals for the poor and backward classes and providing curative services. State Government should take initiative to promote, strengthen and bring efficiency among the NGOs in forming collaboration in Family Welfare Programme. The Health department may develop grant-on-aid schemes for NGOs, invite their participation, fund the activities, monitor or organize monitoring them and provide for/or arrange technical support.

In the Direct Selection Model, State government may play the following important roles:

(i) Promoting collaboration of NGOs by:

- Maintaining continuous dialogue with NGOs;
- Facilitate the identification of grant-in-aid schemes which would meet the objectives of both state government and NGOs;
- Invite proposals for schemes from NGOs and fund them.

(ii) Strengthening functioning of NGOs by:

- Monitoring their programme activities;
- Providing technical assistance in pre-project, project and post project phases;
- Training manpower of NGOs.

(iii) Co-ordinate NGOs work with government by:

- Bringing collaboration with the work of government machinery;
- Avoiding duplication of efforts

(iv) Facilitate exchange of experiences among NGOs and mutual learning between governmental agencies and NGOs in the health and family welfare field.

General Principle

In this model, state government will identify appropriate NGOs applying certain rating mechanism and invite them to participate in PPP project. The relationship should be based on mutual trust and sense of complement, where both sides admit, understand and appreciate each other's strengths and limitations.

General Issues

- The process of identification and selection of the NGOs for funding should be transparent and impartial. An independent forum of experts of wide representation, including from NGOs can be created to assist state government in identification and selection process. Certain rating mechanism should be adopted for identification and selection of NGOs.
- The minimum period of partnership should be for two years, as the first year will be required to complete the formalities, understand the mechanism, start the work and evaluate the performance, and the second year will be the year of taking corrective measures and improving performance.
- The area handed over to the NGO can be part (one or more sub-centers) or whole of a PHC, or equivalent area; the NGO can be fully involved in the choice of the area, the size being determined by the ability of the NGO to handle work load.
- All levels of personnel in the government hierarchy who are likely to come in contact with the work of the NGO from time to time be specifically informed about the special importance of the partnership venture and encouraged to minimize impediments and maximize positive response in dealing with the NGO. This may mean inter- sectoral meetings at high levels that can then percolate such positive attitudes in the hierarchy down to the grassroots levels.
- While specific measures to help the NGOs cope with various elaborate and cumbersome procedure of the government need to be instituted, it would be a useful practice to provide for deputation of various levels of technical and non-technical staff with the working of the government system, and are interested in helping such NGOs. This has, of course, to be done with mutual consent and should provide a welcome break to sincere and committed officers in the government, while being

of immense benefit to the NGOs. Such deputations should naturally be non-transferable for at least the period of the partnership projects.

- In the interest of voluntarism and independence of the NGO, it may be desirable to limit funding to less than 100 per cent of the estimated cost of the project.

Manpower

As mentioned above, necessary technical and non-technical government staff can be placed on deputation to the NGO for the period of the project. Besides health workers and administrative personnel, doctors from various relevant departments of medical colleges would be extremely valuable. This, of course, should be done with complete acceptance of all parties concerned.

While deputation may fill some requirements of the NGOs, it is likely that in other cases, there may be no candidates for some posts fulfilling all the criteria for qualifying for government job. It would be desirable under these circumstances to permit recruitment of less qualified staff by the NGOs, with appropriate relaxation of recruitment criteria.

The staff employed or deputed would be totally answerable and responsible to NGO alone for the period of the project.

When staffs are employed by the NGO, their basic training for the job, as for example for health workers, may need to be arranged at government or government-approved training centres, with acceptance of appropriate relaxation.

Reporting and Monitoring

The system for monitoring the partnership project should be separate and independent of the system that monitors other PHCs. This may be the single most important measure to ensure successful completion of the project.

An apex committee may be constituted by experts, NGO representatives, and government officers familiar with simplified accounting procedures etc. The committee can be empowered to deal with virtually all aspects of the partnership, beginning from identification of the NGO, through finalizing the terms of the project, to ongoing monitoring of all aspects of the progress.

In general, reporting should be simplified as far as possible. State Government should help the NGOs with reporting procedures where required, by giving timely inputs and anticipating problems, if necessary by deputing appropriate accounts officers for the purpose.

The NGOs should be permitted considerable freedom as regards recording procedures. This should be to encourage simplified procedures that should work in the atmosphere of trust that is expected in an NGO.

Grants and Finances

The release of the grant should be from state to avoid bottlenecks from the bureaucracy at lower levels, which can be very upsetting to the work of an NGO unwilling to compromise on basic values. Recognizing the limited resources of NGOs and lack of working funds, some amount can be released in advance. Later releases can be quarterly following submission of simple financial statements.

The accounting system being followed by the NGO should be allowed to be retained, and should be based on norms generally acceptable to the auditing system. Financial aspects may be monitored by the apex committee formed at the State level for monitoring all other aspects of partnership.

In the last, it has been observed that in every subsequent plan period the impetus on tribal development has been gaining momentum. Various tribal groups in India are at different levels of development. The level of development depends upon a large number of variables, the most important of them being the level of contact with the outside world and the extent of change that has occurred in pertinent cultural elements. The situation of tribals in the Himalayan region is very different from that in the plain region. The gist of all that has been explained above is not that we must go back to traditional warp of life and living or become superstitious. However, the merits of these ethnic groups belief make it incumbent to study the traditions scientifically and put forth their rational wisdom while introducing certain changes. Inputs from this research study and improvisation in the traditional system of medicine and health care has become necessary considering the exorbitant cost of modern medicine. Hence, people orientated, equalitarian, comprehensive health care system has been suggested in the thesis.

To conclude, it is essential to adopt holistic approach for improvement of health status of Kinnauri. Now, the time has come to develop the community based mechanism with active participation of all key stakeholders for overall development of Himalayan region in general and Kinnaur in specific. The multi - sectoral approach should be developed to improve inter- sectoral coordination between various key factors (i.e education, health, women empowerment, water and sanitation, employment etc.) of human development. For successful planning, knowledge about the social, economic and demographic conditions becomes essential.

Bibliography

1. Aanodt, Agnes M. 1978. "The Care Component in a Health and Healing System," in R.N. Eleanor E. Bauwens *The Anthropology of Health*, Missouri, The CV Mosby Company.

2. Abernethy, Virginia and Ray Yip. 1990. "Parent Characteristic and Sex Differential Infant Mortality: The Case in Tennessee," *Human Biology*, Vol. 62, No- 2: 279- 290.

3. Acharya, Rajib. 1998. "Gender Disparity, Development, Fertility Transition in India: An Inter-State Analysis," *Journal of Social Science*, Vol. 2, No. 4: 253- 263.

4. Ackerknecht, Erwin H. 1942. "Primitive Medicine and Cultural Pattern," *Bulletin of the History of Medicine* Vol. 12: 545-521.

5. Adak, Dipak Kumar.1994. "The Khasi of Shillong (Meghalaya): A Note on Infant and Early Childhood Mortality," Man *In India*, Vol. 74 No. 3: 287- 295

6. Adak, Dipak Kumar and Chaudhari, Sarit Kumar. 1996. "Birth Rate and Death Rate of Tripura (1971-1989): A Study of Time Series," *J. Hum.Ecol.*,Vol. 7, No. 3: 169-173.

7. Agarwala, S.N 1966. "Raising the Marriage Age of Women: A Means to Lower the Birth Rate," *Eco. Pol. Weekly*, Vol. 1: 797.

8. Agnihotri, B S. 1952-58. "Kinnaur Desh," *Air Force Trekking Society Records*, Shimla, Civil And Military Press: 45-54.

9. Agnihotri, B S. 1959. "The Legend of Kinnaur Desh," *Hindustan Times*, Sept 20, 27 and 4 October 1959.

10. Ahir, D. C. 1971. "Temple of Kanum: The Pride of Himachal Pradesh," Calcutta, *Mahabodhi*, Vol. 79, No. 12: 442-443.

11. Ahir, D. C. 1993. *Himalayan Buddhism Past and Present* Bibliotheca Indo-Buddhica Series-122, Mahapandit Rahul Sanskrityayan Century Volume, Delhi, Sri Satyguru Publication.

12. Ahluwalia, M.S. 1988. *History of Himachal Pradesh*. Delhi, Intellectual Publishing House.

13. Ahluwalia, M.S. 1998. *Social, Cultural and Economic History of Himachal Pradesh*. Delhi, Indus Publishing House.

14. Al- Krenawi, Alean and John R. Graham. 1999. "Gender and Bio Medical/Traditional Mental Health Utilisation Among the Bedouin-Arabs of the Negev," *Culture, Medicine and Psychiatry*, Vol. 23: 219-243.

15. Alain, D.1964. *Hindu Polytheism*. London.

16. Albrecht, Glenn., Sonia Freeman and Nick Higginbotham. 1998. "Complexity and Human Health: A Case for a Tran Disciplinary Paradigm," *Culture, Medicine and Psychiatry*, Vol. 22: 55-92.

17. Aldrich, Sarah and Chris Eccleston. 2000. "Making Sense of Everyday Pain," *Social Science and Medicine*, Vol. 50: 1631-1641.

18. Alland, Alexander Jr.1970. *Adaptation Cultural Evolution: An Approach to Medical Anthropology*, New York Columbia University Press.

19. Amooti-Kaguna, B. and F. Nuwaha. 2000. "Factors Influencing Choice of Delivery Sites on Rakai District of Uganda," *Social Science and Medicine*, Vol. 50: 203-213.

20. Andersson, Roland and Staffan Bergstrom. 1998. "Is Maternal Malnutrition Associated with a Low Sex Ratio at Birth?," *Human Biology*, Vol. 70, No. 6: 1101-1106.

21. Annual Tribal Sub-Plan. 1999-2000. *Tribal Development Department*, Government of Himachal Pradesh, Shimla.

22. Annual Tribal Sub-Plan. 2000-2001, *Tribal Development Department*, Government of Himachal Pradesh, Shimla.

23. Anonymous 1955. *Adivasis*. Publication Division, Govt. of India, pp. 110.

24. Anonymous 1958. "Kinner Conference," Delhi, *Vanyajati*, Vol. VI, No. 10: 190-193.

25. Anonymous .1977. "Kinner Kailesh: Home of Inspiration," *Indian Express*, August 8, 1977 p. 3, c 2-4.

26. Archar, (Major) Edward C.1833. *Tours in Upper India and in Parts of the Himalaya Mountains with Accounts of the Courts of the Native Prince*. London, Richard Bentley, 2 Vols.

27. Aristotle, 1959, *Politics and Anthenian Constitution*. London, J.M. Dent.

28. Arora, Sushil 1993 "Concept of Ageing and Problem of the Aged: Some Observation," *Man in India*, Vol. 73, No. 3: 251-257.

29. Arora, Sushil and A.K. Haldar 1994. "Economy of the Nomadic Communities of India." *Man in India*, Vol. 74 No. 2:181- 191.

30. Aryan, K. C. and Subhashini Aryan. 1985. *Rural Art of Western Himalaya*. Delhi, Rekha Prakashan.

31. Atkinson, Edward, T 1976. *Religion in the Himalayas*. Delhi, Cosmo Publications.

32. Atkinson, Edward T. 1882. *The Himalayan Districts of the North-Western Provinces of India*, Three Volumes, Allahabad, N.W Provinces and Oudh Press (Reprinted in 1973 as Himalayan Gazetteer, Vol I-Kinnaur, Delhi, Cosmo Publications).

33. Atkinson, Edward T. 1884. "Notes on the History of Religion in the Himalaya of the North-Western Provinces," *Journal of Asiatic Society of Bengal*, Vol. LIII Part 1: 39-103.

34. B.D. Sharma; 1976. *Environmental Context and the Personnel System and its Implications for Tribal Areas*. Ministry of Home Affairs, Govt of India.

35. Baboo, Balgovind. 1992. "Development and Displacement: A Comparative Study of Rehabilitation of Dam Bustees in Two Suburban Villages of Orissa," *Man in India*, Vol. 72 No. 1: 1- 14.

36. Badaruddoza and Mohammad Afzal. 1992. "Inbreeding in the Human Population," *Man in India*, Vol. 72, No. 4: 431-453.

37. Bag, H. and Kapoor, A. K. 2004. Religious Beliefs and Practices Among the Kondhs: A Primitive Tribe of Kalhandi District, Orissa. *Man and Life*, 30 (3&4): 71-80.

38. Bagade, A. C., N. N. Bhalerao, S L Kate 1995. "Overview of Tribal Health," in Jain Navinchndra & Tribhuwan Robin (eds) *An overview of Tribal Research Studies* pp. 46-60.

39. Bahadur, K. P. 1978. *Caste, Tribe and Culture of India-Haryana, H. P, Punjab, J & K and Sikkim* Vol VI, Delhi, Ess Ess Publications.

40. Bailey, Rev. T. Grahame 1910. "Kanauri Vocabulary in Two Parts: English Kanauri and Kanauri –English," *The Journal of Royal Asiatic Society* Vol 89 Part -1.

41. Bailey. Rev. T. Grahame 1909. "A Brief Grammar of the Kanauri Language," *Zeitschrift der Deutschen Morgen Landischen Gesellschaft*, Vol 63: 661-687.

42. Bailey, Rev. T. Grahame 1975a. "Linguistic Studies from the Himalayas Being Studies in the Grammar of Fifteen Himalayan Dialects," *Royal Asiatic Society Monograph*.

43. Bailey, Rev.T. Grahame 1975b. *Linguistic Studies from the Himalayas*. Delhi, Asian Publication Services.

44. Bajpai, S. C. 1970. *Northern Frontiers of India, Central and Western Sector*. Delhi, Allied Publishers (p- 15 Bashahr).

45. Bajpai, S. C. 1981. *Kinnaur in the Himalayas—Mythology to Modernity*. New Delhi, Concept Publishing Company.

46. Bajpai, S. C. 1987. *Kinnaur: Remote Land in the Himalayas*. Delhi, Indus Publishing Company.

47. Bajpai, S. C. 1991. *Kinnaur: Restricted Land in the Himalaya*. Delhi, Indus Publishing Company.

48. Baker, Paul T. 1996. "Adventures in Human Population Biology," *Annual Review of Anthropology*, Vol. 25: 1-18.

49. Balgir, R.S. 1994. "Age at Sexual Maturity Among the Eight Endogamous Populations of North Eastern India," *J.Hum.Ecol.*, Vol. 5, No. 2: 91-96.

50. Balokhra, Jag Mohan. 1999. *The Wonderland: Himachal Pradesh: A Survey of Geography, People, History, Administrative History, Art and Architecture, Culture, Tourism and Economy of State*. Delhi, H. G. Publication.

51. Banerjee, B.G. and Bhatia, K. 1988. *Tribal Demography of Gonds*. Delhi, Gian Publishing House.

52. Banerjee, U.K. and Shubhra Banerjee. 1997. "The Mystic Kinnaur in The Himalayas: Its People and Forests" in K C Mahanta (ed) *People of the Himalayas: Ecology, Culture, Development and Change*. pp. 221-224, New Delhi, Kamla Raj Enterprises.

53. Banerjee, D.and Mukherjee, S.P. 1961. "The Menarche in Bengali Hindu Girls," *J. Indian Medical Association*, Vol. 37: 261-270.

54. Banyal, S S. 1980. "Sangla Valley Kinnaur: Land of Lotus Eaters," *The Hindustan Times Weekly*, No. 9 p. Vol. 6, c 4-8.

55. Barillas, Edgar. 1999. "Transformation of National Health System," *Development*, Vol. 42, No. 4: 76-77.

56. Barua, Indira and Roopa Phukan. 1998. "Socio-Religious Aspect of Health Among Sonowal Kachari," *The Eastern Anthropologist*, Vol. 51, No. 4: 351-362.

57. Basu, Ashok Ranjan 1985. *Tribal Development Programmes and Administration in India with Special Reference to H.P.* Delhi, National Book Organisation.

58. Basu, A. 1990. "Anthropological Approach to Tribal Health," in Ashish Bose, Tiplut Nagbri & Nikhlesh Kumar (ed.) *Tribal Demography and Development in North-East India*, pp. 131-142. Delhi, B.R. Publishing Corporation.

59. Basu, Ashok Ranjan 1998. "Ecology," in Ashok Ranjan Basu and Padam Nabh Gautum (edt) *Natural Heritage of India: Essays on Environment Management*. Delhi, H K Publishers & Distributor.

60. Basu, Ashok Ranjan and Padam Nabh Gautum (ed) 1998. *Natural Heritage of India: Essays on Environment Management*. Delhi, H K Publishers & Distributor.

61. Basu, M.P. 1967. "A Demographic Profile of Irula," *Bull. Anthrop. Surv. India*. Vol. V-XVI, No 3 & 4: 267-289.

62. Basu, S. and Kshatriya, G. K. 1989. Fertility and Mortality in Tribal Populations of Bastar District, Madbya Pradesh, India. *Biology and Society*, Vol. 6: 100-112.

63. Basu, S. K. (ed). 1993. *Tribal Health in India*. Delhi, Manak Publishers.

64. Basu, S.K. 1992. "Health and Culture Among the Underprivileged Groups in India," in Alok Mukhopadhyay (ed) *State of India'a Health*. Delhi, Voluntary Health Association of India pp. 175-186.

65. Basu, S.K. 1993. "Tribal Health," *Rural Health*. New Delhi, NIHFW.

66. Basu, S.K., Jindal, A., Kshatriya, G.K.1990. "The Determinants of Health Seeking Behaviour Among Tribal Populations of Bastar District, Madhya Pradesh". *South Asian Anthropologist* Vol. 11, No. 1: 1-6.

67. Basu, S.K., Kshatriya, G.K.1990. "Growth Trends and Adolescent Spurts among Kutia-Kondha—A Primitive Tribal Group of Phulbani district, Orissa," *Acta Medica Auxologica*. Vol. 22. No. 3: 153-164.

68. Basu, Salil 1996. "Need for Action Research For Health Development Among Tribal Communities of India," *South Asian Anthropologist*, Vol. 17, No. 2: 73-80.

69. Basu, S.K and Kshatriya, G.K. 1989, "Fertility and Mortality in Tribal Populations of Bastar District, Madhya Pradesh,". *J.Biol.and. Soc.*, Vol. 6: 110-112.

70. Basu, S. K., Jindal, A. and Kshatriya, G.K. 1994 "Perception of Health and Pattern of Health Seeking Behaviour among the Selected Tribal Population Groups of Madhya Pradesh and Orissa," in S.K.Basu (eds) *Tribal Health in India*. Delhi, Manak Publication Ltd.

71. Basu, S. 1999 *Health Need Assessment of Tribal Populations of District Midnapur, West Bengal*. Ph.D. Thesis, Unpublished, Delhi Uni., Delhi..

72. Basu, S., Kapoor, A.K. and Basu, S.K. 2004. Knowledge, Attitude and Practice of Family Planning Among Tribals. *Journal of Family Welfare*, 50 (1): 24-30.

73. Baura, Indira, 1996. "Menarche in North-East Indian Communities: Some Bio-social Aspects," South *Asian Anthropologist*, Vol. 17, No. 1: 65-72.

74. Beaglehole, Robert and Anthony J. McMichael 1999. "The Future of Public Health in a Changing Global Context," *Development*, Vol. 42 No. 4: 12-16.

75. Beall, C.M. 1983. "Ages at Menopause and Menarche in a High Altitude Himalayan Population," *Ann.Hum.Biol.*, Vol. 10:365.

76. Behera, B.K. 1995. "Socio-cultural and Environment Factors of Health: A Case Study of the Kondhs," *South Asian Anthropologist*, Vol. 16, No. 2: 73- 77.

77. Behera, D. K. 1996. "Development, Environment and the People: An Anthropological Perspective," *South Asian Anthropologist*, Vol. 17, No. 2: 95-100.

78. Bell, Charles *The People of Tibet* pp. 43.

79. Bell, Charles, 1968. *Tibet: Past and Present*. London, Oxford University Press.

80. Belshaw, C.S, 1972. "Development: The Contribution of Anthropology," *International Social Science Journal*, Vol. 24: 83-94.

81. Benerjee, B. G, 1988. *Tribal Demography of Gonds*. Delhi, Gian Publishing House.

82. Benerjee, D and S. Mukherjee, 1961. "The Menarche in Bengal Hindi Girls," *Journal of Indian Medical Association*. Vol. 37: 261.

83. Berlinguer, Giovanni, 1999. "Health and Equity as a Primary Global Goal," *Development*, Vol. 42, No. 4: 17-21.

84. Bernier, Ronald M, 1989. *Himalayan Tower: Temples and Palaces of Himachal Pradesh*. Delhi, S. Chand and Company Ltd.

85. Berreman, Gerald. D, 1960. "Cultural Variability and Drift in the Himalayan Hills," *American Anthropologist*, Vol. 62: 774-794 .

86. Berreman, Gerald D, 1962a. *Hindus of the Himalayas*. Bombay, Oxford University Press.

87. Berreman, G.D, 1962b. "Sib and Clan Among the Pahari of North India," *Ethnology*, Vol. 1: 524-528.

88. Berreman, G.D, 1962c, "Pahari Polyandry: A comparison," *American Anthropologist*, Vol. 64 No. 1: 60-75.

89. Berreman, G.D, 1962d. "Caste and Economy in the Himalayas," *Economic Development and Cultural Change*, Vol. 10.
90. Berreman, Gerald. D, 1963. "People and Culture of the Himalayas," *Asian Survey*, Vol. 3 No. 6: 289-304.
91. Berreman, G.D, 1972a. "Self Situation and Escape from Stigmatised Ethnic Identity," in K. S. Mathur and S.C Verma (ed), *Man and Society*, Lucknow, Ethnographic and Folk Culture Society.
92. Berreman, Gerald D, 1972b. *Hindus of the Himalayas: Ethnography and Change*. Bombay, Oxford University Press.
93. Berreman, G.D, 1975. "Himalayan Polyandry and the Domestic Cycles," *American Ethnologist*, Vol. 2.
94. Berreman, G. D, 1977. "Demography Domestic Economy and Change in the Western Himalayas," *The Eastern Anthropologist*, Vol. 30, No. 2.
95. Berreman, G.D, 1978. "Ecology, Demography and Domestic Strategies in the Western Himalayas," *Journal of Anthropological Research*, Vol. 24, No. 3.
96. Berreman, G.D, 1980. "Polyandry: Exotic Customs vs. Analytic Concept," *Journal of Comparative Family Studies*, Vol. 11, No. 3.
97. Berreman, Gerald D (ed). 1981. *Social Inequality: Comparative and Development Approaches*. New York, Academic Press.
98. Bhagwat, Sunanda and Chirmuly, Deepti, 1997. "Traditional Healing Practises in Rural Karnataka," *Man in India*, Vol. 77, No. 1: 71-82.
99. Bhalla, V, 1958. "The Ethnic Significance and Evaluation of R.B.C. Quantity and Haemoglobin Intensity Among the Kinnar of Chini," *The Anthropologist*, Vol V, No. 1& 2: 9-16.
100. Bhanja, Kamlesh, 1948. *Mystic Tibet and The Himalayas* Darjeeling, Gilbert and Company.
101. Bhanu, B.A. and Saheb, S.Y, 1983 *Tribe in Contemporary India: The Irular of Tamil Nadu* Calcutta, Anthropological Survey of India.

102. Bharadwaj, Shyama and Pradeep Kumar Vaid, 1998. "Management of Environment Pollution in India," in Ashok Ranjan Basu and Padam Nabh Gautum (edt) *Natural Heritage of India: Essays on Environment Management*, Delhi, H K Publishers & Distributor.

103. Bhasin, Veena, 1997. "Medical Pluralism and Health Services in Ladak," *Journal of Human Ecology*, Vol 1 No. 1: 43- 69.

104. Bhatia, P.S, 1990. "Health Care Utilisation by the Tribal Dang District," in A Bose, V.P. Sinha and R.P. Tyagi (ed) *Demography of Tribal Development*. Delhi, B.R. Publishing Corporation.

105. Bhatnagar, V. M, 1973. "The Land of the Gods: A Journey in the Chamba, Lahoul-Spiti and Kinnaur Valleys of Himachal Pradesh," *The Trishul: Western Air Command Journal*, Vol. VIII No. 1 January 1973: 35-47.

106. Bhatt, G.C, 1981. "Polyandry in Western: Some Notes and Observations," *Journal of Social Research*, Special Number on Himalayan Studies, Vol. 24, No. 3.

107. Bhowmick, P.K, 1963. *The Lodhas of West Bengal: A Socio-Economic Study*. Calutta, Punthi Pustak.

108. Birch, Stephen, 2000. "Health Economics: An Important Contribution to Social Science and Medicine," *Social Science and Medicine*, Vol. 50: 307-308.

109. Bisotra, R. L, 1998. "New Strategies in Tribal Development Administration: A Case Study of Himachal Pradesh," in Gupta, S. K., V. P. Sharma, N. K. Sharda (ed), *Tribal Development: Appraisal and Alternatives* Delhi, Indus Publishing House.

110. Biswas, N, 1967. "Age at Menarche Among Punjabi Girls of Delhi," *The Anthropologist*, Vol. 14:153.

111. Biswas, R. K., Patra P. K. and Kapoor, A. K. 2001. "Demographic Profile of Kamar—A primitive Tribe of Madhya Pradesh," *Bulletin of the Tribal Research Insititute*, Vol. XXVIII (1 & 2): 67-77.

112. Biswas, Ranjan Kumar, 2002.Demographic Study of Saharaia—A Primitive Tribe of Madhya Pradesh. Unpublished Ph.D. Thesis, University of Delhi, Delhi

113. Biswas, Ranjan Kumar and Kapoor, A. K. 2003. "Fertility Profile of a Primitive Tribe, Madhya Pradesh," *Anthropologist*, Vol. 5 No. 3: 161-167.

114. Biswas, Ranjan Kumar and Kapoor, A. K. 2003. "A Study on Mortality Among Saharaia—A Primitive Tribe of Madhya Pradesh," *Anthropologist*, Vol. 5 No. 4: 283-290.

115. Biswas, R.K. and Kapoor, A.K. 2003a. Fertility Profile of a Primitive Tribe, Madhya Pradesh. *The Anthropologist*, 5 (3): 161-167.

116. Biswas, R.K. and Kapoor, A. K. 2003b. Education and its Effect on Fertility and Mortality Differentials Among a Primitive Tribe of Madhya Pradesh. *Indian Journal of Population Education*, 23 (Dec.): 36-45.

117. Biswas, R.K. and Kapoor, A. K. 2003c. A Study on Mortality Among Saharia—A Primitive Tribe of Madhya Pradesh. *The Anthropologist*, 5 (4): 283-290.

118. Biswas, R.K. and Kapoor, A. K. 2003. Fertility Profile of a Primitive Tribe, Madhya Pradesh. *The Anthropologist*, 5 (3): 161-167.

119. Biswas, R.K. and Kapoor, A. K. 2004. Socio-Cultural Impact on Maternal Care Among Saharia Primitive Tribe of Madhya Pradesh. *J. Region, Health and Health Care*, 9 (1&2): 1-10.

120. Biswas, R. K. and Kapoor, A. K. 2005. Health Seeking Behaviour Among Saharia—A Primitive Tribe of Madhya Pradesh. *Indian Journal of Social Research*, 46 (3): 249-260.

121. Biswas, R. K. and Kapoor, A. K., 2005. Age at Menarche and Menopause Among Saharia Women—A Primitive Tribe of Madhya Pradesh. *The Anthropologist*, 7 (2): 105-109.

122. Blais, Regis and Aboubacrine Maiga. 1999. "Do Ethnic Groups Use Health Services like the Majority of the Population? A Study from Quebec, Canada," *Social Science and Medicine*, Vol 48:1237-1245.

123. Blumberg, Baruch S. and Jana E. Hesser, "Anthropology and Infectious Disease," pp. 260-293.

124. Bose, A.B. 1970. "Problems of Education Development of Scheduled Tribes," *Man in India* Vol. L No. 1: 26-51.

125. *Brief Facts of Himachal Pradesh*, 1999. Economic And Statistics Department, Himachal Pradesh.

126. Brochure on *Reservation Instruction for Members of Scheduled Caste/Scheduled Tribes/Backward Classes Etc in Service*, Department of Personnel (A-II), Government of Himachal Pradesh.

127. Brown, Daniel E, 1996. "The Use of Isolates and Migrant Populations in Human Biology Research Design," *Reviews in Anthropology*, Vol. 24: 229- 237.

128. Bruice, C.G, 1910. *Twenty Years the Himalaya*. London, Edward Arnold.

129. Buch, Edward J. 1904. *Shimla: Past and Present*. Calcutta, Thacker Spink and Company (Revised Edition Published by The Times Press, Bombay, 1925).

130. Buchanan, W.E. 1930. "In the Footsteps of the Gerards,". *The Himalayan Journal*, Vol II: 73.

131. Burman, B.K. Roy, 1997. "Tribal and Indigenous People—A Global Overview," *The Eastern Anthropologist*, Vol. 50, No. 1: 17-25.

132. Buvinic, M.,Valenzuela, J.P., Molina, T. and Gonzalez, E. 1972. "The Fortunes of Adolescent Mothers and their Children: The Transmission of Poverty in Santiago, Chile," *J.Pupoul.Dev.Rev*.Vol. 18, No. 2: 269-297.

133. Caldwell, John., Pat Caldwell and Bruce Caldwell. 1987. "Anthropology and Demography: The Mutual Reinforcement of Speculation and Research," *Current Anthropology*, Vol. 28 No. 1: 25-43.

134. Caldwell, John.C. 1979. "Education as a Factor in Mortality Decline: An Examination of Nigerian Data". *Opul. Studies*, Vol. 33, No. 3: 395-413.

135. Cameron, N. and I. Nagdee. 1996. "Menarcheal Age in Two Generation of South African Indians," *Annals of Human Biology*, Vol. 23, No. 3: 113-119.

136. Campbell, Benjamin. C and Paul W. Leslie. 1995. "Reproduction Ecology of Human Males," *Year Book of Physical Anthropology*, Vol. 38: 1-26.

137. Census of India, 1960. Vol. XX H.P Part VII b, *Fairs and Festivals* by R.C. Pal Singh, Suptd. of Census Operation, H.P.

138. Census of India. 1961. *Kanum: A Village Survey*. Vol. XX Part VI No. 12 (1965).

139. Census of India, 1961. *Kothi: A Village Survey*. Vol XX Part VI No. 1 (1963).

140. Census of India, 1961. *Nachar: A Village Survey*.Vol XX Part VI No. 16 (1964).

141. Census of India, 1961. *India* Vol. I, Part II- C (ii), Language Tables, p- CLXVI.

142. Census of India, 1971. *Sirwi-7* H.P. Paper 10f 1972 Final Population Table.

143. Census of India, 1991. *Himachal Pradesh District Profile*, Delhi, Registrar General of India,.

144. Chachra, Sushmita Paul and M. K. Bhasin. 1998. "Anthropo-Demographic Study Among the Caste and Tribal Groups of Central Himalayas: 1. Population Structure," *Journal of Human Ecology*, Vol. 9, No. 5: 405-416.

145. Chadha, S.K. 1989. *Himalayan Ecology* Delhi, Ashish Publishing House.

146. Chakrabarti, Dilip K. 1977. "The Archaeological Research in the Western Himalayas," *The Anthropologist*, Vol. XXIV, No. 1 & 2: 72-85.

147. Chakravarti, Indrani. 1998. "Musical Heritage of Kinnauri'", in Gupta, S.K., V.P. Sharma, N.K. Sharda (edt), *Tribal Development: Appraisal and Alternatives* Delhi, Indus Publishing House.

148. Chakravarti, Nirmal Kumar and Shubhasree Subedi. 1995. "Origin of Untouchability in Hinduism," *Man In India*, Vol. 75, No. 2: 139-161.

149. Chakravarti, Prithwis Chandra. 1971. *Evolution of India's Northern Borders*, Bombay, Asia Publishing House. p. 90-93.

150. Chambers Compact, "English Dictionary", p. 683.

151. Chandra Kumar, 1964. "Kinnaur: the Adobe of Kinnars," *Trend*, November, p. 14-15.

152. Chandra, Ramesh, 1981. "Ecology and Religion of the Kinner: Mountain Dwellers of North-Western Himalayas," in R.S Mann (edt) *Nature- Man-Spirit Complex in the Tribal India* Delhi, Concept Publishing Company. pp. 276.

153. Chandra, Ramesh, 1987. "Polyandry in the North-Western Himalayas: Some Changing Trends," in M .K Raha et al (eds) *Polyandry in India.* Delhi, Gian Publishing House.

154. Chandra, Ramesh, 1988. "Marriage Forms in the North-Western Himalayas in the Perspective of Evolution," in M .C Goswamy et al (eds) *Marriage System in India*. Calcutta, Anthropological Survey of India.

155. Chandra, Ramesh, 1992. *Highlanders of North Western Himalayas*. Delhi, Inter-India Publication.

156. Chandra, Ramesh. 1972. "Socio- Economic study of Kinnaures of Himachal Pradesh," Report Submitted to Anthropological Survey of India, Calcutta.

157. Chandra, Ramesh, 1972. "The Notion of Paternity Among the Polyandrous Kanet of Kinnaur, H.P.," *Bulletin of Anthropological Survey of India*, Vol. XXI No. 1-2: 80- 87.

158. Chandra, Ramesh, 1973. "Types and Forms of Marriage in a Kinnaur Village," Ranchi, *Man In India*, Vol. 53 No. 2: 176-187.

159. Chandra, Ramesh, 1974. "A Study of Polyandry in a Himalayan Village," *Bulletin of Anthropological Survey of India*.

160. Chandra, Ramesh, 1976. "Wife Sharing and Economic Security Among the Sirmurese of H. P," *The Proceedings of the 2nd Annual Conference of Ethnography and Folk Culture Society*, Lucknow.

161. Chandra, Ramesh, 1977. "Socialization and Polyandry: A Study in Cultural Continuity among Sirmurese," *The Proceedings of the 3rd Annual Conference of Ethnography and Folk Culture Society*, Lucknow.

162. Chandra, Ramesh, 1979. "Need for Urgent Research on Vanishing Trait of Polyandry," *Review Ethnology*, Vol. 7: 1-9 .

163. Chandra, Ramesh, 1981. "Sex Role Arrangement to Achieve Economic Security in North Himalayas," in Christoph Von-Furer Haimendorf (ed) *Asia Highland Societies in Anthropological Perspective*, New Delhi, Sterling Publishers. pp. 203-213.

164. Chandra, Ramesh, 1982. "Myth of Surplus Women in Polyandry," *Man in India*, Vol. 62, No. 4: 414-419.

165. Chandra, Ramesh, 1987. "Quest for Environment Adaptation in a Highland Society: Case of Kinners," *Man in India*, Vol. 67 No. 3: 177-195.

166. Channa, Subhadra, M. 1991. "Caste, 'Jati' and Ethnicity—Some Reflection Based on a Case Study in an Bhobis," *Indian Anthropologist*, Vol. 21, No. 2: 39-55.

167. Charak, Sukhdev Singh, 1978. *Himachal Pradesh*—Three Volumes, Delhi, Light and Life Publishers.

168. Chatterjee, Sikha, 1994. "Menarche, Menopause and Fertility in the Paschatya Vaidik Brahmean Women of West Bengal," *Man in India*, Vol. 74, No. 3: 271-278.

169. Chattopadhyay, Madhubala and Prasad, B.V. 1995. "Nutritional Status of the Nicobarese Tribal Children of Harminder Bay, Little Andaman". *J.Hum. Ecol.*, Vol. 6, No. 1: 59-61.

170. Chattopadhyay, P.K. and Khullar, S, 1969. "The Age at. Menarche in Indian Girls," *Acta Medica Auxologica*.Vol 1: 58-62.

171. Chattopadhyaya, K.P, 1957. "On Polyandry," *Journal of Asiatic Society of Bengal*, Vol. 22, No. 2.

172. Chaudhary, B. 1986. *Tribal Health*. New Delhi, Sunil Printers.

173. Chaudhary, R.K.Naik, Mamatamayee and Mohanty, Manmath Kumar, 1994. "A Note on Menarcheal Age of Mirdha Women of Orissa," *J. Hum.Ecol.*,Vol. 5, No. 2: 123-126.

174. Chaudhary, Sukanta Kumar, 1990. "Tribal Development: Dimensions of Planning and Implementation," *Indian Anthropologist*, Vol. 20, No. 1 & 2: 49-65.

175. Chaudhuri, Buddhadeb, 1990. "Social and Environmental Dimensions of Tribal Health," in B. Chaudhuri (ed) *Cultural and Environmental Dimensions of Health*. Delhi, Inter-India Publications.

176. Chaudhury, P. C. Roy, 1981. *Temples and Legends of Himachal Pradesh*. Bombay, Bharatiya Vidya Bhavan.

177. Chauhan, Abha. 1990, *Tribal Women and Social Change in India*. Etawah, A.C. Brothers.

178. Chauhan, Dr. Narain Singh, 1999. *Medicinal and Aromatic Plants of Himachal Pradesh*," Delhi, Indus Publishing Company.

179. Chauhan, Ramesh, 1998. *Himachal Pradesh: A Perspective*. Shimla, Minerva Book House.

180. Chawla, I. N. 1981, *Kinnar Kailesh*. Parvesh Prakashan, Bakshi Radho Majra.

181. Chetlapalli et al, 1991. "Estimates of Fertility and Mortality in Kutia Kondhs of Phulbani District. Orissa," *J. Hum. Ecology*, Vol. 2, No. 1: 117-120.

182. Chib, S S. 1982, "Some Aspects of the Population Geography of Kinnaur District (HP): A Case Study of Tribal Area," Doctoral Thesis, Punjab University, Chandigarh.

183. Chib, S. S. 1984, *Caste, Culture and Economy of Kanauras of Trans Himalayas*. Delhi, Ess Ess Publication.

184. Chib, S.S. 1984. "Changing Face of Polyandry in Tribal Kinnaur Himalaya," *Indian Journal of Landscape System and Ecological Studies*, Vol. II, No. 8.

185. Chib, S.S, 1998. "Demographic Dynamics in the Trans-Himalayan Tribal Tract of Kinnaur: Perspective on Socio-Economic Implications," in Gupta, S. K., V. P. Sharma, N.K. Sharda (ed) *Tribal Development: Appraisal and Alternatives*. Delhi, Indus Publishing House.

186. Chib, Sukhdev Singh, 1977. *This Beautiful India: Himachal Pradesh*—Two Volumes, Delhi, Light and Life Publishers.

187. Chitre, R.G., Dixit, M., Agate, V., Vailekar. V. 1976. "The Concept of Essential Amino Acid in Human Nutrition—A Need for Reassessment," *Ind. J. Nutr. Dieter*. Vol. 13:101.

188. Chopra, Pran, 1964. *On an Indian Border* Bombay, Asia Publishing House.

189. Choudhury, Jaya, 1992. "Marriage Conception Interval," *Man in India*, Vol. 72, No. 1: 91-95.

190. Choudhury, N.R. and R. Kumar, 1976. "Demographic Profiles of the Bhils," *Eastern Anthropologist*, Vol. 29, No. 3: 273-280.

191. Choudhury, R.K.,Naik, Mamatamayee and Mohanty, Manmath Kumar, 1994. "Fertility and Mortality Among Mirdhas: A Tribal Community of Orissa," *J. Hum. Ecol.*, Vol. 5, No. 3:191.

192. Choudhury, N.C, 1963. "Munda Settlement in Lower Bengal: A Study in Tribal Migration," *Bulletin of the Cultural Research Institute*. Vol. 2: 23-32.

193. Cleare, John, 1978-79. "Himalchuli-78," *The Himalayan Journal*, Vol. 36.

194. Communication Action Research Centre, 1964-68. *A Final Report on Standard Fertility Survey in Calcutta City, Family Planning Communication Action Research Project*, Family Planning Research Unit, Indian Statistical Institute, Calcutta. p. 34

195. Conder, Josiah, 1829. *Modern Travellers: A Description of the Various Countries of the Glove*, London, Printed for Thomas Tegg, 73, Cheapside.

196. Conference Statement, 1999. "Equity in Health in the Age of Globalization," *Development*, Vol. 42, No. 4: 5-7.

197. Conway, C W W S, 1942. *Sunlit Waters* Bombay, Thacker.

198. Cordahi, Alexandre and Lawrence Tshuma, 1999. "Law, Globalization and Sustainable Development," *Development*, Vol. 42, No. 2: 39-46.

199. Coyle, Joanne, 1999. "Exploring the Meaning of 'Dissatisfaction' with Health Care: The Importance of 'Personal Identity Threat," *Sociology of Health and Illness*, Vol. 21, No. 1: 95-124.

200. Crooke, W, 1975. *The Tribes and Caste of North Western India*—Four Volumes, Delhi, Cosmo Publication.

201. Cunningham, Alexander, 1854. *Ladak, Physical Statistical and Historical with Notices of the Surrounding Countries*," London (Reprinted by Sagar Publication, Delhi, 1970).

202. Cunningham, Alexander, 1884. "Notes on Moorcrafts Travels in Ladhak and on the Gerrand's Account of the Kanawar Including the General Description of the Latter's District," *Journal of Asiatic Society of Bengal*, Vol. 13, No. 1.

203. Curtis, Sarah, Wil Gesler, Glenn Smith and Sarah Washburn, 2000. "Approaches to Sampling and Case Selection in Qualitative Research: Examples in the Geography of Health," *Social Science and Medicine*, Vol. 50: 1001-1014.

204. Dandekar, V.M. and Dandekar, K, 1953. *Survey of Fertility and Mortality in Poona District* Poona, Gokhale Institute of Politics and Economics.

205. Dandekar,K, 1959. *Demographic Survey of Six Rural Communities*. Bombay, Asia Publishing House.

206. Dandekar,V.M. and Dandekar, K, 1953. *Survey of Fertility and Mortality in Poona District* Poona, Gokhale Institute of Politics and Economics.

207. Dang, B.S, 1980. "Technology Strategy for Tribal Development," *Indian Anthropologist* Vol. 20, No. 1&2: 115-124.

208. Das, S. R. 1954. "A Note on the Kinnaras," *Man in India*, Vol. 34, No. 1.

209. Das, S.K. et al, 1982. "Demography and Demographical Genetics of Two Isolated Mountain Villages of Northern Sikkim, Eastern Himalayas," *J. Ind. Anthropol. Soc.* Vol. 17 No. 2: 155-162.

210. Das, T.C. 1973. *Social Organisation* in the Tribal People of India, Publication Division, Ministry of Information and Broadcasting, Government of India, New Delhi.

211. Datta, C. L, 1973. *Ladakh and Western Himalayan Politics*, Delhi, Munshi Ram Manohar Lal.

212. Datta, C. L, 1997. *The Raj and the Simla Hill States: Socio-Economic Problems, Agrarian Disturbances and Paramountcy*. Jalandhar, ABS Publications.

213. Daulaire, Nils. 1999. "Globalization and Health," *Development*, Vol. 42, No. 4: 22-24.

214. Davies, W, 1962. *Trade Report* Lahore.

215. Davis, K and W.E. Moore. 1945. "Some Principles of Stratification," *American Sociological Review*, Vol. X No. 2: 242-49.

216. De. K.B, 1998. "Equilibrium in Ecosystem," in B.D. Sharma and Tej Kumari (ed), *Himalayan Natural Resources: Eco Threats and Restoration Study*, Delhi, Indus Publishing House.

217. Dension, Edward, F. Accounting for United States Economic Growth, 1962-69, Washington: Brookings Institution, 1974.

218. Deuster, R.H, 1996. *Kanawar* Shimla, H. P. Academy of Arts, Culture and Language.

219. Dhamija, Jasleen, 1970. *Indian Folk Arts and Crafts*. Delhi, National Book Trust; pp 70- 71.

220. Dhebar, U.N, 1961. *Report of Scheduled Areas and Scheduled Tribes Commission*. New Delhi.

221. Dhir, D. N, 1990. "Tribes on the North Western Borders of India (West Himalayas)," in K. Suresh Singh (ed) *The Tribal Situation in India*, Shimla, IIAS.

222. District and States Gazetteers of the Undivided Punjab (prior to Independence) Vol. IV, (Reprinted by B.R. Publishing Corporation, Delhi in 1985).

223. District Census Handbook, 1981. *Village and Town Directory*, Kinnaur District, Director of Census Operations, HP.

224. District Census Handbook, 1991. *Village and Town Directory*, Kinnaur District, Director of Census Operations, HP.

225. *District Kinnaur at Glance*—1998. Kinnaur, District Statistical Office.

226. *District Kinnaur at Glance*—1999 Kinnaur, District Statistical Office.

227. *District Kinnaur at Glance*—2000 Kinnaur, District Statistical Office.

228. *District Kinnaur Vikas Khand Suchank*—1994. Kinnaur, District Statistical Office.

229. *District Kinnaur Vikas Khand Suchank*—1998. Kinnaur, District Statistical Office.

230. *District Kinnaur Vikas Khand Suchank*—1999. Kinnaur, District Statistical Office.

231. Donald Bogue, 1969. *Principles of Demography*. New York, John Wiley and Sons Ltd.

232. Dubey, Suman, 1966. "Kinnaur," *The Himalayan Journal*, Vol XXVII: 128-35.

233. Dunlop, R.H, W. 1860. *Hunting in the Himalayas*.

234. Dunn, James R. and Michael V. Hayes, 2000. "Social Inequality, Population Health and Housing: A Study of Two Vancouver Neighbourhoods," *Social Science and Medicine*, Vol. 51: 563-587.

235. Dutta,Bimal Chandra. 1999. "Some Observations on the Diseases and Traditional Medicine of the Singphos," *Man in India*, Vol. 79, No. 1&2:179-189.

236. Earickson, Robert, 2000 "Health Geography: Style and Paradigms," *Social Science and Medicine*, Vol 50:457-458.

237. *Economic Review: Himachal Pradesh*, 1999. Economic and Statistics Department, Himachal Pradesh.

238. Eickstedt, Baron. E. Von. 1926. "The Races and Types of the Western and Central Himalayas," *Man in India*, Vol. VI, No. 4: 237-276.

239. Fabrega, Jr. Horacio, 1971. "Medical Anthropology," *Biennial Review of Anthropology*, p. 167-229.

240. Fernandes, Walter, 1993. "Forest and Tribals: Informal Economy, Dependence and Management Traditions" in Mrinal Miri (ed.) *Continuity and Change in Tribal Society*, New Delhi, Elegant Printers.

241. Fix, Alan G, 1979. "Anthropological Genetics of Small Population," *Annual Reviews of Anthropology*, Vol. 8: 207-30.

242. Fortes, M, 1975 "Introduction" in J. B. Loudon (ed). *Social Anthropology and Medicine*. London, Academic Press.

243. Francke, A.H, 1907. *A History of Western Tibet* London, S.W. Partridge.

244. Francke, A. H (ed), 1913-26. *Antiquities of Indian Tibet*, ASI (Reprinted by S. Chand, Delhi in 1972) Two Volumes (Vol. 1-Kinnaur).

245. Fraser, James Baillie, 1820. *Journal of Tour Through Parts of the Snowy Range of the Himalaya Mountains and to the Sources of the Jamuna and Ganges*," London, Roadwell and Martin Band Street.

246. Fraser, James Baillie, 1882. *The Himala Mountains*. Delhi, Neeraj Publishing House.

247. Fuchs, Victor R, 1974. *Who Shall Live? Health Economics and Social Choice*, New York, Basic Books.
248. Gage, Timothv B, 1985. "Demographic Estimation from Anthropological Data: New Methods," *Current Anthropology*, pp. 644-646.
249. Gaulin, Steven. J.C and James S. Boster, 1992. "Human Marriage System and Sexual Dimorphism in Stature," *American Journal of Physical Anthropology*, Vol 89: 167-175.
250. Gautam, Padam Nabh, 1995. *Basic Needs of the Hill People: A Study of Social and Administrative Response in Himachal Pradesh*. Delhi, Indus Publishing House.
251. Georgiadis, E., C.S. Mantzoros, C. Evagelopoulou and D. Spentzos, 1997. "Adult Height and Menarcheal Age of Young Women in Greece," *Annals of Human Biology*, Vol 24, No. 1: 55-59.
252. Gerand, Alex, 1842. "Narrative of a Journey from Subathoo to Shipkee in Chinese Tertiary," *Journal of Asiatic Society of Bengal*, Vol. II: 363-91.
253. Gerard, Alexander, 1841. *An Account of Koonawur in the Himalaya and Lloyd, George, James Madden and Co*. London (Reprinted by Indus Publishing Company, Delhi in 1993.
254. Gerards, Alexander, 1842. "A Vocabulary of the Koonawar Language," *Journal of the Asiatic Society of Bengai*, Vol. XI: 479.
255. Gerards, Alexander, 1996. *Tours in the Himalaya: Account of an Attempt to Penetrate by Bekhur to Garoo and the Lake Manasarowara*. Delhi, Indus Publishing Company.
256. Germain, Adrienne, 1999. "Reproductive Health: The Continuing Challenge," *Development*, Vol. 42, No. 1: 38-40.
257. Ghosal, Samit, 1983. "A Frontier Tribe and its Transformation into Peasantry in Himalayan State," *Man in India*, Vol. 63, No. 4: 363-369.
258. Ghosh, A.K, 1970. "Selection Intensity in the Kota of Nilgiri Hills. Madras,". *Social Biology*, Vol. 17, 224-225.
259. Ghosh,A.K. and Kumari, S, 1973, "Effect of Menarcheal Age on Fertility," *J.Ind.Anthrop. Soc.*, Vol. 8, 165-172.

260. Ghoshmaulik, S.K. and R.K.K. Mohapatra, 1995. "Health Situation of the Juangs of Orrisa," *South Asian Anthropologist*, Vol. 16, No. 2: 79-83.

261. Ghurya, G.S, 1924. "The Ethnic Theory of Caste," *Man in India*, Vol. IV, No. 3-4: 209-273.

262. Gianchand and Manohar Puri, 1991. *Explore Himachal*. Delhi, International Publisher.

263. Gibbons, F.X, 1999. "Social Comparison as Mediator of Response Shift," *Social Science and Medicine*, Vol. 48, 1517-1530.

264. Gibson, Jack, 1976. *As I Saw it*. Delhi, Mukul.

265. Glover, H.M, 1930. "Round the Kinnaur Kailesh," *The Himalayan Journal*, Vol II, 1-12.

266. Gogoi, J.K, 1990. "Tribal Demography in North-East India. Some Preliminary Observation," in Ashish Bose, Tiplut Nangbri & Nikhlesh Kumar(ed) *Tribal Demography and Development in North-East India*. Delhi, BR. Publishing Corporation. pp. 85-94.

267. Gogoi, D, 1972. "Menarche and Menopause Among Women of an Ahom Village in Upper Assam," *Bull*. Vol 1, 18.

268. Goldstein, M.C, 1978. "Pahari and Tibetan Polyandry Revisited," *Ethnology*, Vol. XVII, No. 3.

269. Gopalan, T, 1987. "National Status of Some Selected Tribes of Western and Central India," *Society of India*. Vol. 33, 76-93.

270. Gorrie, R. Maclagan, 1929. "Two Easy Passes in Kanawar," *The Himalayan Journal*, Vol. 1, No. 1: 75-77.

271. Gunasundaramma, Kambham, 1980. *Fertility and Family Planning Among the Balijas of Tirupati, Andhra Pradesh*. Unpublished Ph.D.Thesis. Department of Physical Anthropology and Prehistoric Archaeology.

272. Gupt, Bharat, 1997. "What is Ethnic," *The Eastern Anthropologist*, Vol. 50, No. 2: 139-146.

273. Gupta, Krishnamurthi, 1965. "Kinnaur Ka Desh Ma," *Vanyajati*, Vol. XIII, No. 4: 148-54, 159.

274. Gupta, Neena and I.J.S. Jaiswal, 1992. "Age at Menarche among Three Caste Groups of Jammu City," *Man in India*, Vol. 72, No. 4: 463-467.

275. Gupta, S.K, 1998. "Tribal Areas and Tribes of Himachal Pradesh: Some Myths and Realities," in Gupta, S.K., V.P. Sharma, N.K. Sharda (ed), *Tribal Development: Appraisal and Alternatives*, Delhi, Indus Publishing House.

276. Gupta, S. K. and S. P. Bansal, 1998. " Environmental and Ecological Aspects of Hydro- Electric Power Projects in Tribal Areas of Himachal Pradesh," in Gupta, S. K., V. P. Sharma, N. K. Sharda (ed), *Tribal Development: Appraisal and Alternatives*, Delhi, Indus Publishing House.

277. Gupta, S. K., V. P. Sharma, N. K. Sharda (edt), 1998. *Tribal Development: Appraisal and Alternatives*, Delhi, Indus Publishing House.

278. Gupta, Sunil, 1998. "Handicraft of Kinnaur District in Himachal Pradesh," in Gupta, S. K., V. P. Sharma, N. K. Sharda (edt), *Tribal Development: Appraisal and Alternatives*, Delhi, Indus Publishing House.

279. Handa, O.C, 1987. *Buddhist Monasteries in Himachal Pradesh*, Delhi, Indus Publishing House.

280. Handa, O.C, 1994. *Buddhist Art and Antiquities of Himachal Pradesh: Upto 8th Century AD*, Delhi, Indus Publishing Company.

281. Haque, M, 1990. "Height, Weight and Nutrition Among the Six Tribes of India," in Chaudhuri, B. (eds), *Cultural and Environmental Dimension on Health*, Delhi, Inter-India Publication, pp. 192-206.

282. Haridas, M.M, 1997. *Kailesh: Impression and Expressions*, Delhi, Gyan Publishing House .

283. Harnor, S.P, 1997. *Yatra: Kinnaur, Spiti, Lahul and Maniemahesh*, Shimla, Minerva Book House.

284. Harnot, S.R, and Raj Pal Verma, 2000. *Himachal at a Glance: More than 3000 Facts*, Shimla, Minerva Book House.

285. Harrison, G.A, 1966. "Human Adaptability with Reference to the Proposals for High Altitude Research," in P.T. Baker and J. S Weiner (ed), *The Biology of Human Adaptability* Oxford. Clarendon.

286. Hasan and Prasad, 1959. "A Note on the Contribution of Anthropology to Medical Science" *Journal of the Indian Medical Association*. Vol. 33: 182-190.

287. Hasan, K.A, 1981. "The Ecology of Health and Disease: Some Biological and Cultural Considerations," *The Mankind Quarterly*, Vol. XXI, No. 4: 315-325.

288. Hassan, Fekri A, 1979. "Demography and Archaeology," *Annual Reviews of Anthropology*, Vol, 8, 137-60.

289. Haug, Marie R, 1988. "Medical Technology and Quality of Life," *Indian Anthropologist*, Vol. 18, No. 2: 15-26.

290. Hedin, Sven, 1909. *Trans Himalayan Discoveries and Adventures in Tibet*, London, Macmillan, 3 Vols. (Vol. II—Kinnaur p. 360-414, Vol. III p. 415-23).

291. Hedin, Sven, 1985. *Adventures in Tibet*, Delhi, Manas Publications.

292. Heggenhougen, H.K, 2000. "More Than Just " Interesting!" Anthropology, Health and Human Rights," *Social Science and Medicine*, Vol. 50, 1171-1175.

293. Herbert, J. D, 1825. " An Account of a Tour to Lay Down the Course and Bends of the River Sutlej or Satudara as far as Traceable within the Limits of the British Authority Performed in 1819 A.D.," *Asiatic Researches*, Vol XV (reprinted by Cosmo Publishers, Delhi 1980) p. 339-428.

294. Himachal Pradesh, Directorate of Industries, 1965. *Survey Report on Handicrafts of Kinnaur District, Himachal Pradesh*, Shimla, Directorate of Industries, H.P.

295. Hoben, Allan, 1982. "Anthropologist and Development," *Annual Review of Anthropology*, Vol. 11, 349-75.

296. Honigmann, John.J, 1976. *The Development of Anthropological Ideas*, Illinois, The Dorsey Press.

297. Horne, C, 1876. "Notes on Villages in the Himalayas, in Kumaon and Garhwal, and on the Sutlej," *The Indian Antiquary*, Vol. 5, 161-167.

298. Howell, Nancy, 1986. "Demographic Anthropology," *Annual Review of Anthropology*, Vol. 15, 219-46.

299. Huard, Pierre, 1969. "Western Medicine and Afro-Asian Ethnic Medicine" in FNL Poynter (ed.) *Medicine and Culture*. London, Welcome Institute of the History Medicine.

300. Hughes, Charles, 1968. "Ethnomedicine," in International Encyclopaedia of the Social Sciences, New York, The Free Press.

301. Huijbers, P. M. J. F., J. L. M. Hendriks, W. J. M. Gerver, P. J. De Jong and K. DE Meer, 1996. "Nutritional Status and Mortality of Highland Children in Nepal: Impact of Socio-cultural Factors," *American Journal of Physical Anthropology*, Vol 101:137-144.

302. Hussain, I. Z, 1970. *An Urban Fertility Field: A Report on City of Lucknow*, Lucknow, Demographic Research Centre.

303. Hussain, I.Z, 1970. "Educational Status and Differential Fertility in India," *Social Biology*, Vol. 17: 132.

304. Hutton, Thomas, 1938. *Journal of a Trip Through Kunawur Hungrung and Spiti*, pp. 10-16.

305. Hyde, Jim, 1999. "Health System Reform and Social Capital," *Development*, Vol. 42, No. 4: 49-53.

306. Jacquemont, Victor, 1934. *Letters from India Describing a Journey in the British Dominions of India, Tibet, Lahore and Cashmere During the Years 1829, 1830, 1831, tr. From the French*, London, Edward Churton.

307. Jacquemont, Victor, 1936. *Letters from India 1829- 1832, Being a Selection from the Correspondence of Victor Jacquemont, tr. From the French with an intro from Catherine Alison Phillips*, London, Macmillan.

308. Jagat, Bandu, 1961. "Kinnaur Jila Ki Basha," *Himprastha*, Vol. 6, No.10: 22, 35.

309. Jagat, Bandu, 1961. "' Kinnur Dharti ," Shimla, *Himprastha*, Vol. 6, No. 11:5-7.

310. Jagmohan and J. N Panda, 1985. "For Her: Factors Controlling Human Fertility," December.

311. James, S., Robert Martin and David Pilbeam (ed), 1992. *The Cambridge Encyclopaedia of Human Evolution*, Cambridge University Press. pp. 287-291.

312. Janes. Craig R, 1999. "The Health Transition, Global Modernity and the Crisis of Traditional Medicine: the Tibetan case," *Social Science and Medicine*, Vol. 48, 1803-1820.

313. Jayal, Amenia, 1979. "Kinnaur Remembered: Life in a Lovely Himachal District," *Times of India Annual*, p. 87-90.

314. Jerath, Ashok, 1995. *The Splendour of Himalayan Art*, Delhi, Indus Publishing House.

315. Jha, Ugranath, 1966. "The Origin of Panji System," *The Eastern Anthropologist*, Vol. XIX, No. 3: 190-204.

316. Johri, Sita Ram, 1964. *Our Boderland*, Lucknow, Himalayan Publication.

317. Joshi, L.D, 1984. "Tribal People of the Himalayas: A Study of the Khasas," pp. 77-89.

318. Joshi, Pt. Tikka Ram. 1909. "A Grammar and Dictionary of Kanawari, The Language of Kanawar in the Bashahr State, Punjab," ed. By H. A. Rose, *Journal of Asiatic Society of Bengal*, Vol. 5.

319. Joshi, Pt. Tikka Ram, 1911. "Ethnography of Bushahr State," ed. By H.A. Rose, *Journal of Asiatic Society of Bengal*.

320. Kalita, Mondira and Sengupta, Sarthak, 1997. "Age at Menarche and Menopause Among the Sonowal Women of Dibrugarh, Assam," *J.Ecol.*, Vol: 8, No. 6:485-486.

321. Kantikar, T, 1979. "Development of Maternal and Child Health Services in India," in K. Srinivasan, P.C Saxena and Tara Kantikar (eds), *Child in India*. Delhi, Himalaya Publishing House.

322. Kapadia, Harish, 1985-86. "A Note of Kinnaur," *The Himalayan Journal*, Vol. 93.

323. Kapadia, Harish. 1993. *High Himalaya Unknown Valleys*, Delhi, Indus Publishing Company.

324. Kapadia, Harish, 1999. *Across Peaks and Passes in Himachal Pradesh*, Delhi, Indus Publishing Company.

325. Kapoor, A. K, 1992. "Ecology and Development in Himalayas' in S. K Chada, (ed). *Fragile Environment*, New Delhi, Anmol Publication.

326. Kapoor, A.K, 1993a. "The Environmental and Settlement Patterns of the Bhotias in the Central Himalayas," in S.K Biswas (ed), *Central Himalayan Panorama Vol I,* pp 169-179. Calcutta Institute of Social Research and Applied Anthropology.

327. Kapoor, A.K, 1993b. "Ecology and Development in Himalayas," in S.K. Chadha (ed), *Fragile Environment*. Delhi, Anmol Publications. pp. 118-125.

328. Kapoor, A.K, 1994. "Ecology and Tribal in Central Himalayas," in A.K Kapoor and Satwanti Kapoor (ed.) '*Ecology and Man in the Himalayas,*" Delhi, M.D. Publication Pvt. Ltd. pp. 69-90.

329. Kapoor A.K, 1996. ""Tribe, Anthropologist & Future," in R.S. Mann (Eds), *Tribes of India: Ongoing Challenge,* pp. 67-80 New Delhi, M.D. Publications.

330. Kapoor, A. K, 1996. "Ecology, Demographic Profile and Socio-Economic Development of a Tribe ot Central Himalayas," in P. K Samal (ed), "*Tribal Development: Options,*" (Proceedings of a National Seminar), pp 118-138. Nainital, Gynoday Prakashan.

331. Kapoor A.K, 1998a. "Role of NGO's in Human Development: A Domain of Anthropology," *J.Ind. Anthrop. Soc.,* Vol. 33: 283:300.

332. Kapoor. A.K, 1998b. "Law and Scheduled Tribes: A Socio-Legal Perspective," in V.K. Pant & B.S. Bisht. (Eds), *Backward Communities: Identity, Development and Transformation*. New Delhi, Gyan Publishing House. pp. 175-182.

333. Kapoor. A.K, 1998c. Genetic Diversity Among Indian Population Groups pp.216-224. In: *Physical Anthropology & Human Genetics: Contemporary Perspectives*. I.J.S. Jaswal (eds.). Asia Visions, Ludhiana.

334. Kapoor. A. K, 2000a Genetic Affinity Among Caste and Tribal Populations of India. pp. 125-138. In: *Studies on Man: Issues and Challenges*. M.K. Bhasin (eds.), KRE, Delhi.

335. Kapoor. A.K, 2000b. Environment, Health and Development in Sahariyas: A Primitive Tribes of Rajasthan. pp. 105-134. In: *Environment, Health & Development: An Anthropological Perspective*. P. Dash Sharma (eds.), S.C. Roy Institute of Anthropological Studies, Ranchi.

336. Kapoor. A. K, 2000c. Ecology and Religious Practices Among the Bhotias in Himalayas. pp. 221-235. *In: Tribal Religion: Change and Continuity.* M.C. Behra (ed.). Commonwealth, New Delhi.

337. Kapoor, A.K, 2000d. "Environment Technology and Development—An Anthropological Study of a Non Government Organisation (NGO) in Himachal Pradesh," *Man and Life*, Vol. 26, No. 3 and 4: 137-159.

338. Kapoor, A.K, 2003a. Tribes of Andaman & Nicobar Islands: Then and Now. *Dialogue Quarterly*, 4 (4): 59.

339. Kapoor, A.K, 2003b, Self-Hostage for Jarawa of Andaman & Nicobar Islands: Dimension of Development. pp. 1-34. In: *Tribal Development in Andaman Islands.* A.N. Sharma (ed.). Sarup & Sons, New Delhi.

340. Kapoor, A.K and D. Singh (eds). 1996. *Man and Development in the Himalayas*, Delhi, Academic Foundation.

341. Kapoor, A. K. and D. Singh, 1997. *Rural Development through NGOs*," Jaipur, Rawat Publications.

342. Kapoor, A. K. and Kshatriya, G.K, 2000. Fertility and Mortality Differentials Among the Selected Tribal Population Groups of North-Western and Eastern India. *J.Biosoc. Sci.(U.K.)*., 32: 253-264.

343. Kapoor, A. K. and R.R Prasad, 1986. "Relationship Between Environment and Economic Structure: Study of a Himalayan Community' in L.P. Vidyarthi and M Jha (eds), *Ecology, Economy and Religion of Himalayas*, Delhi, Orient Publications. pp. 50-58.

344. Kapoor, A.K. and S. Kapoor, 1985. "The Secular Growth Trend Among Himalayan Populations," *Anthrop. Kozl.* Vol. 29: 85-88.

345. Kapoor, A.K. and S. Kapoor, 1986. "The Effects of High Altitiude on Age at Mearche and Menopause," *International Journal Biometeor* . Vol. 30: 21-26.

346. Kapoor, A.K. and Satwanti, 1986. "Some of the Urgent Measures for the Development of the Non-Literate Societies in the Himalayas' in L. P. Vidyarthi and M Jha (eds), *Ecology, Economy and Religion of Himalayas*, Delhi, Orient Publications. pp. 101-112.

347. Kapoor, B.L, 1972. "Kinnaur: Ak Paricharcha," Shimla, *Himprastha*, Vol. 16, No. 10: 46-50.

348. Kapoor, D.D, 1958. "The Kinship system of the Non-Polyandrous Kanets of Mahasu," *The Anthropologist*, Vol. V, No. 1 & 2: 19-31.

349. Kapoor, D.D, 1990. "The People of Kinnaur," in K. Suresh Singh (ed), *The Tribal Situation in India*, Shimla, IIAS, p. 158-168.

350. Kapoor, S, 1996. "Research, Planning and Development in the Himalayan Region," in A.K. Kapoor and D. Singh (eds), *Man and Development in the Himalayas*, Delhi, Academic Foundation, pp. 197-211.

351. Kapoor, Suresh, 1993. *Schedule Tribes of H.P: Marriage and Divorce Customs*, Delhi, Navrang.

352. Kapoor, A. K. Kshatriya, G. K. and Kapoor, S, 2003. Fertility and Mortality Differentials Among the Population Groups of the Himalayas. *Human Biology*, 75 (5): 729-747.

353. Kapoor, S., Patra, P. K., Kshatriya G. K. and Kapoor, A.K. 2002. Fertility and Mortality Differentials Among Four Caste Groups of Haryana. pp. 188-194. *In: Tribal Welfare and Development: Emerging Role of Anthropological Explorations*. A.N. Sharma (ed.) Sarup & Sons, New Delhi.

354. Kapoor, A.K., Khanna, R., Mishra, P. and Patra, P. K, 2004. Koya. pp. 285-295. *In Ethnographic Atlas of Indian Tribes*. Prakash Chandra Mehta (ed.). Discovery Publishing House, New Delhi.

355. Kapoor, S. and Kapoor, A.K, 2005. Body Structure and Respiratory Efficiency Among High Altitude Himalayan Populations. *Coll. Anthropol.*, 29 (1): 37-43.

356. Kapoor, A.K, 2006. Ecology, Disease Patterns and Healing Aspects in Himalayan Communities. (Send for Publication).

357. Kapur, Suresh, 1998. "Tribal-Non Tribal Interaction and Conflict of laws," in Gupta, S. K., V. P. Sharma, N. K. Sharda (ed), *Tribal Development: Appraisal and Alternatives*, Delhi, Indus Publishing House.

358. Kar, P.C, 1982. *The Garos in Transition*. Cosmo Publications, New Delhi.

359. Kar, R.K, 1993. "Health Behaviour Among the Tribes of North-East India: A Profile," *Journal of Indian Anthropologist*, Vol. 28, No. 2: 115-124.

360. Kar, R. K. and Juri Gogoi, "Health Culture and Tribal Life: A Case study Among the Noctes of Arunachal Pradesh" in R. C. Swarankar (ed), *Indian Tribes: Health Ecology and Social Structure* pp. 17-41.

361. Kar.R.K. and Gogoi, Juri, 1993. "Health Culture and Tribal Life: A Case Study Among the Noctes of Arunachal Pradesh". *Man and Life*, Vol. 19, No. 1&2: 29-53.

362. Karmakar,Tanuja,D. Sampathkumar,V. Jeyalakshmi,S. and Abel.R, 1995. "Nutritional Status of Tribal Women in Bihar," *Man in India*, Vol. 75, No. 2: 209-214.

363. Kaushal, R. K, 1962. *Kamnaa Kinnaur*, Solan, Kumarson.

364. Kaushal, R.K, 1962. "Kinnaur and Kinnar," *Himprastha*, Vol. 8, No. 3: 11-13; Vol 8, No. 4: 9-11; Vol. 8, No. 5: 46-55; Vol. 8, No. 6:2-5.

365. Kaushal, R.K, 1963. "Kinnauri Lok Gito ki Bhavlok," Shimla, *Himprastha*, Vol. 14, No. 5.

366. Kaushal, R.K, 1988. *Himachal Pradesh: Socio Economic, Geographical And Historical Survey*, Delhi, Reliance Publishing House.

367. Kaushal, R.K, 1965. *Himachal Pradesh, A Survey of the History of the Land and its People* Delhi, Reliance Publishing House.

368. Kesby, Mike, 2000. "Participatory Diagramming as a Means to Improve Communication about the Sex in Rural Zibabwe: A Pilot Study," *Social Science and Medicine*, Vol. 50: 1723-1741.

369. Khan, A.R, 1996. *Man, Environment and Development in Himachal Pradesh*, Delhi, Indus Publishing House.

370. Khan,N.C, 1976. "Some Aspects of Marital Tend Among the Tribals of West Bengal," in Kanti B. Pakrasi , Amulay R. Banerjee and Amal K. Das(ed), *Biosocial Studies in India*, Calcutta Editions Indian. pp. 143-152.

371. Khanna, H.R, 1971. "Trip to Kalpa Valley," *Times of India*, 28 Feb.

372. Khanna. R and A.K.Kapoor, 2000. "Ethnographic Profile of Kinnauri: A Scheduled Tribe of Himachal Pradesh' in P.C. Mehta (Ed), *Ethnographic Atlas of Tribals*, Udaipur Shiva Publications Distributors.

373. Khanna. R and A.K.Kapoor, 2004. "Public—Private Partnership: A Strategy for Improvement of Health in Scheduled Tribe" in AK Kalla and PC Joshi (etd), *Tribal Health and Medicines*, Delhi, Concept Publishing House.

374. Khare, R.S, 1957. "A Plea for Cultural Dimension in Medicine," *The Eastern Anthropologist*, Vol. XII, No. 3: 196-201.

375. Khongsdier, R and Ghosh, A.K, 1995. "Bio Demographic Study Among the War Khasi of Meghalaya," *J.Ind.Anthrop.Soc.*, Vol. 29, No. 1&2: 195-202.

376. Khongsdier, R, 1992. "Some Demographic Traits Among the PNAR of Sutnga And Moopala in Jaintia Hills District of Meghalaya," *Man in India*, Vol. 72, No. 4: 491-495.

377. Khongsdier, R, 1995. "Prenatal and Postnatal Mortality in the War Khasi of Meghalaya" *J.Hum.Ecol.*,Vol. 5, No. 4: 307-310.

378. Khongsdier, R. and A.K Ghosh, 1995. "Bio-Demographic Study Among the War Khasi of Meghalaya" J. Ind. Anthrop. Soc., Vol. 29 No. 1 & 2: 195-202.

379. Khosla, Romi, 1979. *Buddhist Monasteries in the Western Himalaya*, Bibliotheca Himalayica Series III Vol. 13, Nepal, Ratan Pustak Bhandar.

380. Khullar, D. K. 1973. "Kinnaur Kailes," *The Himalayan Journal*, Vol. XXXII: 105-112.

381. Kinnar, Khem Singh, 1956. "Kinnar Geet," Shimla, *Himprastha*, Vol. 1, No. 12: 40-41.

382. Kinnar, Khem Singh, 1957. "Kinnar lok Geet," *Vanyajati*, Vol. 5, No. 1: 35-36.

383. Kinnar, Khem Singh. 1957."Vyas Moni Ka Ak Kinnar Lok Geet," *Vanyajati*, Vol. 5, No. 10:69.

384. Kirkpatrick, C. S, 1878. "Polyandry in the Punjab," *The Indian Antiquary*, Vol. 7: 86.

385. Klarman, Herbert E, 1965. *The Economics of Health*, New York, Columbia University Press.

386. Kohli, Capt. M. S, 1983. *The Himalayas: Playground of the Gods: Trekking, Climbing, Adventure*, Delhi, Vikas Publishing House Pvt. Ltd.

387. Konow, S, 1905. "Some Facts Connected with the Tibeto-Burman Dialect Spoken in Kanawar," *Zeitschrift der Deutschen Morgenlandischen Gesellschaft*, Vol. 59: 117-125.

388. Kopparty, S.N.M, 1992. "Caste and Utilisation of Health Resources," *The Eastern Anthropologist*, Vol. 45 No. 4: 365-383.

389. Kramer, Karen L. and Garnett P. McMillan, 1999. "Women Labour, Fertility and The Introduction of Modern Technology in Rural Maya Village," *Journal of Anthropological Research*, Vol. 55, No. 4:499-520.

390. Kroeber, A. L, 1946. *Anthropology*, Bombay, Oxford and I.B.H. Publishing.

391. Krohn-Hansen, Christian, 1994. "The Anthropology of Violent Interaction," *Journal of Anthropological Research*, Vol. 50: 367-381.

392. Kshastriya, G.K, 2000. "Ecology and Health with Special Reference to Indian Tribes," in M.K. Bhasin and V. Bhasin (Eds), *Man—Environment Relationship*. Delhi, Kamla-Raj Publications.

393. Kshastriya, G. K., Salil Basu and Anil Jindal, 1994. "Perception of Health and Pattern of Health Seeking Behaviour Among the Selected Tribal Population Groups of Madhya Pradesh and Orissa' in Salil Basu (Ed), *Tribal health in India*, Delhi, Manak Publications Pvt. Ltd.

394. Kshatriya, G.K, 1992. *Health as Parameter for Women's Development National Workshop on Education and Women's Development*. Delhi, National Institute of Education Planning and Administration.

395. Kshatriya, G. K, Paramjit Singh and S. K. Basu, 1993. "Fertility and Mortality in Bison Horn Madias of Dantewara Tehsil of Bastar District, Madhya Pradesh," *Journal of Human Ecology*, Vol. 4, No. 2: 93- 96.

396. Kshatriya, G.K, Paramjit Singh and S.K. Basu, 1997. "Anthropo-Demographic Features and Health Care Practices Among the Jaunsaris of Jaunsar-Bawar, Dehradun, Uttar Pradesh ," *Journal of Human Ecology*, Vol. 8, No. 5: 347-354.

397. Kshatriya. G. K. and Basu, Salil, 1993. *Fertility and Mortality Trends in Dudh Kharia Tribal Population of Sundergarh District, Orissa. Genetic and Variation*. Hyderabad, Indian Society of Human Genetic.

398. Kshatriya, G.K. and Kapoor, A.K, 2003a. Population Characteristics of the Bhil of Rajasthan. *Indian Anthropologist*, 33 (1): 1-16.

399. Kshatriya, G.K. and Kapoor, A. K. 2003b. Progressive Dhodias of South Gujarat pp. 600-624. *In: Understanding People of India: An Anthropological Insight*. (eds.) Dept. of Anthropology, University of Delhi, Delhi.

400. Kshatriya, G.K. and Kapoor, A. K. 2005. Demographic Structure and Health Care Practices of Dhodia Tribal Population of District Valsad, Gujarat. pp. 107-135. *In: Social Dimensions of Health*. Ajit K. Dalal and Subha Roy (eds.). Rawat Publications, Jaipur.

401. Kumar, K.I, 1979. *Expedition Kinner-Kailesh*, New Delhi, Vision Books.

402. Kumar, N and A.K. Miitra, 1975. "Reproductive Performance of Tharu Women East," *Anthrop*. Vol. 28: 349-357.

403. Kumar, R. K. and Kapoor, A.K, 2004. Religious Festivals of Parhaiya: A Tribal Community of Jharkhand. *Vanyajati*, XLXII (3): 1-7.

404. Kumar, R. K. and Kapoor, A.K, 2005. Some Aspects of Health and Role of Public Health Services in a Primitive Tribe of Jharkhand. pp. 552-567 In: *Reproductive and Child Health Problems in India*. K. K. N. Sharma (ed.). Academic Excellence, Delhi.

405. Kumar, R.K. and Kapoor, A.K, 2006. Health Profile of Pahariya: A Primitive Tribe of Jharkhand. (Send for Publication).

406. Kumar, S. Vijaya and C. Chakrapani, 1994. "Rural Health Delivery System: A Case Study," *Trends in Social Science Research*, Vol. 1 No. 2:41-48.

407. Lamba, Rajni and Shalina Mehta, 1995. "Priorities in Indian Medicine: A Tribal Perspective," *Indian Anthropologist*, Vol. 25, No. 2: 1-12.

408. Lambert, Helan, 1997. "Illness, Inauspiciousness and Modes of Healing in Rajasthan," *Contributions to Indian Sociology,* Vol. 31, No. 2: 253-71.

409. Laughlin, William, S, 1963. "Primitive Theory of Medicine: Empirical Knowledge," in Lago Galdson (ed) *Mans Image in Medicine and Anthropology,* New York, International University Press.

410. Leighton, Alexander H and Dorothea C Leighton, 1941. "Element of Psychotherapy in Navaho Religion," *Psychiatry* Vol. 4: 515-523.

411. Levinson, David and Melvin Ember, 1996. *Demography-Encyclopaedia of Cultural Anthropology, Vol. 3,* New York, A Henery Holt Reference Book, Henry Holt & Co. pp. 319-323.

412. Levinson, David and Melvin Ember. 1996. *Rural Health Care-Encyclopaedia of Cultural Anthropology,* Vol. 3, New York, A Henery Holt Reference Book, Henry Holt & Co. pp. 1126-1131.

413. Lieban, R.N, "1977. "The Field of Medical Anthropology," in David Landy (ed) *Culture Disease and Healing,* New York, Mac Millian Publication Inc. pp.13-31.

414. Living Stone, Frank B. 1958. "Anthropological Implications of Sickle Cell Gene Distribution in West Africa", *American Anthropologist* Vol. 60: 533-56.

415. Loesch, D.Z.R, Huggins, E. Rogucka, N.H. Hoang and J.L Hopper. 1995. "Genetic Correlates of Menarcheal Age: A Multivariate Twin Study," *Annals of Human Biology,* Vol. 22, No. 6: 479-490.

416. Logan, Michael and Edward E Hunt. 1978. *Health and Human Conditions,* Massachusetts, Daxbusty Press.

417. Loukid, M., A. Balli and M. K. Hilali, 1996. "Secular Trend in Age at Menarche in Marrakesh (Morocco)," *Annals of Human Biology,* Vol. 23, No. 4: 333-335.

418. Luft, Harold S, 1978. *Poverty and Health: Economic Causes and Consequences of Health Problems,* Massachusetts, Ballinger.

419. Luther, Indira, 1958. "The People of Kinnaur District," *Social Welfare,* Vol. V, No. 4: 13.

420. Macfarlane, Alan, 1981. "Death, Disease and Curing in a Himalayan Village," in Christoph Von-Furer Haimendorf (ed), *Asia Highland Societies in Anthropological Perspective*, New Delhi, Sterling Publishers. pp. 79-130.

421. Madan, T.N, 1951. "Education of Tribal India," *Eastern Anthropologist* Vol. V, No. 4: 179-82.

422. Madrigal, L, 1995. "Differential Fertility of Mothers of Twin and Mothers of Singletons: Study in Limon, Costa Rica," *Human Biology*, Vol. 67, No. 5:779-789.

423. Mahanta, K. C, 1993. "Socio-Economic Status of the Aged: A Case Study," *Man In India*, Vol. 73, No. 3: 201- 213.

424. Mahapatro.Meerambika, Sachdeva, M.P. and Kalla, A.K, 1999. "Knowledge, Attitude and Practice of Birth Control Devices Among Bhattara Tribals of Orissa," *J.Ecol.*, Vol. 10, No. 1: 7-13.

425. Mahato, S.N and M. K Raha, 1975. "The Kinnaurese of Western Himalaya," Calcutta, *Bulletin of the Cultural Research Institute*, Vol. XI, No. 1 & 2.

426. Maheo, Lorho Mary and Kalla,A.K, 2001, "Knowledge, Attitude and Practice(KAP)of Birth Control Measures Among the Mao Nagas of Senapati District, Manipur," *Anthropologist*, Vol. 3, No. 1: 33-42.

427. Maitra, Kiranshankar. 1989. *Himalayan Dreamland—Journey to Kinnarlok*, Delhi, Mittal Publications.

428. Majumdar, Paul, Gopa, 2001. "Some Aspects of Fertility of the Mahotas of Midnapore, West Bengal," *J.Hum.Ecol.*, Vol. 12, No. 5: 379-382.

429. Majumdar, D.N, 1953. "Children in a Polyandrous Society," *The Eastern Anthropologist*, Vol, 6: 177-189.

430. Majumdar, D.N, 1955a. "Family and Marriage in a Polyandrous Society," *The Eastern Anthropologist*, Vol. 8: 85-110.

431. Majumdar, D.N, 1955b. "Demographic Structure in a Polyandrous Village," *The Eastern Anthropologist*, Vol. 8: 161-172.

432. Majumdar, D.N, 1961. *Races and Culture of India*, Bombay, Asia Publishing House.

433. Majumdar, D.N, 1962. *Himalayan Polyandry*, Bombay, Asia Publishing House.

434. Majumdar,D.N., and Madan, T.N., 1986. *An Introduction to Social Anthropology*. Uttar Pradesh, Mayoor Paperbacks,.

435. Majumdar, Murari, 1962. *Vital Rates National Sample Survey, 54*, Government of India: 24.

436. Majumder, Partha P, 1991. "Recent Developments in Population Genetics," *Annual Review of Anthropology*, Vol. 20: 97-117.

437. Malik, D.S, 1975. "Land The Ogress stalked: Kinnaur," *ITBP Bulletin*, July-Sept.

438. Malik, Harji, 1975. "A Remote Village Tucked in the Himalayas—Sangla feels the Impact of the Change," *Sunday Standard Magazine*, 3 August, p. 7, c 6-7.

439. Malik, S.L. and Hauspie, R.C, 1986. "Age at Menarche Among High Altitude Bodhs of Ladakh (India),"*Hum.Bio* Vol. 58:541.

440. Mamgain M.D, 1971. *Gazetteer of India: Himachal Pradesh: Kinnar*, State Editor, District Gazetteer H.P.

441. Mandelbaum, David, G, 1972. *Society in India*. 2 Vols. Bombay, Popular Prakashan.

442. Mann, K, 1987. *Tribal Women in a Chaging Society*. Delhi, Mittal Publications.

443. Mastana, S.S, 1987. *The Genetic Structure and Affinities of the Lobanas of Punjab(India)*. Dissertation, University of Cambridge, Cambridge.

444. Mathai, Achamma. J, 1963. "Kinnaure: Belle of the Himalayas," *Social Welfare*, Vol. 10, No. 5, August, p. 4.

445. Mayer, Jonathan D, 2000. "Geography, Ecology and Emerging Infectious Diseases," *Social Science and Medicine*, Vol. 50: 937-952.

446. Mc Cracken, Robert D, 1971. "Lactase Deficiency: An Example of Dietary Evolution" *Current Anthropology* Vol. 12: 479-517.

447. McKenna, James J, 1996 "Sudden Infant Death Syndrome in Cross-Cultural Perspective: Is Infant-Parent Cosleeping Protective?," *Annual Review of Anthropology*, Vol. 25: 201-16.

448. McKeown, T, 1987. "Looking at Disease in the Light of Human Development," *World Health Forum*, Vol. 6.

449. Mcleish, Alexander, 1983. *The Frontier People of India*, Delhi, Mittal Publication.

450. Mehra, Anil, 1974. "Kinnaur Trekker's Delight," *NPA Tourism and Travel*, Vol. 1. No. 5: 6-7.

451. Mehra, P. N, 1961. "Ramniya Kinnaur," *Himprastha*, Vol. 6, No. 10: 48-55.

452. Mehta, Shaline, 1992. "Industrialisation of Tribal Belt: Some Observations," *Man In India*, Vol. 72, No. 3: 271-280.

453. Mehta, Swarjit Kaur, 1968. "Kinnaur: Its Geographical Setting," *Everyday Science*, Vol. XIII, No. 2: 32-34.

454. Melkania, N.P. and Uma Melkania, 1988. "The Human Environment in Himalaya," *Man in India*, Vol. 68, No. 2.

455. Menon, Geeta, 1987. "Tribal Women. Victims of the Development Process," *Social Action*, Vol. 37.

456. Menon, Geeta, 1991. "Ecological Transitions and the Changing Context of Women's Work in Tribal India," *Purusartha*, Vol. 14: 291-314.

457. Metzgua, Duane and Gerald Williams, 1963. "Tenejapa Medicine 1: The Curer," *South Western Journal of Anthropology* Vol. 19: 216-234.

458. Mishra, Braja Kishori, 1993. "Feeding Practises of Preschool Children in Western Orissa. I. Demographic Character, Prelacteal Feed and Colostrum Rejection," *J. Hum.Ecol.*, Vol. 4, No. 1: 85-92.

459. Misra, B.D, 1982. *An Introduction to the Study of Population*. New Delhi, South Asian Publisher Pvt. Ltd.

460. Misra, P. and Kapoor, A.K, 2002a. Patterns of Health Care and Attitude Towards Healing Practices in a Primitive Tribe of Rajasthan. *J. Region, Health and Health Care*: 7 (1&2): 28-40.

461. Misra, P. and Kapoor, A.K, 2002b. Environments, Disease Pattern and Health Status in a Primitive Tribe of Rajasthan. pp. 33-51. *In: Eco Degradation and Population Health: The Challenges and Managements in New Millennium*. B. N. Pandey (ed.). Daya Publishing House, New Delhi.

462. Misra, P. and Kapoor, A.K, 2002c. Educating Tribal Girls. *Social Welfare*, 49 (6): 23-29.

463. Misra, P. and Kapoor, A.K, 2003. Ecology, Economy and Culture: An Anthropological Perspective of the Meena—A Scheduled Tribe of Rajasthan. *Social Change*, 32 (1&2): 1-26.

464. Misra, P. and Kapoor, A.K, 2003-2004 Interaction Between Traditional and Modern System of Medicine in a Primitive Tribe of Rajasthan. *Artha Journal of Social Sciences*, 2 (2): 31-47.

465. Misra, P. and Kapoor, A. K, 2005a. Anthropo-Demographic Profile of Saharia: A Primitive Tribe of Rajasthan. In: Contemporary Studies in Primitive Tribes. S.K. Chaudhary and S.S. Chaudhari (eds.). Mittal Publications, New Delhi.

466. Misra, P. and Kapoor, A. K. 2005b. Ecology and Economy of a Primitive Tribe in a Semi-and Zone pp. 121-158. *In Primitive Tribes in Contemporary India*, Vol. II. S. K. Chaudhari and S. S. Chaudhari (eds.). Mittal Publication, New Delhi.

467. Misra, P. and Kapoor, A.K, 2005c. Socio-cultural and Environmental Factors of Health Among Sahariya: A Primitive Tribe of Rajasthan. pp. 217-275. *In: Contemporary Society: Tribal Studies*, Vol. 6. Deepak Kumar Behera and George Pfeffer (eds.). Concept Publishing company, New Delhi.

468. Misra, P. Kapoor, and A.K. 2006. Health Culture and Health Seeking Behaviour in a Primitive Tribe of Desert Zone. pp. 291-322 *In: Anthropology of Primitive Tribes in India*. P. Dash Sharma (ed.). Serials Publications, New Delhi.

469. Misra, P. and Kapoor, A.K, 2006. 'Utilization of ANC Services in a Primitive Tribe of Rajasthan. (Sent for Publication).

470. Mitchell, Duncan G.A, 1972. *New Dictionary of Sociology*, London, Routledge and Kegan Paul.

471. Mitra, Koumari., Koushambhi Basu and Salil Basu. 1998. "An Eco Health Study on Urban Slums of Delhi," *Journal of Human Ecology*, Vol. 9 No. 2: 153-157.

472. Mitto, Hari Krishan. 1978. *Himachal Pradesh*, Delhi, National Book Trust.

473. Mohan, Krishna. 1998. "Scheduled Population of Tribal Areas in Himachal Pradesh: A Spatial Analysis," in Gupta, S. K., V.P. Sharma, N. K. Sharda (ed), *Tribal Development: Appraisal and Alternatives*, Delhi, Indus Publishing House.

474. Monga, O.P, 1998. "Dynamics of Family Structure Among Kinnauras of Himachal Pradesh: Some Reflection," in Gupta, S.K., V.P. Sharma, N.K. Sharda (ed), *Tribal Development: Appraisal and Alternatives,* Delhi, Indus Publishing House.

475. Mooney, Cathleen., Jack Zwanziger, Ciaran S. Phibbs and Susan Schmitt, 2000. "Is Travel Distance a Barrier to Veterans' Use of VA Hospitals for Medical Surgical Care?," *Social Science and Medicine,* Vol. 50: 1743-1755.

476. Mukherjee, D.P, 1974. *Fourth Annual Report on the Genetic Studies in Relation to Fertility,* ICMR Project.

477. Mukherji, R.K, 1951. *The Original Inhabitants of India.*

478. Mukherji, S.K, 1998. "Managing Protected Areas," in Ashok Ranjan Basu and Padam Nabh Gautum (ed), *Natural Heritage of India: Essays on Environment Management,* Delhi, H.K. Publishers & Distributor .

479. Mukherji, D.P, 1972. "Some Recent Trends in Population Genetics in India. Genetics and Our Health," ICMR Technical Report Series. Vol. 20: 234-243.

480. Murdock, George, Suzanne F. Wilson and Violetta Frederick. 1978. "World Distribution of Theories of Illness," *Ethnology,* Vol. XVII, No. 4: 449-470.

481. Murray, Ansley J.C, 1882. *An Account of Three Months Tour from Shimla Through Bussahir Kunowar and Spiti to Lahaul,* Calcutta, Thacker Spink and Co.

482. Murthy Venka, G.B, 1987. "The Soligas of B.R. Hills: A Demographic Study," *Journal of Family Welfare,* Vol. XXXIV 154-58.

483. Murty, J.S. and A. Ramesh, 1978. "Selection Intensities Among the Tribal Population of Adilabad District. Andhra Pradesh," *Social Biology,* Vol. 25: 302-305.

484. Mushkin, Selma J. 1962."Health as Investment,"*Journal of Political Economy,* Vol. 70, No. 5.

485. Mutharayappa, R. 1993. "Socio-Cultural Factors and Marriage Among Jenukuruba and Kadukuruba Tribes of Karnataka," *Man in India,* Vol. 79, No. 1: 17-27.

486. Nadda, A. L, 1987. *Economics of Apple: A Case Study of Himachal Pradesh*, Delhi, B.R. Publishing Corporation.

487. Nag, M, 1974. "Socio Cultural Pattern, Family Cycle and Fertility," in Paper of the World Population Conference-3. Bruchest, *The Population Debate: Dimensions and Perspectives*, United Nations, New York. pp. 289.

488. Nag, M, 1962. *Factors Affecting Human Fertility in Non-Industrial Societies: A Cross Cultural Study*. USA, Yale Univ. Publications in Anthropology. No. 66.

489. Nag, Narendra Gopal, 1981. "Some aspects of Ethnomusicology of Kinnara with Special Reference to Issues Relating to Primitiveness," *Man in India*, Vol. 61, No. 4: 305- 326.

490. Nag, Narendra Gopal, 1981. "Kinnara: Some Aspects of Ethnomusicology of Kinnara with Special Reference to Issues Relating to Primitiveness," *Man in India*, Vol. 61, No. 4: 306-308.

491. Nanda, Satyajeet and S. Niranjan, 1999. "Utilisation of Antenatal Care(ANC) Service by the Schedule Tribes in India," *Journal of Human Ecology*, Vol. 10, No. 2: 99-104.

492. Nas, Peter. J. M and Margriet Veenma, 1997. "Towards Sustainable Cities: Urban Community and Environment in the Third World," *Journal of Human Ecology*, Vol. 1, No.1: 29-41.

493. *National Health Policy 2002*, Ministry of Health and Family Welfare, Government of India.

494. *National Policy on Scheduled Tribes 2004*, Ministry of Tribal Affairs, Government of India.

495. *National Population Policy 2000*, Ministry of Health and Family Welfare, Government of India.

496. Nath, Krishan, 1999. *Kinnaur Dharmlok*, Bikanar, Vaghadevi Pocket Books.

497. Nayak,A.N. and Babu,B.V, 2001, "Utilization of Services Related to Safe Motherhood Among the Scheduled Caste and Scheduled Tribe Population of Orissa: An Overview," *South Asian Anthropologist*, Vol. 1, No. 2: 117-122.

498. Negi, D.B, 1987. "Changing Pattern of Political Leadership Among the Tribals of Kinnaur" in M.K. Raha, *The Himalayan Heritage*, Delhi, Gian Publishing House, pp. 371-386.

499. Negi, D. B, 1990. "Jajmani Relations in Tribal World: A Study of Kinnaura Tribes," *Man in India*, Vol. 70, No. 2: 131-143.

500. Negi, D.B, 1998. "Tribal Development and Social Change: Some Issues," in Gupta, S.K., V.P. Sharma, N.K. Sharda (ed), *Tribal Development: Appraisal and Alternatives*, Delhi, Indus Publishing House.

501. Negi, G.R, 1998. " The Moravian Missionaries: Their Contribution to the Cause of the Tribal People of Himachal Pradesh," in Gupta, S.K., V.P. Sharma, N. K. Sharda (ed), *Tribal Development: Appraisal and Alternatives*, Delhi, Indus Publishing House.

502. Negi, Gopal Chand, 1960. *Kinnaur and Kinnar Desh*, Shimla, Bhart Sewak Samaj.

503. Negi, Hans Raj and Madhav Gadgil, 1997. "Conserving Livestock Genetic Resources: A Case Study of Kinnaur in Himachal Pradesh," in K.C Mahanta (ed), *People of the Himalayas: Ecology, Culture, Development and Change*, New Delhi, Kamla Raj Enterprises, pp. 317-324.

504. Negi, J.M. S, 1995. *Himalayan Heritage: A Socio-Economic Cultural and Tourism Analysis*, Delhi, Gitanjali Publishing House.

505. Negi, R.S, A.C Srivastava and B.R. Bhatnagar, 1972. "Distribution of ABO Blood Groups in Central and Western Himalayan Population," *Bulletin of the Anthropological Survey of India*, Vol. XXI: 57-76.

506. Negi, S.S, 1990. *A Hand Book of the Himalaya*, Delhi, Indus Publishing Company.

507. Negi, S.S, 1991. *Himalayan: Rivers, Lakes and Glaciers*, Delhi, Indus Publishing Company.

508. Negi, S.S, 1995. *Cold Desert of India*, Delhi, Indus Publishing Company.

509. Negi, S.S, 1998, *Discovering the Himalaya*, Delhi, Indus Publishing Company, 2 Vol.

510. Negi, S. S, 1993. *Himachal Pradesh: The Land and People*, Delhi, Indus Publishing Company.

511. Negi, T.S, 1998. "Tribes of Himachal Pradesh: Past, Present and Future," in Gupta, S.K., V.P. Sharma, N. K. Sharda (ed), *Tribal Development: Appraisal and Alternatives*, Delhi, Indus Publishing House.

512. Negi, T. S, 1971. "Land of Ancient Kinners," *Cultural Forum*, Vol. XIII, No. 3-4: 52-56.

513. Negi, T.S, 1976. *Schedule Tribe of Himachal Pradesh: A Profile*, Meerut, Raj Printers.

514. Negi, T.S, 1976. "The Kanauras: Transformation of a Traditional Society," Hindustan Times, 28 June, p. 9, Col. 1-5.

515. Negi, T.S, 1990. "The Tribal Situation in Himachal Pradesh: Some Socio-Economic Consideration," in K. Suresh Singh (ed), *The Tribal Situation in India*, Shimla, IIAS.

516. Nemet, Gregory F. and Adrian J. Bailey, 2000. "Distance and Health Care Utilisation among the Rural Elderly," *Social Science and Medicine*, Vol. 50: 1197-1208.

517. Newall, D. J.F, 1882-87. *Highlands of India*, London, Harrison and Sons, 2 Vols. (Vol. 2- Kinnaur) (Reprinted by B.R. Publishing Corporation, Delhi in 1984.

518. Noponen, Helzi, 1999. "Participatory Internal Learning for Grassroots NGOs in Micro-Credit, Livelihoods and Environment Regeneration," *Development*, Vol. 42, No. 2: 27-34.

519. Notiyal, Virander Dutt, 1961. "Kinner Lok: A Survey," *Vanyajati*, Vol. IX No. 4: 151-52.

520. Nurge, Etiel, 1985. "Etiology of Illness in Guinnangdam" *American Anthropologist*, Vol. 60:1158-1172.

521. Ohri, Vishwa Chandra.(ed), 1975. *Arts of Himachal-Studies in Arts of Himachal Pradesh*, Shimla. Deptt. of Language and Cultural Affairs, State Museum.

522. Okonofua, Friday E., Diana Harris, Adetanwa Odebiyi, 1997. "'The Social Meaning of Infertility in Southwest Nigeria," *Health Transition Review*, Vol. 7: 205-220.

523. Palakshappa, T.C, 1978. *Tibetans in India: A Case Study of Mund God Tibetans*, Delhi, Sterling Publishers Pvt. Ltd.

524. Pallis, Marco, 1933. "Gangotri and Leo Pargial," *The Himalayan Journal*, Vol VI:106.

525. Pallis, Marco, 1939. *Peaks and Lamas*, London, Readers Union.

526. Panchani, C.S, 1994. *The Himalayan Tribes*, Delhi, Konark Publishers Pvt. Ltd.

527. Panda, Madhumita and Rupa Satpathi, 1996. "Population Genetics of Kondh Tribe of Orrisa," *South Asian Anthropologist*, Vol. 17, No. 2: 101-104.

528. Pandey, B.N., P.K.L. Das, A.K. Jha and A.K. Ojha, 1999. "Impact of Industrialization on Socio-Cultural and Health Aspects of Tribal Groups of South Bihar, India," *Journal of Human Ecology*, Vol. 10, No. 1: 15-28.

529. Pandey, G.D., and Saxena, B.N, 1988. "District Level Planning for Health and Family Welfare for Tribals," *Demography in India*, Vol. 17, No. 2: 216.

530. Pandey, G.D and R. S. Tiwary, 1996. "Fertility in Hill Korwas-A Primitive Tribe of Madhya Pradesh," *Man In India*, Vol. 76, No. 4: 325-329.

531. Pandey, G.D, 1994. "Demographic Characteristic of Tribal and Non Tribal Female—A Comparative Study," *Man in India*, Vol. 74, No.1: 39-47.

532. Pandey, G.D. and Goel A.K, 1999. "Some Demographic Characteristics of Abujhmaria of Madhya Pradesh," *J. Hum. Ecol.*, Vol. 20, No. 2: 85-88.

533. Pandey, G.D. and Tiwari, R.S, 1994. "Literacy and Demographic Characteristics: A Study of Association Among the Tribals of Madhya Pradesh," in S.K. Tiwary (ed) *Tribal Situation and Development in Central India*. Delhi, M.D. Publication.

534. Pandey, G.D., Verma, Arvind and Tiwari, R.S. 2001. "Some Correlates of Infant Mortality in a Primitive Tribe of Madhya Pradesh," *South Asian Anthropologist*, Vol. 1, No. 1:17-20.

535. Pandey, G.D., Roy, J. and Tiwary, R.S, 2001. "Socio-Cultural Aspects and Health Care in Pando Tribe of Madhya Pradesh," *J. Hum.Ecol.*, Vol. 12, No. 5: 391-394.

536. Pant, S.D, 1935. *The Social Economy of the Himalayas*, London.

537. Papiha S.S., S.M.S. Chahal, D.F. Roberts, and I.P. Singh, 1980. "Genetic Studies Among Kanet and Koli of Kinnaur District in Himachal Pradesh, India," *Man in India*.

538. Parkash, Mohinder and S. L Malik, 1990. "Differences in Fertility in Highlander and Lowlander Bods of Himachal Pradesh," *Journal of Human Ecology*, Vol. 1, No. 2:175-180.

539. Parmar, Dinesh and L .K. Sengupta, 1994. "Marital Distance and Village Endogamy in Bhilala and Barela Tribes of West Nimar in Central India," *Journal of Human Ecology*, Vol. 5, No. 2: 127-130.

540. Parmar, H.S, 1992. *Tribal Development in Himachal Pradesh: Policy, Programme and Performance*, New Delhi, Mittal Publications.

541. Parmar, Y.S, 1972. "Tribal Economy of Himachal Pradesh" *Indian Express*, 17 April, Vol. 7: 6.

542. Parmar, Y.S, 1975. *Polyandry in the Himalayas*, Delhi, Vikas Publishing House Pvt. Ltd.

543. Pasricha, Ramesh. 1959. "Chini and Baspa Vadi Ki Yatra," *Himprastha*, Vol. 5, No.4: 20-23.

544. Pasrsons, Talcott, 1954. " A Revised Analytical Approach to Theory of Social Stratification," in R. Bendix and S. M. Lipset (ed), *Class Status and Power*, London, Routledge and Kegan Paul. p. 92-108.

545. Patel, Shirsha. 1985. *Ecology, Ethnology and Nutrition: A Study of Khondh Tribals and Tibetan Refugees*. Delhi, Mittal Publication.

546. Pathy, J., et al, 1976. "Tribal Studies in India: An Appraisal," *Eastern Anthropologist*. Vol. 29, No. 4: 399-417.

547. Patki, G.S, 1959. "In the Land of Kinnaurs," *All India Congress Committee—Economic Review*, Vol X , No.16-18, 9 January, p. 131-35.

548. Patnaik, N, 1989. *The Saorn: Tribal and Harijan*, Research-cum-Training Institute, Bhubaneshwar.

549. Patnaik, Soumendra Mohan, 1990. "Relevance of Case Study Method in Anthropology of Development," *Indian Anthropologist*, Vol. 20, No.1&2: 31-38.

550. Patra, Prasanna Kumar, 2001.Demographic Profile with Reference to Development Programmes Among the Rajas: A Primitive Tribe of Uttaranchal. Unpublished Ph. D Thesis, University of Delhi, Delhi

551. Peery, Jonathan P, 1979. *Caste and Kinship in Kangra*, Delhi, Vikas Publishing Pvt. Ltd .

552. Peter, Prince, 1955. "Polyandry and the Kinship," *Man*, Vol. 55, No. 197-98: 179-181.

553. Peter, Prince, 1963. *A Study of Polyandry*, Mouton & Co. The Hague.

554. Phillips, W.S.K. 1967. "Social Distance: A Study of the Attitudes of the Upper Caste Towards the Lower Caste," *The Eastern Anthropologist*, Vol. XX, No. 2: 177-196.

555. Piplai,Chumki, 1993. "Effects of Family Planning Practises and Income on Fertility and Mortality among Two Tea-Garden in the Duars Area of Jalapiguri District, West Bengal," *J.Hum.Ecol.*, Vol. 4, No. 2: 105-110.

556. Pitanguy, Jacqueline, 1999. "Reproductive Rights are Human Rights," *Development*, Vol. 42, No. 1: 11-14.

557. Polgar Steven, 1962. "Health and Human Behaviour: Areas of Interest Common to the Social and Medical Sciences." *Current Anthropology* Vol. 3: 159-205.

558. Prabhakar, V.P, 1977."Kinnaur: A Vista of Nature's Beauty," *The Tribune*, Vol. XXVII: 94-106.

559. Prakash, M. and S.L. Malik, 1990. "Differences in Fertility in High Landers and Lowlanders Bods of Himachal Pradesh," *Journal of Human Ecology* Vol. 1, No.2: 175-180.

560. Prasad, N. Purendra and Rao, P.Venkata, 1996. "Patterns of Migration from a Drought-prone Area in Andhra Pradesh: A Case Study". *J.Hum. Ecol.*,Vol. 7, No. 3:135-162.

561. Prasad,K. Raghava and Nagaraj, K, 2001. "Socio-Demographic Correlates of Place of Delivery and Person Conducting the Delivery in a Rural Community of Andhra Pradesh," *J.Hum.Ecol.*,Vol. 12, No. 5: 357-362.

562. Prem, Agyaram, 1974. "Hastha Gata Nachta Kinnaur," *Himprastha*, Vol. 19, No. 12: 9-11.

563. Prema, L and Thomas, F, 1992. "Nutrition and Health Problems Faced by Kanikkar Women," in P.D. Tiwari and R.S. Tripathi (ed) *Dimensions of Scheduled Tribes Development in India*. Delhi, Uppal Publishing House.

564. Punjab District Gazetteers. 1904. Vol. VIII *A. Gazetteer of the Shimla District 1904* (Reprinted by Indus Publishing Company in 1997).

565. Punjab States Gazetteer. 1910. Vol. VIII, *Gazetteers of Shimla Hills States*, Government of Punjab, Lahore. (Reprinted by Indus Publishing Company in 1995).

566. Purohit, A.N. 1994. "The Sequence of Change-Growth, Development and Progress," *Man in India*, Vol. 74 No. 2: 129-140.

567. Raghuram, Shobha and Manashi Ray. 1999. "The State and Civil Society: Meeting health Needs, Reaching Equity," *Development*, Vol. 42 No. 4: 54-58.

568. Raha, M K, 1987. "The Khandan system of the Kinnaura of Western Himalayas," in M.K. Raha (ed), *The Himalayan Heritage*,Delhi, Gian Publishing House, pp. 329-371.

569. Raha, M. K, J. C Das and R.S. Negi, 1978. *Tribal Women of the Western and Central Himalayas in Tribal Women in India,"* Calcutta, Indian Anthropological Society.

570. Raha, M. K, J. C Das and S. C. Mondal, 1980. "Village Structure in Kinnaur," *Journal of Social Research*, Vol. XXIII, No. 1.

571. Raha, M.K, 1981. "Urgent Anthropology of the Himalayas," *Journal of Social Research*, Special Number on Himalayan Studies, Vol. XXI (11).

572. Raha, M.K, (ed) 1987. *Polyandry in India*, Delhi, Gian Publishing House.

573. Raha, M.K., and Palash Chandra Coomar, 1987. "Polyandry in High Himalayan Society: Persistence and Change," in M. K. Raha (ed), *Polyandry in India* Delhi, Gian Publishing House, pp. 62-129.

574. Raha, M.K. and S.N. Mahto, 1985. *The Kinnaurese of the Himalayas*, Calcutta, Anthropological Survey of India.

575. Raha, M. K, 1974. "The Highlanders of Western and Central Himalayas," *Vanyajati*, Vol. XXII, No.3: 88-95.

576. Raha, M.K, 1982. "Economic Strategies, Religious Pluralism and Change in Diverse Ecological Setting: A Case from Western Himalaya," in M.K. Raha (ed), *The Himalayas and the Himalayans: Bio-Cultural Aspects*, Calcutta, Anthropological Survey of India.

577. Raha, M.K., S. N. Mahato and R. S. Negi, 1976. "The Kinnaurese Kinship System, A Terminological System," *Bulletin of Anthropological Survey of India*, Vol. XXV, No.3-4: 85-102.

578. Raha, Manis Kumar, 1978. "Stratification and Religion in a Himalayan Society," in James F. Fisher (ed), *Himalayan Anthropology: The Indo Tibetan Interface*, Paris, Mouton Publishers, pp. 83-102.

579. Rahul, Ram, 1970. *The Himalayan Borderland*, New Delhi, Vikas Publishing House.

580. Raina, Jawahar Lal, 1989. "Tourism and Its impact on the Eco-System of the Himalayas," in S.K. Chadha (ed), *Himalayan Ecology*, Delhi, Ashish Publishing House, pp. 161-169.

581. Raja, Ram, 1942. "Through the Himalayas from Shimla to Tibet," *The Punjab Geographical Review*, Vol. 1, No. 2.

582. Rajpramukh,K.E, 1998. "Tribal health In Visakhapatnam District of Andhra Pradesh," *Journal of Human Ecology*, Vol. 9, No. 2:191-193.

583. Rajyalakshmi, P. and Geervani, P, 1992. "Food Nutrition and Morbidity: Some Observations Among Four Tribal Groups of South India," *J. Hum.Ecol.*, Vol. 3, No. 2: 115-214.

584. Rana, Kulwant and M.K Sharma, 1988. *Industrialisation of Hill States in India*, Delhi, Deep & Deep Publication.

585. Ranchan, Som P. and H.R. Justa, 1981. *Folk Tales of Himachal Pradesh*, Bombay, Bharatiya Vidya Bhavan.

586. Randhawa, M.S, 1974. *Travels in the Western Himalayas in Search of Painting*, Delhi, Thomson Press (India) Ltd.

587. Rao, B. Dharma, G Jaikishan and B.R. Busi, 1998. "Growth Progression and Nutritional Sataus Among Some Tribal Children of Visakhapatnam District, Andhra Pradesh," *Journal of Human Ecology*, Vol. 9, No. 4: 373-378.

588. Rao, K. Visweswara, 1995. "'Anthropometry for the Assessment of Various Forms of Malnutrition—Available Approaches and Their Relative Merits," *Man in India*, Vol. 75, No. 2: 195-207.

589. Rao, T. J, 1993. "Sampling Methodology in Anthropology," *Journal of Indian Anthropologist*, Vol. 28, No. 2: 115-124.

590. Ray, A and A. Roth, 1991. "Indian Tribals Fertility Patterns from Orissa." *Man in India* (special) Vol. 71, No. 1: 235-239.

591. Ray, Subha, 1998. "Health Seeking Behaviour of Two Squatter Groups of Calcutta City: A " Case Study Approach ," *Journal of Human Ecology*, Vol. 2, No. 4: 273-279.

592. Reddy, P.H and B. Modell, 1995. "Consanguinity and Reproductive Behaviour in a Tribal Population the Baiga in Madhya Pradesh," *Annals of Human Biology*, Vol. 22, No.1: 235-246.

593. Reeves, H.C, 1913. "Notes on the Mountains of Bussahir and Spiti," *Alpine Journal*, Vol. XXVII.

594. Relethford, John. H, 1992. "Cross-Cultural Analysis of Migration Rates: Effects of Geographic Distance and Population Size," *American Journal of Physical Anthropology*, Vol. 89: 459-466.

595. Report of Symposium on *Social and Economic Problems of Hilly Area*. 1973, Directorate of Economics and Statistics, Shimla.

596. Risley, H Sir, 1915. *The People of India*, Delhi, Orient Book Reprint Corporation (Reprinted 1969).

597. Rose, H.A, 1970 Reprint. *A Glossary of the Tribes and Castes of the Punjab and North West Frontier Province*, Lahore, Vol. II.

598. Rosen, G.A, 1976. *A History of Public Health*. New York, MD Publications.

599. Roy,Burman, 1961. *Ethnographic Study of Soligas, Census of India, Part V-B, No 6*, Monograph Series, Karnataka.

600. Roy, S.B, 1996. "Social Indicators Towards Institutionalisation of Development Programme: A Case Study of Joint Forest Management," *South Asian Anthropologist*, Vol. 17 No. 2: 81-87.

601. Roy,Debesh, 1995. "Marriage Rules and Marriage Rituals Among Kokna Tribes of Maharashtra," *Man in India*, Vol. 75, No. 1:25-35.

602. Rubin, Vera, 1961. "The Anthropology of Development," *Biennial Review of Anthropology*, p. 120-172.

603. Rund, Nadine H. and Lynn L. Krause, 1978. "The Health Attitudes and Your Health Programme" in R.N. Eleanor E. Bauwens (ed), *The Anthropology of Health*, Missouri, The CV Mosby Company.

604. Sabat, Kalpana Rani and Dash, Nirmal Chandra, 1996. "Socio-Economic and Demographic Profile of a Kandh Village of Eastern Ghats, Orissa," *Man in India*, Vol. 76, No. 2: 127-140.

605. Sabat, Kalpana Rani,Dash Nirmal Chandra and Das, Bijayalaxmi, 1997. "Habitat and Nutritional Status of a Kandh Village of Eastern Ghats, Orissa," *Man in India*, Vol. 76, No. 2: 127-140.

606. Sachchidananda, 1978. "Social Structure, Status and Mobility Patterns the Case of Tribal Women," *Man in India*, Vol. 58, No. 1.

607. Sachchidananda, 1994. "Women, Environment and Development," *Trends in Social Science Research*, Vol. 1, No. 1: 71-79.

608. Sahlins, Marshall D, 1969. "Culture and Environment: The Study of Cultural Ecology," in Manners and Kaplan (Ed), *Theory in Anthropology* London. Routledge and Kegan Paul Ltd.

609. Sahu,P.N, 1983. "Demographic and Genetic Constitution of a Small Population (The Malia)," *J. Indian Anthrop. Soc*, Vol. 18: 55-59.

610. Saklani, Bina, 1995. "Mechanical Strategies in Folk Medicine in Rawain," *Indian Anthropologist*, Vol. 25 No. 2: 25-33.

611. Saklani, Dinesh Prasad, 1998. *Ancient Communities of the Himalaya*, Delhi, Indus Publishing Company.

612. Samal, Prasanna K, 1993. "'The Status of Women in Central Himalaya: A Cultural interpretation," *Man In India*, Vol. 73, No.1: 87-95.

613. Samal P.K (ed), 1996, *Tribal Development: Options*, (Proceedings of a National Seminar), pp. 118-138. Nainital, Gynoday Prakashan.

614. Sanam, Deepak and Dhanu Swadi, 1998. *Exploring Kinnaur and Spiti in the Trans-Himalaya*, Delhi, Indus Publishing Company.

615. Sandhu, Balwant, 1979. "Kinnaur-1978," *The Himalayan Journal*, Vol. XXXV: 224-28.

616. Sandhu, J. N, 1972. "New Horizons Open for First Settlers of Himachal Pradesh (Kinnauras)," *Indian Express*, 23 July, Vol. IV, Col. 1.

617. Sankrityanan, Rahul, 1957. *Rigvedic Arya*, Allahabad and Delhi, Kitab Mahal.

618. Sanskritayan, Rahul, 1957. *Himalaya Parichaya*, Delhi, Vani Prakashan.

619. Sanskritayan, Rahul, 1956. *Kinnar Desh*, Allahabad, Kitab Mahal.

620. Saraswat, H.C, 1970. *Himachal Pradesh*, Publication Division, Ministry of Information and Broadcasting, Govt. of India.

621. Sarkar, R. M, 1992. "From Nomadism to Sedentism: Adaptation and Response: As Exemplified by the Birhors of Chotanagpur," *Man in India*, Vol. 72, No. 3: 259-270.

622. Sastry, V.N.V.K, 1993. "Chenchu Food Gatherers and Hunters of Nallamala Forests—Problems of Transition," *Man and Life*, Vol. 19, No. 1&2: 83-96.

623. Satwanti, R. Balla, A.K. Kapoor amd Indra P. Singh, 1987. "Varitaions in the Age at Menarche due to Physical Execise and Altitude," Z. *Morph. Anthrop*, Vol. 73, No. 3: 323-332.

624. Sawalha, Leah, 1999. "Barriers of Silence: Reproductive Rights for Women in Jordan," *Development*, Vol. 42, No.1: 41-46.

625. Sawant, S.D, 1963. "The Kinnaur Valley," *Illustrated Weekly of India*, February 10.

626. Saxena, D.N, 1990. "Family Building, Fertility and Family Welfare Among Two Tribal Community of UP," in A. Bose, U.P. Sinha and R.P. Tyagi (ed) *Demography of Tribal Development*. pp. 249-269.

627. Schneider, Harold K, 1975. "Economic Development and Anthropology" *Annual Review of Anthropology*, pp. 271-292.

628. Schwartz, Carolyn E. and Mirjam A G. Sprangers, 1999. "Methodological Approach for Assessing Response Shift in Longitudinal Health-Related Quality of Research," *Social Science and Medicine*, Vol. 48: 1531-1548.

629. Schwartz, Carolyn E. and Rabbi Meir Sendor. 1999. "Helping Others Helps Onself: Response Shift Effects in Peer Support," *Social Science and Medicine*, Vol. 48: 1563-1575.

630. Sebastian,A, 1986. "Migrants in North Eastern Region of India," in Data Ray (ed). *The Patterns and Problems of Population in North-east India* Delhi, Uppal Publishing House. pp. 63-68.

631. Semwal, P.N, 1963. "Kinnaur and Bhot Samrajaha," *Himprastha*, Vol. 8, No. 10: 10-12.

632. Semwal, P.N, 1965. "Traces of Ancient Republics in the Himalayan Region," Shimla, *Himalayan Kalpadrum*, Vol. 1, No. 4: 65-79.

633. Sen, Biswajit, 1966. "The Himalayans and Their Occupation Patterns," *The Eastern Anthropologist*, Vol. XIX No. 3: 177-189.

634. Sen, Biswajit, 1969. "Problem of Western Himalaya Borderland," in *Urgent Research in Social Anthropology*, Shimla, Indian Institute of Advanced Study, Vol. 10, pp. 1-5.

635. Sen, Biswajit, 1979. *Caste, Class and Leadership in a Himalayan District* Shimla, Indian Institute of Advanced Study, pp. 529-30.

636. Sen, Biswajit, 1996. "The Process of Development and Change in a Tribal District: A Case Study of Kinnaur in Himalayas," in A.K. Kapoor and Dharamvir Singh (ed), *Man and Development in the Himalayas*, Delhi, Academic Foundation, pp. 118-125.

637. Sen, D.K, 1956. "Some Notes on the Fertility of Jaunsari Women," *Eastern Anthropologist* Vol. 10, No. 1: 60-67.

638. Sen, Tulika, 1994. "A Historical Study of Ages At Menarche, Marriage, Menopause and Family Size in High-Caste Bengalees of Calcutta," *Man In India*, Vol. 74, No. 3: 241-250.

639. Sengupta, Sarthak and Chakravarty, Kanta, 1995. "Family Type, Fertility and Mortality: A Study Among the Ahoms of Assam," *J.Hum.Ecol.* Vol. 6, No. 3: 197-200.

640. Sengupta, Sarthak and Rajkhowa, Mina, 1996. "Menarche and Menopause Among the Ahom Women of Dibrugarh, Assam," *J.Ecol.*, Vol. 7, No. 3: 211-213.

641. Shah,A.K, 1958. "The Age of Menarche in Gujrati College Girls," *J. Ind. Med.Assoc.*, Vol. 30:347.

642. Sharma, Avinash K, 1998. "Conservation, Management and Development of Non Wood Forest Resources of the Himalaya," in B.D. Sharma and Tej Kumari (ed), *Himalayan Natural Resources: Eco threats and Restoration Study*, Delhi, Indus Publishing House.

643. Sharma, B.D, 1998. "Eco-Conservation of Hill Areas: The Challenges and Proposals" in B.D. Sharma and Tej Kumari (ed), *Himalayan Natural Resources: Eco threats and Restoration Study*, Delhi, Indus Publishing House.

644. Sharma, B.D and Tej Kumari (ed), 1998. *Himalayan Natural Resources: Eco Threats and Restoration Study*, Delhi, Indus Publishing House.

645. Sharma, B.R, 1998. "Tribal Myths and Legends and Their Role in Development in Himachal Pradesh," in Gupta, S. K., V.P. Sharma, N. K. Sharda (ed), *Tribal Development: Appraisal and Alternatives*, Delhi, Indus Publishing House.

646. Sharma, Bashi Ram, 1976. *Kinner Lok Sahitya*, H.P, Lalit Prakashan.

647. Sharma, Bashi Ram. "Kinnaur Shatriya Vivha Prathyaa," Shimla, *Himprastha*, Vol. 21, No. 3, June 1975: 22-25; Vol. 21, No. 4, July 1975: 41-44; Vol. 21, No. 5, August 1975: 52-56; Vol. 21, No. 6, September 1975: 16-19; Vol. 21, No. 7, October 1975: 32-37; Vol. 21, No. 8, November 1975: 17-20.

648. Sharma, D.D, 1988. *A Descriptive Grammar of Kinnauri*, Delhi, Mittal Publication.

649. Sharma, Jaidev, " Kinnaur Ki Gatai," Shimla, *Himprastha*, Vol. 5, No. 2, May 1959: 8-9; vol. 5, No. 3, June 1959: 3-4; Vol. 5, No. 4, July 1959: 3-4; Vol. 5, No. 5, August 1959: 49-52; Vol. 5, No. 6, September 1959: 15-17; Vol. 5, No. 7, October May 1959: 15-17; Vol. 5, No. 8, November 1959: 6-7; Vol. 5, No. 9, December 1959: 20-21; Vol. 5, No. 10, January 1960: 33-35; Vol. 6, No. 1, April 1960: 19-20, 35; Vol. 6, No. 2, May 1960: 21-22, 45; Vol. 5, No. 4, July 1960: 33-34.

650. Sharma, K.K.N and Abdul Samad Khan, 1990. "Fertility and Mortality Trends of the Khairwar Tribal Women of Madhya Pradesh," *Man in India*, Vol. 70, No.4: 446-453.

651. Sharma, K.C, 1994. *Poverty, Unemployment and Inequalities in Tribal India with Special Reference to Himachal Pradesh*, Delhi, Reliance Publishing.

652. Sharma, K.K.N. and A.S. Khan, 1990. "Fertility and Mortality Trends of the Khairawar Tribal Women of Madhya Pradesh," *Man in India*, Vol. 70, No. 4: 446-453.

653. Sharma, K.K.N. and B. M. Mukherjee, 1995. "Nutritional Status Among the Tribal Hill Korwa Children of Madhya Pradesh," *South Asian Anthropologist*, Vol. 16, No. 2: 95-98.

654. Sharma, Krishan, 1979. *The Kondhs of Orissa: An Anthropometric Study*. Delhi, Concept Publishing Company.

655. Sharma, L.R, 1987. *The Economy of Himachal Pradesh*, Delhi, Mittal Publication.

656. Sharma, L.R, 1998. "Redesigning the Strategy for Tribal Development in Himachal Pradesh," in Gupta, S.K., V.P. Sharma, N.K. Sharda (ed), *Tribal Development: Appraisal and Alternatives*, Delhi, Indus Publishing House.

657. Sharma, Ranbir, 1977. *Party Politics in Himalayan State*, Delhi, National Publishing House.

658. Sharma, S.P.and Sharma J.B, 1999. *Tribal Demography*, Delhi Radha Publications.

659. Sharma, V.S, 1958. "The Scheduled Tribes of Himachal Pradesh and Their Problem," *Vanyajati*, Vol. VI, No. 7: 117-122.

660. Sharma, A.N, 1998. "Socio-Cultural and Demographic Profile of Bharias of Patalkot," in Chaturbhuj Sahu (ed), *Tribal Culture and Identity*. Delhi, Sarup and Sons Publication pp. 146-187.

661. Sharma, Anima, 2001. "Health and Hygiene Among the Tribals—A Case study of Gonds," *Anthropologist*, Vol. 3, No. 3: 191-194.

662. Sharma, K, 1978. "Fertility and Mortality in Khonds: A Tribal Community of Orissa," *Man in India*, Vol. 58, No. 1:77-88.

663. Sharma, M. B. and Chowdhury, Dipesh, 1995. "Menarche and Fertility: A Correlation Study Among the Gind Tribe of Maharashtra," *J Hum.Ecol.*, Vol. 6, No. 3: 209-212.

664. Shashi, S.S, 1978. *The Tribal Women of India*, Delhi, Sundeep Prakashan.

665. Shashi, S.S, 1979. *The Nomads of the Himalayas*, Delhi, Sundeep Prakashan.

666. Shashi, S.S, 1971. *Himachal Nature's Peaceful Paradise*, New Delhi, Sundeep Prakashan.

667. Shastri, Badari Datt, 1970. " Kinnauri Pahali," Shimla, *Himprastha*, Vol. 16, No. 3.

668. Shastri, Dharm Dev, 1957. "Delhi Sa Chini (Kinnar Desh)," *Vanyajati*,Vol. V, No. 10: 167-172.

669. Shastri, Dharm Dev, 1963. " Sri Gavar Bhai Ki Kinnaur Yatra, Hamara Simant Prahari Adam Jati- Kinnar," *Vanyajati*, Vol. XI, No. 3: 109-20.

670. Shay, Ronald, 1977. *Himalayan Bhutan, Sikkim and Tibet*, Delhi, Ess Ess Publication.

671. Shnirelman, Victor.A, 1994. "Hunters and Gatherers in the Modern Context," *Current Anthropology*, Vol. 35, No. 3: 298-301.

672. Sidhu, S and Sidhu, L.S, 1985. "Age at Menarche and Menopause in Sansi Female of Punjab," *Ind.J.Phus.Anthrop.Hum.Genet.*, Vol. 11: 33-37.

673. Sigerist HR, 1941. *Medicine and Human Welfare*, New Haven, Yale University Press.

674. Singh, A, 1994. "Geographical Pathology of Gastroduodenal Ulcer Disease in Ambah Town, Madhya Pradesh," *Journal of Human Ecology*, Vol. 5, No. 2: 149-152.

675. Singh, Bageshwar, 1983. "Role of Anthropologist in Development: Tribal or Otherwise," *Spectra of Anthropological Progress*, Vol. 5: 13-20.

676. Singh, Bhupinder, 1977. "Tribal Development At Cross-Roads: A Critique and a Plea," *Man in India*, Vol. 57, No. 3: 229-243.

677. Singh, Bulbul, 1975. "Kinnaur," *Illustrated Weekly of India*, March 23.

678. Singh, Chetan, 1998. *Natural Premises: Ecology and Peasant Life in the Western Himalaya*, Delhi, Oxford University Press.

679. Singh, J.P. Vyas N.N. and Mann R.S, 1988. *Tribal Women and Development*. Jaipur, Rawat Publications.

680. Singh, K. S.(ed), 1994. *The Schedule Tribe, Vol. III, People of India*, Calcutta, Anthropological Survey of India.

681. Singh, K.S. (edt), 1996. *People of India: Himachal Pradesh Vol. XXIV*, Anthropological Survey of India, Manohar Publishers & Distributors.

682. Singh, K. Suresh, (ed) 1990. *The Tribal Situation in India*, Shimla, IIAS.

683. Singh, K.S, 1988. "Tribal Women: An Anthropological Perspective," in Singh, Vyas and Mann (ed), *Tribal Women and Development*, Jaipur, Rawat Publications.

684. Singh, Mian Goverdhan, 1964. "The Origin of the Man and the Himalaya," *Himalayan Kalpadruma*, Vol. I, No.1.

685. Singh, Mian Goverdhan, 1982. *History of Himachal Pradesh*, Delhi, Yogbodh Publishing House.

686. Singh, Mian Goverdhan, 1983. *Art and Architecture of Himachal Pradesh*, Delhi, B.R. Publishing Corporation.

687. Singh, Mian Goverdhan, 1985. *Descriptive Bibliography of Himachal Pradesh*, Shimla, Himachal Academy of Art, Culture and Language.

688. Singh, Mian Goverdhan, 1985. *Social, Cultural and Economic Survey of Himachal Pradesh*, Shimla, Minerva Book House.

689. Singh, Mian Goverdhan, 1992. *Festivals, Fairs and Customs of Himachal Pradesh*, Delhi, Indus Publishing House.

690. Singh, Mian Goverdhan, 1994. *Himachal Pradesh: History, Culture and Economy*, Shimla, Minerva Book House.

691. Singh, Mian Goverdhan, 1999. *Wooden Temples of Himachal Pradesh*, Delhi, Indus Publishing House.

692. Singh, Mutum Bokul, 1995. "Some Aspects of Socio-Demographic Profile of Taro," *Man In India*, Vol. 75, No.1: 79-95.

693. Singh, Nishi, 1995. "Family and Marriage Among Khonds of Koraput District, Orissa—A Comparative Study" *Man in India*, Vol. 75, No. 1: 105-110.

694. Singh, Prabhat K, 1994. "Ethno-Peasantry System: A Case of the Kurmis," *Man In India*, Vol. 74, No. 2: 169-179.

695. Singh, Udai Pratap and B.R.K. Shukla, 1992. "Trend of Menarche in Five-Endogamous Groups of Tharu Tribal Females of Uttar Pradesh," *Man in India*, Vol. 72, No.3: 343-352.

696. Singh, Mutum Bokul, 1995. "Some Aspects of Socio-Demographic Profile of Tarao," *Man in India*, Vol. 75, No.1: 79-95.

697. Singh. Gopal S, 1999. "Tribal Women of Western Himalaya: A Socio-Cultural Interpretation," in Deepak Kumar Behera and Georg Pfeffer (ed), *Contemporary Society: Tribal Studies*, Vol. III, Delhi, Concept Publishing Company, pp. 203-214.

698. Singhal, Praveen,Gupta Sagarika and Kaur, Baljinder, 1994. "Genetic and Socio-Economic Influence on Menarche," *J.Hum.Ecol.*, Vol. 5, No. 4: 271-275.

699. Sinha, Durganand, 1960. "Caste Dynamic: A Psychological Analysis," *The Eastern Anthropologist*, Vol. XIII, No. 4: 159-171.

700. Sinha, Nirmal Chandra, 1966. "The Lama," *Man in India*, Vol. 46, No. 4: 345-351.

701. Sinha, Surajit, 1965. "Tribe-Caste and Tribe-Peasant Continua in Central India," *Man In India*, Vol. 45: 57-83.

702. Sinha, U.P, 1986. *Ethno Demographic Study of Tribal Population in India*. Bombay International Institute for Population Studies,

703. Sinha, U.P, 1990. "Demographic Profile of Tribal Population in India," in Bose (et al.) *Demography of Tribal Development* Delhi, B.R. Publishing Corporation.

704. Sirajuddin.S.M. and Basu,A, 1984. "Population Structure, Consanguity and its Effect on Fertility and Mortality among Chenchu of Andhra Pradesh," *Hum.Sci.*, Vol. 33:179-200.

705. Sishaudhia, V.K, 1981. "Demographic Structure of a Tribal Village: A Preliminary Study," *Vanyajati* Vol. XXIX, No. 3: 3-11.

706. Sleen-Van- Dar, W.G.N, 1924. *Four Months Camping in the Himalayas*, London, Philip Allan and Co.

707. Slewen, D.G.P.M, 1929. "The Way to the Baspa," *The Himalayan Journal*, Vol. 1: 67-69.

708. Smith, Andrea L, 1994. "Colonialism and the Poisoning of Europe: Towards an Anthropology of Colonists," *Journal of Anthropological Research*, Vol. 50: 383-392.

709. Snellgrove, D.L, 1957. *Buddhist Himalaya*, Oxford, Bruno Cassirer.

710. Soklani, Dinesh Prasad, 1998. *Ancient Communities of the Himalaya*, Delhi, Indus Publishing Company.

711. Som, Sujit, 1993. "Demographic Profile of an Orissa Village," *Man in India*, Vol. 73, No. 1: 49-63.

712. Som, Sujit, 1993. "Demographic Profile of an Orissa Village," *Man In India*, Vol. 73, No. 1: 49-63.

713. Sorkin, Alan L, 1976. *Health Economics in Developing Countries*, Massachusetts, D.L. Health.

714. Spitulnik, Debra, 1993. "Anthropology and Mass Media," *Annual Review of Anthropology*, Vol. 22: 293-315.

715. Sprangers, Mirjam A.G. and Carolyn E. Schwartz, 1999. "Integrating Response Shift into Health Related Quality of Life Research: A Theoretical," *Social Science and Medicine*, Vol. 48: 1507-1515.

716. Srivastava, A.C, 1972. "Mid-Phalangeal Hair Variability in Some Himalayan Populations," *Bulletin of the Anthropological Survey of India*, Vol. XXI: 139-150.

717. Srivastava, A.R.N, 1990. "Rise of Ecological Studies in Cultural Anthropology," *Man in India*, Vol. 70, No. 3: 278-287.

718. Srivastava, V.K, 1986. "Culture and Development," *Man In India*, Vol. 66, No. 1: 67-80.

719. Stacton, David, 1954. *Ride on a Tiger: The Curious Travel of Victor Jacquemont*, London, Museum Press.

720. *Statistical Outline Dist. Kinnaur, 1998-99*, District Statistical Office, Kinnaur.

721. *Statistical Outline, Himachal Pradesh, 1996*. Economic and Statistics Department, Himachal Pradesh.

722. *Statistical Outline, Himachal Pradesh, 1997*. Economic and Statistics Department, Himachal Pradesh.

723. *Statistical Outline, Himachal Pradesh, 1998*. Economic and Statistics Department, Himachal Pradesh.

724. Steward, Julian H, 1955. *Theory of Culture Change*, Urbana. University of Illinois Press.

725. Stulpnagel,1878. "Polyandry in the Himalayas," *The Indian Antiquary*, Vol. VII: 132-135.

726. Sudhakar, G., B.V. Babu and V. Padma, 1997. "A Demographic Study of Chakali Caste From Andhra Pradesh," *South Asian Anthropologist*, Vol. 18, No.1: 37-39.

727. Susanne, Charles, 1982. "Anthropology and Health," *Spectra of Anthropological Progress*, Vol. 4: 39-42.

728. Swain, S, S.C. Jena and P. Singh, 1990. "Morbidity Status of the Kondha Tribes of Phulbani (Orissa)," in Buddhadeb Chaudhuri(ed), *Cultural and Environmental Dimensions on Health* Delhi, Inter-India Publications.

729. Swedlund, Alan C. 1978. "Historical Demography as Population Ecology," *Annual Review of Anthropology*, Vol. 7: 137-73.

730. T.L, Subha Sri and Shashikala Puttaraj, 1999. "Nutritional Need Assessment of Member Families of a Family Helper Project with Women as Indicators of Household Nutrition," *Journal of Human Ecology*, Vol. 10, No. 2: 115-120.

731. *The Himachal Pradesh Panchayati Raj Act, 1994 (Act No. 4 of 1994)*, Rural Development and Panchayati Raj Department, Government of Himachal Pradesh.

732. *The Himachal Pradesh Panchayati Raj Manual (Part- 1)*, Rural Development and Panchayati Raj Department, Government of Himachal Pradesh, Shimla.

733. Thomas, T, 1952. *A Narrative of a Journey Through the Mountains of Northern India During the Years 1847-48*, Convengarden, Reeva and Co. (reprinted as 'Western Himalayas and Tibet: A Narrative on Ladakh and Mountains of Northern India," by Cosmo Publication, Delhi in 1978).

734. Thorton, Edward, 1854. *A Gazetteer of the Territories Under the Govt. of the East India Company and of the Native States on the Continent of India* p. 518.

735. Tiwari, D.N. 1984. *Primitive Tribes of Madhya Pradesh.* Government of India, Ministry of Home Affairs, Tribal Development Division, New Delhi.

736. Tiwari,R.P, 1998. "Tribal Habitat, Economy and Society—A Case Study of Kawar Tribe of Madhya Pradesh," in Chatarbhuj Sah (ed). *Tribal Culture and Identity,* Delhi, Sarup and Sons Publication.

737. Tribhuwan, Robin. D and Ram D. Gambhir, 1996. "Ethnomedical Pathway: A Conceptual Model," *The Eastern Anthropologist,* Vol. 49, No. 2: 139-163.

738. Tungdim, M. G., Kapoor, S. and Kapoor, A. K, 2007. Tribes, Tuberculosis and Treatment: A Study in North-East India. (Send for Publication).

739. Turmen, Tomris. 1999. "Making Globalization Work for Better Health," *Development,* Vol. 42, No.4: 8-11.

740. Upreti, R.P, 1969. *Butias of Uttarakhand: A Study in Cultural Geography* Ph.D. Thesis, Agra University: 256-267.

741. Vaidya, K.L, 1975. "Kinnaur and Spiti before the Tragedy," *Patriot Magazine,* 16 February p. 2, c 1-4.

742. Vaidya, K. L, 1977. *The Cultural Heritage of the Himalayas,* Delhi, National Publishing House.

743. Vashistt, Sudharshan, 1996. *Rang Badalta Parbhat,* Delhi, Sanmarg Parkashan.

744. Vashistt, Sudharshan, 1997. *Himachal,* Delhi, Atam Ram and Sons.

745. Vedanthan, Rajesh and Suneeta Krishnan, 1999. "The Swasthya Community Health Partnership: Gender-Based Health Care in Rural South India," *Development,* Vol. 42, No.1: 95-96.

746. Venugopala Sarma, A and K. Nagaraj, 1999. "A Study on Pattern of Utilisation of Health Services in an Integrated Child Development Services Project in Andhra Pradesh," *Journal of Human Ecology,* Vol. 10, No. 1: 1-6.

747. Verrier, Elwin, " A New Deal for Tribal India, p. 1

748. Vidyarthi, L.P, 1983. *Tribes of India. In Peoples of India: Some Genetical Aspects*. XV International Congress of Genetics, Dec. 12-21, Delhi, ICMR.

749. Vidyarthi, L.P. and Rai, B.K, 1977. *The Tribal Culture of India*, Delhi, Concept Publishing Company.

750. Vidyarthi,L.P, 1971. "The Ranchi Tribals in Andaman and Nicobar Islands," *Journal of Social Research*, Vol. 14: 50-59.

751. Vidyarthi,L.P, 1976. "Cultural Diversities in the Andaman and Nicobar Islands: A Preliminary Report," *Journal of Social Research*, Vol. 19: 1-15.

752. Vij, R.L, 1998. "Forestry Development Programme in Tribal Areas of Himachal Pradesh," in Ashok Ranjan Basu and Padam Nabh Gautum (ed), *Natural Heritage of India: Essays on Environment Management*, Delhi, H K Publishers & Distributor, pp. 191-197.

753. Vijayaraghavan, K, 1987. "Anthropology for Assessment of Nutritional Status," *Indian Journal of Paediatric*, Vol. 54: 511-520.

754. *Village/Town Primary Census Abstract, 1991*. District Statistical Office, Kinnaur.

755. Vithal, C.P, 1992. "Socio-economic Transformation of the Primitive Tribal Group: A Study of Chenchus of Andhra Pradesh," *Man in India*, Vol. 72, No. 2: 189-206.

756. Voland, Eckart. 1998. "Evolutionary Ecology of Human Reproduction," *Annual Review of Anthropology*, Vol. 27: 347-74.

757. Wakefield, (Sir) Edward, 1996. *Past Imperative: My Life in India 1927-47*, London, Chatto and Windus, p. 24-28.

758. Weiss, Kenneth M, 1976. "Demographic Theory and Anthropological Inference," *Annual Review of Anthropology*, Vol. 5: 351-81.

759. Wikan, Unni, 1995. "Sustainable Development in the Mega-City: Can the Concept be Made Applicable?," *Current Anthropology*, Vol. 36, No. 4: 635- 638.

760. Wilson, A, 1885. *The Adobe of Snow*, London. William Blackwood and Sons (reprinted by Ratan Pustak Bhandar, Kathmandu, Nepal in 1979).

761. Wilson, Horace Hayman, 1825. *Travels in the Himalayan Provinces of Hindustan and The Punjab; in Ladakh and Kashmir, in Peshawar, Kabul, Kunduz and Bokhara by William Moorcraft and Grogre Trebeck from 1819 to 1825*, (Reprinted by Sagar Publication, Delhi in 1971) Two Vol.

762. Wisenfeld, S. L, 1967. "Sickle- Cell Trait in Human Biological and Cultural Evolution," *Science* Vol. 157: 1134-1140.

763. Wolanski, Napoleon, 1990. "Origin and Methodology of Human Ecology," *Journal of Human Ecology*, Vol. 1, No. 2: 109-119.

764. Wood, James W, 1990. "Fertility in Anthropological Population," *Annual Review of Anthropology*, Vol. 19: 211-42.

765. World Health Organization, 1948. *Constitution. World Health Organization*, Geneva.

766. World Health Organization, 1977. *Resolution WHA40.43- Technical Cooperation*. Geneva, World Health Organization, May.

767. World Health Organization, 1985 *Targets for Health for All: Targets in Support of the European Strategy for Health for All*. Copenhagen, World Health Organization Regional Office for Europe.

768. World Health Organization, 1998. *Health for All in the Twenty-first Century*, Geneva, World Health Organization.

769. World Health Organization, 1998. *The World Health Report 1998. Life in the 21 st Century: A Vision for All*. Geneva, World Health Organization.

770. Yadav, Kumkum, 1998. "Rahul Sankrityayan's Kinnar Desh Mein," in Gupta, S.K., V.P. Sharma, N.K. Sharda (ed), *Tribal Development: Appraisal and Alternatives*, Delhi, Indus Publishing House.

771. Yadav, Ankur, Sharma, A.N. and Jain Amita, 2001. "Socio-Demographic Characteristics of Semi-Nomadic Lohar-Gadiyan of Malthon Town of Sagar District, Madhya Pradesh," *Anthropologist*, Vol. 3, No. 2: 135-137.

Index

❑❑❑